Birkhäuser Advanced Texts

Series Editors
Steven G. Krantz, Washington University, St. Louis
Shrawan Kumar, University of North Carolina at Chapel Hill
Jan Nekovář, Université Pierre et Marie Curie, Paris

For further volumes:
http://www.springer.com/series/4842

Tomaž Pisanski • Brigitte Servatius

Configurations from a Graphical Viewpoint

Tomaž Pisanski
IMFM
Oddelek za Teoretično Računalništvo
Univerza v Ljubljani
Ljubljana, Slovenia

Brigitte Servatius
Department of Mathematics
Worcester Polytechnic Institute
Worcester, MA, USA

ISBN 978-0-8176-8363-4   ISBN 978-0-8176-8364-1 (eBook)
DOI 10.1007/978-0-8176-8364-1
Springer New York Heidelberg Dordrecht London

Library of Congress Control Number: 2012944998

Mathematics Subject Classification (2010): 05-01, 05Bxx, 05Cxx, 20-01, 51-01

© Springer Science+Business Media New York 2013
This work is subject to copyright. All rights are reserved by the Publisher, whether the whole or part of the material is concerned, specifically the rights of translation, reprinting, reuse of illustrations, recitation, broadcasting, reproduction on microfilms or in any other physical way, and transmission or information storage and retrieval, electronic adaptation, computer software, or by similar or dissimilar methodology now known or hereafter developed. Exempted from this legal reservation are brief excerpts in connection with reviews or scholarly analysis or material supplied specifically for the purpose of being entered and executed on a computer system, for exclusive use by the purchaser of the work. Duplication of this publication or parts thereof is permitted only under the provisions of the Copyright Law of the Publisher's location, in its current version, and permission for use must always be obtained from Springer. Permissions for use may be obtained through RightsLink at the Copyright Clearance Center. Violations are liable to prosecution under the respective Copyright Law.
The use of general descriptive names, registered names, trademarks, service marks, etc. in this publication does not imply, even in the absence of a specific statement, that such names are exempt from the relevant protective laws and regulations and therefore free for general use.
While the advice and information in this book are believed to be true and accurate at the date of publication, neither the authors nor the editors nor the publisher can accept any legal responsibility for any errors or omissions that may be made. The publisher makes no warranty, express or implied, with respect to the material contained herein.

Printed on acid-free paper

Springer is part of Springer Science+Business Media (www.springer.com)

Preface

Geometric configurations are a natural and attractive topic for students, teachers, and researchers. This text serves as a concise introduction to several important topics and the interesting interplay between them. Our unifying theme is graph theory: basic graph theory to introduce graphs themselves; algebraic graph theory to introduce groups; topological graph theory to introduce surfaces; and geometric graph theory to talk about geometries, in particular incidence geometries.

Incidence structures are seemingly uninteresting mathematical objects. They are too general to possess any deep algebraic structure. Lineal, semi-regular incidence structures, alias combinatorial configurations, usually result from an appropriate collection of points and lines in the plane. These arrangements of lines aroused considerable interest among leading mathematicians of the nineteenth century and much less during the rise of discrete mathematics of the twentieth century. As one puts different limitations on incidence structures, their number decreases and they become much more structured. This effect was observed in the theory of designs which became a successful and important piece of modern combinatorics.

In the second half of the twentieth century graph theory boomed. Since incidence geometries are essentially vertex-colored graphs, the tools from graph theory are now available for studying discrete geometrical objects. Symmetries of configurations establish an important link between group theory and combinatorics. There are several pathways between configurations and geometry, and certainly there are deep links between configurations and algebraic topology involving covering spaces and maps on surfaces. However, the reason for a success is the fact that configurations and incidence geometries can be considered as part of graph theory.

Shifting the paradigm from algebra and topology to graph theory puts new balls into the playing field of the discrete mathematician. We devote neither a chapter nor a section to incidence geometry. We use it throughout the book and it provides a wonderful viewpoint from which difficult topics involving graphs, groups, maps, and configurations become accessible.

The first book on configurations was written in 1929 by Friedrich Levi who already pointed out the interdisciplinary nature of configurations. The most recent book on configurations is *Configurations of Points and Lines* [44] by Branko Grünbaum. We would like to thank him for sharing his work with us. We adjusted terminology with that of Grünbaum in all feasible cases. Our term *Grünbaum calculus* is but a small tribute to his profound influence on the subject. The approach in [44] is geometric. We expect that our graph theoretical approach will be a welcome and timely complement.

Our book is nonstandard in the sense that it brings together a wide variety of mathematical concepts. It is not a monograph. It is a graduate textbook that is accessible to advanced undergraduate students. Each chapter contains a set of exercises. Problems denoted by (*) are more difficult and usually require material that was not covered up to this point. Problems denoted by (**) represent research problems or very difficult problems. The exercises are intended to give working knowledge of the subject and enable the reader to get as close to the research frontier as possible.

Parts of the book were tested in various graduate courses at the University of Ljubljana in Slovenia and at the Worcester Polytechnic Institute. Chapters 1–3 can be used as an invitation to algebraic graph theory, and Chaps. 1–4 give much more than an overview of algebraic and topological graph theory. Chapters 1, 2, 5, and 6 can serve as an introduction to the study of incidence geometries and configurations of points and lines.

In the course of writing the book several mathematicians read and commented on various parts of the manuscript. We are sincerely thanking everyone. We are listing those, who tested several exercises and proofread the manuscript in the last year, giving us invaluable suggestions for improvements. Special thanks go to: Nino Bašić, Leah Berman, Marko Boben, Jürgen Bokowski, Marston Conder, Maria del Rio Francos, Gábor Gévay, Branko Grünbaum, Martin Juvan, Matjaž Konvalinka, Aleksander Malnič, Martin Milanič, Marko Petkovšek, Primož Potočnik, Ivona Puljić, Tom Tucker, Gordon Williams, and Arjana Žitnik.

By far the most important contributor to this volume was the Canadian mathematician H. S. M. Coxeter, whose foundational 1950 paper "Self-dual configurations and regular graphs" in the *Bulletin of the American Mathematical Society* was the model for both the style and content of this volume. In that paper Coxeter coined the term "Levi graph" and established the combinatorial approach; described the connection to cages; examined the connection to Petersen and generalized Petersen graphs, their readily computable automorphism groups and their easy realization as unit distance graphs; exploited the symmetrical nature of small configurations, their classical roots, and their geometric connection to regular maps on surfaces; exposed the important combinatorial and algebraic descriptions of classical configurations which yield new perspectives onto their automorphism groups and their level of transitivity, as well as many other essential insights which have ever since guided the combinatorial approach to configurations. The reader who wishes to move beyond the material presented here can do no better than to consult this paper and those papers which cite it with the love and respect it deserves.

Tomaž Pisanski conceived the idea and started this project, while Brigitte Servatius insisted on the last word and is therefore prepared to take full responsibility for all the mathematical shortcomings of this volume.

Ljubljana, Slovenia
Worcester, Massachusetts

Tomaž Pisanski
Brigitte Servatius

Contents

1 **Introduction** 1
  1.1 Hexagrammum Mysticum 1
      1.1.1 The Fano Plane 5
      1.1.2 The Miquel Configuration 6
      1.1.3 The Pappus Configuration 6
      1.1.4 Pappus-Like Configurations 8
      1.1.5 The Tetrahedron 10
      1.1.6 Overview 12
  1.2 Exercises 12

2 **Graphs** 15
  2.1 Basic Definitions 15
  2.2 Examples of Graphs 16
      2.2.1 Paths 17
      2.2.2 Cycles 18
      2.2.3 Complete Bipartite Graphs and Multipartite Graphs 19
      2.2.4 Wheel Graphs 20
      2.2.5 Prism Graphs 21
      2.2.6 Antiprism Graphs 21
      2.2.7 Platonic and Archimedean Graphs 22
      2.2.8 Polyhedral Graphs 23
      2.2.9 Generalized Petersen Graphs 25
      2.2.10 Cages 27
      2.2.11 Planar Graphs 30
  2.3 Regularity 31
      2.3.1 Regular Graphs 31
      2.3.2 Cubic Graphs and LCF Notation 32
      2.3.3 Regularity and Bipartite Graphs 32
      2.3.4 Semiregular Bipartite Graphs 35
      2.3.5 Permutations 35
      2.3.6 Directed Graphs and Multigraphs 37

	2.4	Operations on Graphs	38
		2.4.1 Graph Complement	38
		2.4.2 Graph Union	38
		2.4.3 Graph Join, Cone, and Suspension	38
		2.4.4 One-Point Union and Connectivity	38
		2.4.5 Cartesian Product	39
		2.4.6 Tensor Product	40
		2.4.7 Strong Product	40
		2.4.8 Line Graph	40
		2.4.9 Subdivision Graph	40
		2.4.10 Graph Square	40
	2.5	Graph Colorings	41
		2.5.1 Vertex Colorings	41
		2.5.2 Edge Colorings	42
	2.6	From Geometry to Graphs and Back	43
		2.6.1 Metric Space and Distance Function	43
		2.6.2 Distances in Graphs	44
		2.6.3 Intersection Graphs	44
		2.6.4 Intersection Graphs of a Family of Balls	44
		2.6.5 Convex Sets	45
		2.6.6 Representations and Drawings of Graphs	46
		2.6.7 Generalized Petersen Graphs as Unit Distance Graphs	47
	2.7	Exercises	48
3	**Groups, Actions, and Symmetry**		**55**
	3.1	Groups	55
		3.1.1 Graph Automorphisms	55
		3.1.2 Definition of Groups	55
		3.1.3 Subgroups	58
		3.1.4 Cosets	59
		3.1.5 Group Homomorphisms and Isomorphisms	61
	3.2	Cayley Graphs	62
		3.2.1 Definition of the Cayley Graph	62
		3.2.2 Examples of Cayley Graphs	63
	3.3	Group Actions	67
		3.3.1 Permutation Groups	67
		3.3.2 Actions on Cayley Graphs	69
		3.3.3 Primitive Versus Imprimitive Actions	71
		3.3.4 Burnside's Theorem	72
		3.3.5 The Escher Problem	73
	3.4	Symmetry and Transitivity	76
		3.4.1 Vertex- and Edge-Transitive Graphs	76
		3.4.2 Semisymmetric Graphs	76
		3.4.3 Arc-Transitive Graphs	78
		3.4.4 s-Arc Transitivity	78

		3.4.5	1/2-Arc Transitivity	79
		3.4.6	Automorphisms of Generalized Petersen Graphs	80
	3.5	Voltage Graphs and Covering Graphs		82
		3.5.1	Quotient Graphs	82
		3.5.2	Pregraphs Revisited	83
		3.5.3	Pregraphs on a Single Vertex	84
		3.5.4	Voltage Graphs and Regular Coverings	84
		3.5.5	Voltage Assignments on $B(1;1)$	85
		3.5.6	Generalized Petersen Graphs as Coverings	86
		3.5.7	$\{0,1\}$ Voltage	87
		3.5.8	Permutation Voltage Assignments and Ordinary Coverings	88
		3.5.9	Cages as Covering Graphs	89
	3.6	Automorphisms of the Symmetric Group		90
	3.7	Exercises		97
4	**Maps**			105
	4.1	Geometric Surfaces		105
		4.1.1	Polyhedral Graphs	105
		4.1.2	Polygonal Surfaces	107
		4.1.3	The Topology of Polygonal Surfaces	111
	4.2	Maps and Flags		112
		4.2.1	A Graph Theoretical Approach to Surfaces	112
		4.2.2	Dual Constructions	116
		4.2.3	Flag Orbits and Orientability	118
		4.2.4	Map Projections	118
	4.3	The Classification of Surfaces		119
		4.3.1	Vertex Splitting and Edge Contraction	120
		4.3.2	Reduction to a Unitary Map	121
		4.3.3	Assembling Crosscaps	122
		4.3.4	Assembling Handles	123
		4.3.5	Crosscaps Canceling Handles	123
		4.3.6	Normal Forms	124
	4.4	Operations on Maps		125
		4.4.1	Uniform Flag Operations	125
		4.4.2	The Medial	129
		4.4.3	Truncation	130
		4.4.4	Barycentric Subdivision and Combinatorial Map	131
		4.4.5	Semiuniform Subdivisions	131
		4.4.6	Edge and Vertex Joins	133
	4.5	Map Automorphisms		136
		4.5.1	Regular Maps and Fundamental Regions	136
		4.5.2	Harmonious Maps	141
		4.5.3	Self-dual Maps	145
		4.5.4	Automorphisms of Planar Graphs of Low Connectivity	150
	4.6	Exercises		153

5 Combinatorial Configurations 157
- 5.1 A Combinatorial Approach to Configurations 157
  - 5.1.1 Incidence Structures 157
  - 5.1.2 Coset Incidence Structures 161
  - 5.1.3 Lineal Incidence Structures.................................. 163
  - 5.1.4 Regularity of Incidence Structures 164
  - 5.1.5 Definition of Combinatorial Configurations 164
- 5.2 Combinatorial (v_3) Configurations of Small Order 165
- 5.3 Classical Configurations... 170
  - 5.3.1 The Fano Plane... 170
  - 5.3.2 The Möbius–Kantor Configuration 174
  - 5.3.3 The Pappus Configuration 176
  - 5.3.4 The Desargues Configuration 176
  - 5.3.5 The Cremona–Richmond Configuration 179
  - 5.3.6 The Reye Configuration 181
  - 5.3.7 Möbius (8_4) Incidence Structures 182
- 5.4 Autopolar Combinatorial Configurations............................. 184
- 5.5 Cyclic Haar Graphs and Cyclic Configurations 188
  - 5.5.1 Cyclic Haar Graphs as Cyclic Covering Graphs Over a Dipole 188
  - 5.5.2 Cyclic Haar Graphs as Certain Cayley Graphs for the Dihedral Group....................................... 189
  - 5.5.3 Associating a Cyclic Haar Graph to a Number 189
  - 5.5.4 Isomorphisms of Cyclic Haar Graphs....................... 190
  - 5.5.5 Cyclic Configurations and Cyclic Haar Graphs 191
- 5.6 Exercises 191

6 Geometric Configurations... 197
- 6.1 Geometric Planes 197
  - 6.1.1 From Euclid to Descartes and Beyond 197
  - 6.1.2 The Projective Plane 198
  - 6.1.3 Homogeneous Coordinates 199
  - 6.1.4 Calculations in the Real Projective Plane 201
  - 6.1.5 The Theorems of Pappus and Desargues 205
  - 6.1.6 Proving Desargues from Pappus 206
- 6.2 Finite Projective Planes... 209
  - 6.2.1 Affine and Projective Realizations over Finite Fields 215
- 6.3 Realization of Classical Configurations 219
  - 6.3.1 Fano Plane.. 219
  - 6.3.2 The Möbius–Kantor Configuration 222
  - 6.3.3 The Pappus Configuration 224
  - 6.3.4 The Desargues Configuration 225
  - 6.3.5 The Cremona–Richmond Configuration 226
  - 6.3.6 The Reye Configuration 226

6.4	Representations and Realizations		234
	6.4.1	The Dimension of a Realization	236
	6.4.2	Lifting Configurations	237
	6.4.3	General Lifting	239
6.5	The Grünbaum Incidence Calculus		243
6.6	Constructing Treelike Configurations		251
6.7	Realizing the Gray Graph and Bouwer Graph		252
6.8	The Zindler Degree of Regularity of an Incidence Structure		257
6.9	Polycirculants and Polycyclic Configurations		259
	6.9.1	Cubic Circulants	260
	6.9.2	Bicirculants	260
6.10	Exercises		261

References 265

Index 271

Chapter 1
Introduction

1.1 Hexagrammum Mysticum

Conic sections have excited the curiosity of mathematicians even before their first systematic study was begun in 200 BC by Apollonius of Perga. One would think that the effort of almost two millennia would be sufficient to compel these simple quadratic curves to reveal all their mysteries; however, in 1640, 2 years before inventing the first mechanical calculator, 16-year-old Blaise Pascal published the small pamphlet *Essai pour les coniques* which contained the following stunning new property of conics:

Theorem 1.1. *If a hexagon is inscribed in a conic, then the three points in which pairs of opposite sides meet will lie on a straight line.*

The inscribed hexagon need not be regular or even convex, see Fig. 1.1, and for Pascal, this result was not simply a hitherto unnoticed property but the key to easily proving many known results about conics without having to resort to tedious case analysis. It is reported that Leibniz expressed admiration for Pascal's proof when visiting Paris [18], and we shall have to accept Leibniz's judgment since, unfortunately, the proof was later lost.

The line whose existence is asserted by the above theorem is called the *Pascal line* of the hexagon. Six distinct points can be arranged in cyclic order in $5!/2 = 60$ different ways, so, for a given set of 6 points on a conic, there are 60 hexagons, which in turn give rise to up to 60 Pascal lines. It is possible, for a given hexagon, that some of these Pascal lines coincide, as well as some of the intersection points of the diagonals, for instance, in the case of a regular hexagon inscribed in a circle; see Fig. 1.2. Similarly, given the six points, there are $\binom{6}{4} = 15$ sets of four vertices to be partitioned into opposite sides of a hexagon in three ways, giving up to 45 points. If the six original points of the hexagon are chosen with sufficient generality, then these 60 lines and 45 points are all distinct; see Fig. 1.3. There are four possible ways to complete two segments to be opposite sides of a hexagon on the six points,

Fig. 1.1 Pascal's Hexagrammum Mysticum: The points of intersection of the three pairs of opposite sides of any hexagon inscribed in a conic are collinear

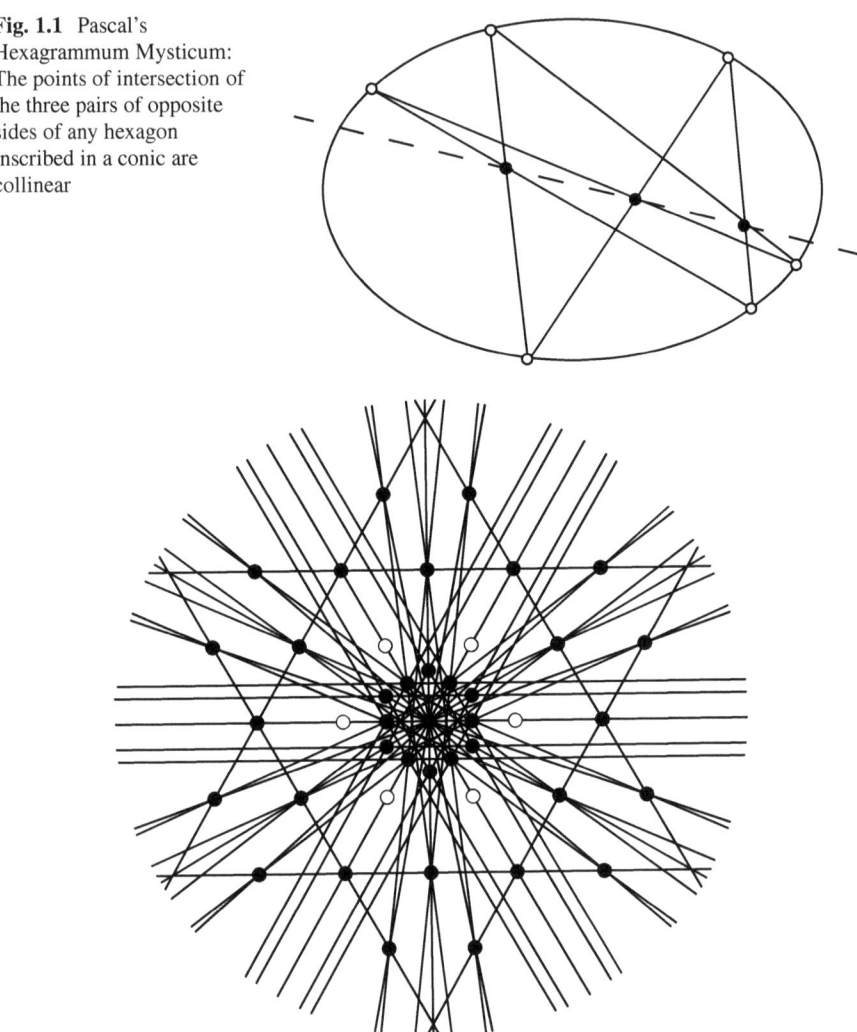

Fig. 1.2 The Pascal configuration on the vertices of a regular hexagon. Some lines have multiplicity, and the line at ∞ is not shown. The point at the origin has multiplicity 3

so each of the 45 points is the intersection point of 4 Pascal lines. We take this fourfold coincidence as evidence that each of these 45 points is geometrically and combinatorially significant. After all, any two nonparallel lines determine an intersection point. It is not until we have three or more such lines that we can be sure that the coincidence is not coincidental. This is the same quality, from the dual point of view, that gives the Pascal line itself its interest, that it joins more than just two points. The $4 \cdot 45 = 180$ point-line incidences are distributed equally among the 60 Pascal lines, 3 to a line, giving in the general case a collection of 45 points and 60 lines, with each of these points and lines significant to the collection.

1.1 Hexagrammum Mysticum

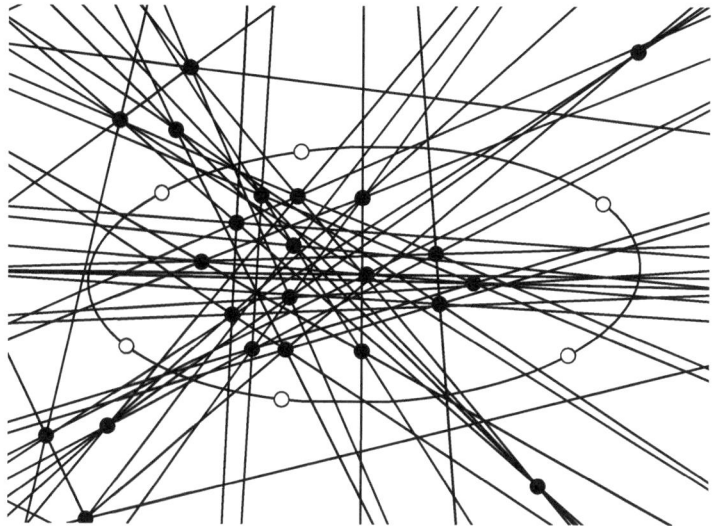

Fig. 1.3 The Pascal configuration corresponding to six generic points on an ellipse

A quick scan of the complicated mass of points and lines in Fig. 1.3 dissuades us from searching for more significant points to add to our collection by hunting with a magnifying glass, real or electronic, to spot three or more lines intersecting at a non-Pascal point. Likewise, especially since it is even more difficult to judge by eye, it seems pointless to hunt for more significant collinearities among the 45 Pascal points. Yet, even if we were to notice new significant points or new significant lines, we might be loth to add them to our collection since that might disturb its *regularity*, the fact that each point is incident to 4 lines and each line passes through 3 points. The attraction of this collection relies on both these qualities, the significance of the points and lines and the regularity of their mutual incidences. This collection of points and lines is called the *Pascal configuration.*

On the other hand, perhaps we ought not to be so sanguine about the lack of additional significant points in the Pascal configuration. If we consider the hexagon $ABCDEF$, the implication of Pascal's theorem that the opposite sides meet in three collinear points says exactly that the triangle formed by the sides AB, CD, and FE and that formed by BC, DE, and FA have their corresponding sides in perspective from a line, see Fig. 1.4, in which that perspective line is close to infinity and not pictured. So, by Desargues' theorem, the corresponding vertices of these triangles are in perspective from a point, S. Moreover, the lines through the corresponding vertices are each Pascal lines, albeit not for the hexagon $ABCDEF$, and the intersection point, baring special positions, is *not* one of the 45 Pascal points. Since the point S is significant, as are all 20 such constructed points, we could now add these new points, called *Steiner points*, to our collection; however, we would no longer have just a single class of points. In the new collection, the Pascal points belong to four lines while the Steiner points belong to just three.

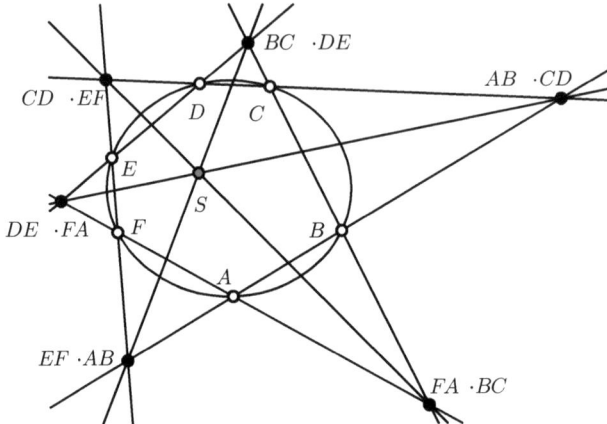

Fig. 1.4 The Steiner point where the Pascal lines for hexagons $ABCFED$, $BCDAFE$, and $CDEBAF$ all meet. The 60 Pascal lines are partitioned into 20 triples which meet in Steiner points

Fig. 1.5 Three hexagons on six points used to construct a Kirkman point

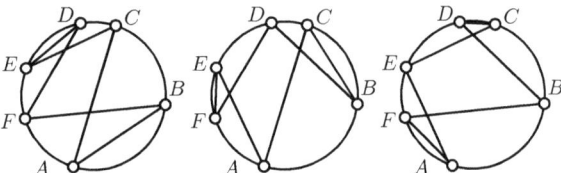

And the hexagrammum mysticum has still not yielded up all her secrets. The Pascal lines corresponding to the three hexagons $ABFDEC$, $DBCAEF$, and $AECDBF$; see Fig. 1.5, form a triangle, depicted in blue in Fig. 1.6, since each pair of hexagons shares a pair of opposite sides. But each of the three hexagons also shares a different pair of opposite sides with the hexagon $ABCDEF$, so each of the sides of the blue triangle extends to meet one of the three points of the Pascal line of hexagon $ABCDEF$. Similarly, the sides of the triangle formed by the non-Pascal lines AB, CD, and EF, depicted in green in Fig. 1.6, whose vertices are Pascal points, meet the Pascal line of hexagon $ABCDEF$ in the same three points; hence, the green and blue triangles are, in the language of Desargues' theorem, in perspective from a line, and consequently the corresponding vertices are in perspective from a point. See Fig. 1.6. The three lines of perspectivity through this new point K, shown in red in Fig. 1.6, can easily be checked to be Pascal lines by noting that they each arise from a hexagon. This new point of coincidence among the Pascal lines is called a *Kirkman point*, and the six points on the conic give rise to 60 Kirkman points.

Nor is mystery yet completely revealed. In addition to 60 Pascal lines, the six generically placed points on a conic give rise to not only 20 Steiner points and 60 Kirkman points but also to 20 *Cayley lines*, 15 *Plücker lines*, and 15 *Salmon points*.

1.1 Hexagrammum Mysticum

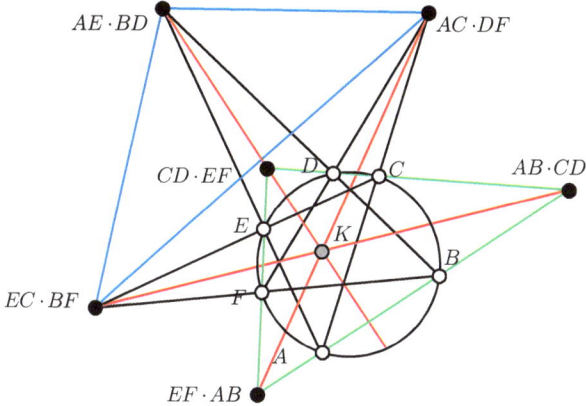

Fig. 1.6 *K* is one of the 60 Kirkman points associated with the Pascal configuration

We could examine each of these geometric incidences in turn, adding each new set of points or lines to our collection and at the same time noting in their names just some of the many geometers who at various times have been brought under the spell of Pascal's hexagrammum mysticum. And to this list, we should add, among others, Brianchon, Hesse, Von Staudt, Veronese, Cremona, and Sylvester.

But this is not the program which we shall follow in this book. The reader wishing to delve further into the mysteries of the hexagrammum mysticum may consult Salmon's "Treatise on Conic Sections" [86] or, for more details, the encyclopedic work by Friedrich Levi [60], the first book on configurations, which treats the subject from a purely geometric point of view.

1.1.1 The Fano Plane

Another classic source of geometric configurations of points and lines is planes over fields other than $\mathbb{R}$. Consider the eight three-dimensional vectors over the field $\mathbb{Z}_2$, $\mathbf{v}_0 = (0, 0, 0)$, $\mathbf{v}_1 = (1, 0, 0)$, $\mathbf{v}_2 = (0, 1, 0)$, $\mathbf{v}_3 = (0, 0, 1)$, $\mathbf{v}_4 = (0, 1, 1)$, $\mathbf{v}_5 = (1, 0, 1)$, $\mathbf{v}_6 = (1, 1, 0)$, and $\mathbf{v}_7 = (1, 1, 1)$. Over $\mathbb{Z}_2$, linear combinations are as simple as possible. Any nontrivial one of them determines a line through the origin consisting only of itself and the origin, and any two determine a plane consisting of exactly four elements, the two basis elements, their sum, and the zero vector. So, we have $\binom{7}{2} / \binom{3}{2} = 7$ planes through the origin, and just as in the real case, we can regard them not as planes containing four points, including the origin, but as projective lines each joining three projective points. So, we have a much smaller configuration than Pascal's configuration, but having the same quality in that each point and line is significant by virtue of three incidences and so having the same regularity. It is called the Fano configuration or, more commonly, the *Fano plane*.

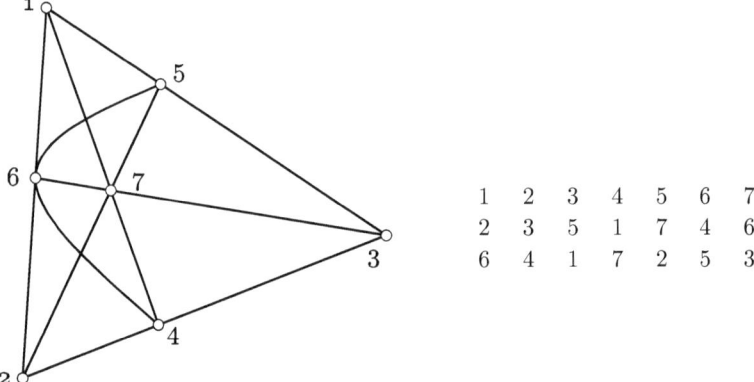

Fig. 1.7 The seven points and seven lines of the Fano plane

The reader may check that the incidences of the seven projective lines are recorded in seven columns of the table in Fig. 1.7 as well as schematically represented in the diagram on the left, if we are generous enough to allow the parabolic arc to stand for a line.

This illustrates the second historical thrust of the theory of configurations, that of *realizability*. Given the incidence data, or, less restrictively, given only the regularity requirements, does there exist a collection of points and lines having that regularity? It is also not difficult to verify that there do not exist seven distinct points in the Euclidean plane which are distributed evenly among seven distinct lines, three lines to a point, three points to a line. See Sects. 5.2 and 6.3.

1.1.2 The Miquel Configuration

For another example, suppose we wish to have eight points and six lines, each point on three lines and each line passing through four points. Consider any particular line. It comprises half the points. Any other distinct line would have to use at least three of the remaining four points, since it could not intersect the first line in two points. Thus, there could not be a third line, and such a collection is not realizable by points and lines in the Euclidean plane. As with the Fano plane, we can still achieve geometric realizability by relaxing the requirements. In this case, there exists a collection of eight points and six circles called the *Miquel* configuration (Fig. 1.8).

1.1.3 The Pappus Configuration

A classic theorem of Pappus states that the intersection points of the opposite sides of a hexagon whose vertices alternate between two lines are collinear. Since the

1.1 Hexagrammum Mysticum

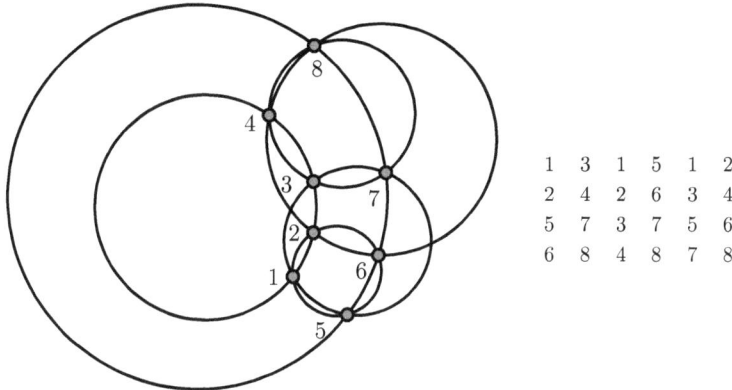

Fig. 1.8 Miquel configuration of *points* and *circles*

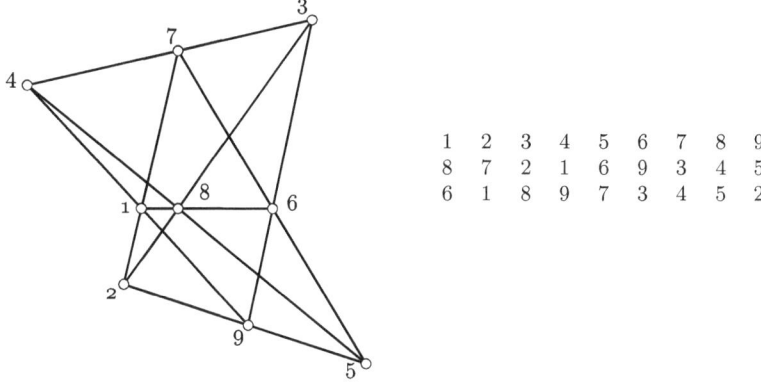

Fig. 1.9 The Pappus configuration and its configuration table

two given lines may be interpreted as a degenerate conic, the theorem of Pappus is a special case of Pascal's theorem. If we consider the points of the hexagon together with the three points of intersection of the opposite sides, together with the two parallel lines, the six lines of the hexagon, and the line whose existence is guaranteed by the theorem, we have an incidence structure with nine points and nine lines in which each point is incident with three lines and each line is incident with three points. This configuration of nine points and nine lines is called the *Pappus configuration*; see Fig. 1.9. It is important to note that the diagram represents *the* Pappus configuration and not simply *a* Pappus configuration. In other words, the Pappus configuration is considered to be any collection of nine points and nine lines constructed in this manner. In Fig. 1.9, the lines 473 and 295 contain the alternating hexagon (493275) and generate via their intersections the three collinear points 1, 8, and 6.

This same collection of lines and points can be constructed via Pappus' theorem in several other ways. We could, for example, have instead used the lines 473 and 186 with hexagon (148367) or, for a more radical departure, lines 127 and

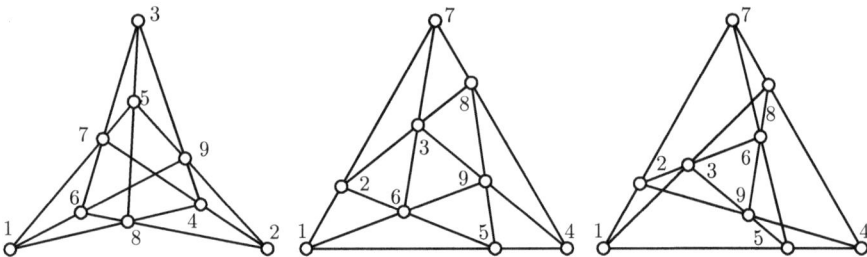

Fig. 1.10 Geometrically symmetric configurations on nine points and nine lines

369 with hexagon (167329). The 27 point-line incidences required for the Pappus configuration are encoded in the *configuration table* to the right of Fig. 1.9. Any collection of points and lines satisfying the incidences prescribed in this table is an example of the Pappus configuration.

This immediately presents two questions: finding all the Pappus configurations in the plane and finding all the ways in which the same configuration may be described as the Pappus configuration. The first is again that of realizability, and the second is the question of automorphisms. The nine points and nine lines of the Pappus configuration are specified by freely choosing five points in the plane to play the roles of $\{1, 2, 3, 4, 5\}$, which determine as well the lines 25 and 34. Points 7 and 9 are constructed as the intersections of lines 12 and 34 and, respectively, lines 14 and 25. Then, points 8 and 6 are constructed as 23 intersect 45 and 39 intersect 57. Thus, the Pappus configuration is specified by the 10 coordinates of the five points $\{1, 2, 3, 4, 5\}$, and we say that it has ten degrees of freedom; however, as we have seen, some of these choices give rise to the same collection of points and lines.

For the second question, we are interested in which permutations of the nine points respect the required incidences. Such a permutation need not arise from a transformation of the plane itself, such as a translation or a dilation, so the configuration of Fig. 1.9 may be more symmetric than its drawing suggests. In fact, it can be shown that for each pair of points there is an incidence preserving permutation mapping one to the other.

1.1.4 Pappus-Like Configurations

Another natural question is whether the different manifestations of the Pappus configuration exhaust all collections of nine points and nine lines with three points on every line and three lines through every point. So, for instance, are the configurations in Fig. 1.10 each *isomorphic* to the Pappus configuration, that is, is there a one-to-one correspondence of points and lines preserving incidence? Since all points in the Pappus configuration, Fig. 1.9, are equivalent under some automorphism, we may, without loss of generality, focus our attention on vertex 1,

1.1 Hexagrammum Mysticum

Fig. 1.11 Constructing an inscribed and circumscribing triangle

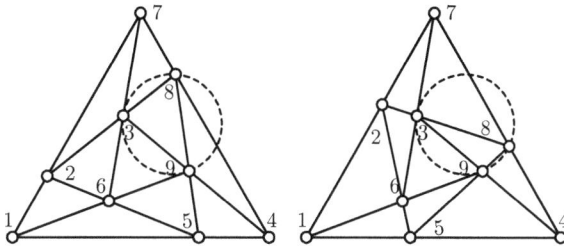

which is contained in the three lines of the Pappus configuration which together are incident with all but two vertices 3 and 5, where 3 and 5 belong to no line of the Pappus configuration. Of the three configurations in Fig. 1.10, only the first has this property at point 1 and in fact at every point. Therefore, only the first configuration remains a candidate for being isomorphic with the Pappus configuration; see Sect. 5.2. In any case, the first configuration is not isomorphic with either of the second or the third.

Whether or not the second and third configurations of Fig. 1.10 are isomorphic is not immediately obvious. Indeed, once we have satisfied ourselves that neither is an example of the Pappus configuration, it is no longer clear that either is a configuration at all. The historical literature is littered with examples in which a suggestive picture, rather than being worth a thousand words, is in fact less than worthless. So we have two problems of realization, and each solution indicates a geometric dependency. In this case, we will use the symmetry of the drawing. Starting with points 1, 4, and 7, the points of an equilateral triangle, the center of rotational symmetry is now fixed, so choosing the slope for the line 19, its rotated images prescribe the points 3, 6, and 9; see Fig. 1.11. The structure can be completed if we can find an equilateral triangle 258 which circumscribes 369 and inscribes 147. To this end, construct a circle with chord 39 such that the chord intercepts an arc of 120°. Clearly, there exist slopes such that this circle intersects the line 47 in a point 8, after which symmetry produces 2 and 5, as well as establishing the collinearity of {1, 2, 7}, {7, 8, 4}, and {4, 5, 1}. In fact, for many choices of slopes, there are two different symmetric drawings. It is left as an exercise for the reader to formulate a similar construction for the third configuration; see Exercise 1.16.

To address now the question as to whether the second and third configurations of Fig. 1.10 are isomorphic, the drawings are clearly different, but there are many bijections to check to see whether or not the third is in fact just the second drawn inside out. In fact, for this question, the realizations of the two configurations are irrelevant, and it is only necessary to consult their respective configuration tables, and the most efficient way to do this is to encode the information in the configuration table in the *incidence graphs* or *Levi graphs*, whose vertices are the points and lines and whose edges record point-line incidences; see Fig. 1.12. So our question reduces to whether the Levi graphs of the second and third configurations are isomorphic, and even before settling that question, we can conclude that graphs and their theory will be of considerable utility in the study of configurations. In fact, this essential interplay between configurations and graphs is the central inspiration for this book.

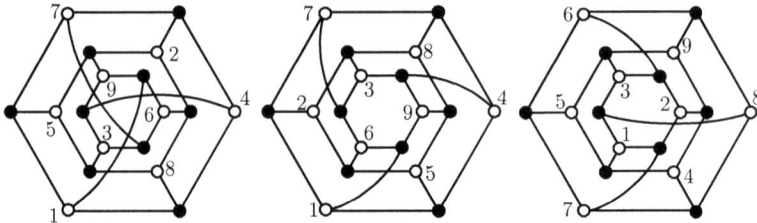

Fig. 1.12 Incidence graphs or Levi graphs of the nine-point and nine-line configurations in Fig. 1.10

Other combinatorial objects will also play a role. Notice that each row of the configuration tables of Figs. 1.7 and 1.9 has the property that each vertex occurs exactly once; in other words, each row is a permutation. Whenever this happens, since the order of the columns in the configuration table is not significant, we can always arrange the table so that the top row is the identity permutation. With this in mind, the configuration table of the middle configuration of Fig. 1.10 may be presented in a particularly nice form:

$$
\begin{array}{c}
1\ 2\ 3\ 4\ 5\ 6\ 7\ 8\ 9 \\
2\ 3\ 4\ 5\ 6\ 7\ 8\ 9\ 1 \\
7\ 8\ 9\ 1\ 2\ 3\ 4\ 5\ 6
\end{array}.
$$

This table makes it clear that in this configuration, just as in the Pappus configuration, any two points are equivalent since cycling through the nine vertices in order preserves incidence. Even more striking, the second permutation is the 9-cycle (123456789), which presents the configuration in the guise of the 9-gon [123456789], although twisting around on itself with each of the nine points also playing the role of an inscribed point, indicated by its position in the third row. But the third row permutation is also the permutation (123456789), so these inscribed points are actually an inscribed 9-gon or, if you prefer, the first two rows describe a 9-gon circumscribing the 9-gon indicated by the third row. This property has been expressed as a 9-gon which *inscribes and circumscribes itself*; see [50]. An n-gon that inscribes and circumscribes itself is a natural geometric realization of a configuration which necessarily has the same number of points and lines, each having three incidences.

1.1.5 The Tetrahedron

As a final motivating example, consider a three-dimensional tetrahedron depicted in Fig. 1.13. Its geometric features include four faces along with its four vertices and six edges. We may consider the pairwise incidences between any pair of objects

1.1 Hexagrammum Mysticum

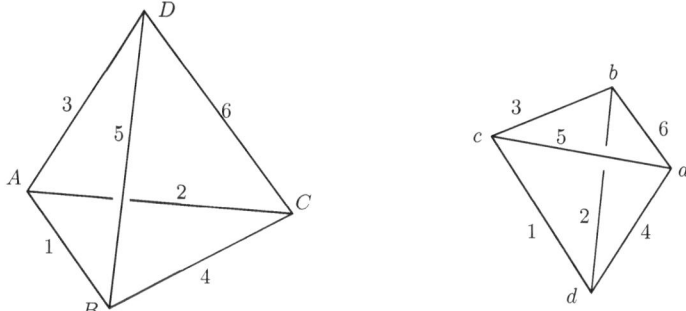

Fig. 1.13 A tetrahedron and its dual polyhedron

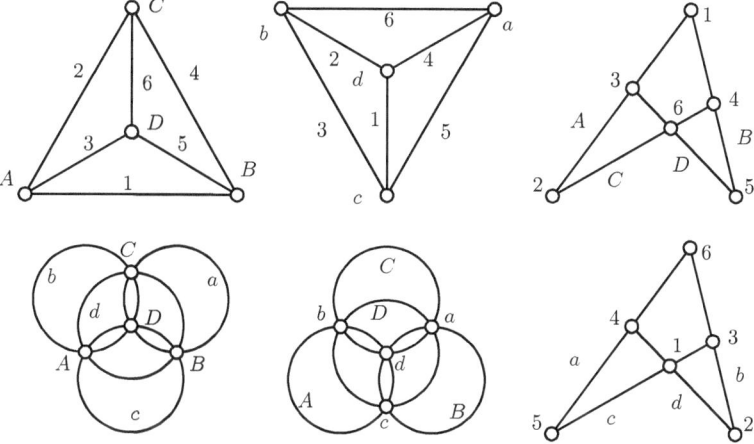

Fig. 1.14 The tetrahedron gives rise to six incidence structures

from different classes. These incidences give six candidates for consideration as the incidences of a configuration since each class of objects may be regarded as the "points" and either of the nonchosen classes regarded as the "lines." The six possibilities are illustrated in Fig. 1.14 with the simplest example having vertices regarded as points and edges regarded as lines, giving an example of a configuration, simple in the fact that the lines only have two endpoints. Such a configuration is simply the drawing of a graph, and we have a drawing of the graph of the tetrahedron, followed by the drawing of the graph of the dual tetrahedron.

Letting the edges of the tetrahedron play the role of the points of a configuration and the vertices the lines, there is a realization as a *complete quadrangle*, a historically and computationally important configuration which is also simple in the sense that the points are the intersections of only two lines. The first and third diagrams, which exhibit the same collection of incidences but with the objects, vertices, and edges, playing different roles as lines and points, are said to be *dual* to one another. Duality will play an important role in the study of configurations. The six configurations of Fig. 1.14 are naturally arranged in three dual pairs. The dual of the complete quadrangle is called the *complete quadrilateral*.

If we regard the vertices as the points and the faces as the lines, there will be no realization in which all the points are not collinear or all the vertices coincide since two distinct points determine at most one line, and two distinct lines determine at most one intersection point. Thus, a planar realization can only be accomplished with other geometric ingredients such as points and circles, as in the Miquel configuration and the Fano configuration. The incidences of these unrealizable configurations have the same regularity as their realizable cousins and so may be studied in the same manner, and we include them in our subject as *incidence structures*. In fact, in this book, as with this example, we regard the combinatorics of the incidence structures as the main object of study from which vantage we consider the questions posed by realizability, taking the geometric configurations as the central source of inspiration and examples.

Notice that the four realizable examples occur in two dual pairs, and we shall see that there is a strong connection between duality and realizability.

1.1.6 Overview

The focus of this book is to draw the connection between configurations, geometric configurations such as the Pascal configuration or more general constructs, and the combinatorial properties of their incidences, particularly as encoded using graphs. The emphasis is thus shifted from nineteenth-century geometry to discrete mathematics and combinatorics. The study will involve not simply graphs but also curves, surfaces, groups, and real and abstract geometry. But graph theory will be the unifying structure and is the topic of the next chapter.

1.2 Exercises

Exercise 1.1. Consider the five points $A_k = (\cos(2k\pi/5), \sin(2k\pi/5))$ with $k \in \{1, 2, \ldots, 5\}$. A_k are the vertices of a regular pentagon on the unit circle. Set A_6 to be $(\cos(\theta), \sin(\theta))$. Describe the Pascal lines for the hexagon $A_1, A_2, A_3, A_4, A_5, A_6$ and how they change as θ varies from 0 to 2π.

Exercise 1.2. The given configuration table for the Fano configuration has the property that each number, that is, each point, occurs once in every row and not more than once in each column. Does that property alone ensure that the incidences form a configuration?

Exercise 1.3. Since the Fano configuration has no realization in the plane as seven straight lines, Fig. 1.7 has rather asymmetrically represented one line as a curve. Can you find a realization with all seven lines represented as circles?

1.2 Exercises 13

Exercise 1.4. Determine the configuration table of the configuration, dual to the Miquel configuration, in which the roles of points and circles are reversed. This means that the entries of the table correspond to the circles while the columns correspond to the points.

Exercise 1.5. What is the minimum and what is the maximum number of lines passing through a pair of distinct points of the Pappus configuration?

Exercise 1.6. Jennie, the brightest little girl in school, is showing a clever puzzle to her classmate Joe. After drawing six small circles on the fence, she said, "Now you can see only two rows with three circles in a line. I want you to mark out one circle and draw it somewhere else on the fence so that there will be four rows of three-in-a-line." *This nineteenth-century puzzle is by Sam Loyd from his 'Cyclopedia of Puzzles'.*

Exercise* 1.7. Plant 9 trees in 10 rows in such a way that there are 3 trees in each of the 10 rows.

Exercise* 1.8. Prove Pascal's theorem for the case when the conic is a circle.

Exercise* 1.9. Derive Pappus' theorem from Pascal's theorem by considering two lines as a degenerate conic.

Exercise 1.10. Write down the configuration tables for the six configurations determined by the tetrahedron.

Exercise 1.11. Determine the configuration table for the point-face configuration of a cube. Show that it is equivalent to the configuration table of the Miquel configuration.

Exercise 1.12. Construct a realization of the Miquel configuration in which the six "lines" are represented by two lines and four circles. Can you construct a realization with fewer circles?

Exercise 1.13. At a Florida robotics competition, there are eight teams. Schedule a sequence of matches such that each match involves three teams and such that no two teams are in more than one match. Show that the solution is essentially unique. Write down the configuration table.

Exercise 1.14. Show that for each pair of points in the Pappus configuration there is an incidence preserving permutation of the points mapping one to the other.
Can you do the same for any two pairs of points?

Exercise 1.15. Prove that the first configuration drawn in Fig. 1.10 is isomorphic to the Pappus configuration.

Exercise 1.16. Show by construction that the third drawing in Fig. 1.10 gives a configuration.
What is the degree of freedom as a threefold symmetric figure?
Can you find a version without a threefold rotation?

Exercise 1.17. Determine whether or not the second and third configurations illustrated in Fig. 1.10 are isomorphic.

Exercise 1.18. Find all combinatorial configurations on seven points which have a description as a heptagon which simultaneously inscribes and circumscribes itself.
Are any of them realizable as point-line configurations in the plane?

Exercise 1.19. Find all combinatorial configurations on eight points which have a description as an octagon which simultaneously inscribes and circumscribes itself.
Are any of them realizable as point-line configurations in the plane?

Exercise 1.20. Describe all configurations of nine points and nine lines which have a description as a polygon which simultaneously inscribe and circumscribe themselves.
Are any of these isomorphic to the Pappus configuration?
Are any of these isomorphic to the third configuration of Fig. 1.10?

Exercise 1.21. It seems that Pascal proved his theorem in June 1639, and according to Coxeter, G. W. Leibniz (1646–1716) admired Pascal's proof.
Research the roots of this folklore tale by first examining the statements in Pascal's pamphlet and how these statements agree with the modern formulation of what we now call the Pascal theorem. Then, trace as many citations of the pamphlet as you can find to see how this early work of Pascal was used.

Exercise 1.22. Determine the total number of incidences among vertices, edges, and faces of the tetrahedron.

Chapter 2
Graphs

2.1 Basic Definitions

The most common and useful structure for encoding discrete information is the *graph*, an abstract structure which is designed to record relationships between objects. For simple undirected graphs, the following definition suffices. A *graph* G is a pair $G = (V, \sim)$, where $V = V(G)$ is the vertex set and $\sim$ is an irreflexive, symmetric relation on $V(G)$, called *adjacency*. We let $E(G)$ denote the edge set, that is, the set of unordered pairs of adjacent vertices of G. If it is more convenient, we will indicate a graph $G = (V, \sim)$ by specifying its vertex and edge set, $G = (V, E)$.

For example, Fig. 2.1 illustrates the graph whose vertex set is the set of twelve numbers $V = \{2, 3, \ldots, 13\}$ in which two numbers are said to be adjacent if they have a common prime divisor. The utility of graphs is their ability to facilitate the discovery or description of properties of the defining relation, and to this end, a large body of common vocabulary has been developed. For instance, a subset I of the vertex set V of a graph is called *independent* if no edge of G has two endpoints in I. Clearly, independent sets of the divisor graph will have significance in number theory.

If $G = (V, \sim)$ with edge set E, the edge corresponding to adjacent vertices a and b is denoted variously by $(a, b) = \{a, b\} = ab = ba$. Vertices a and b are called the *endpoints* (or *endvertices*) of the edge ab. The number of vertices adjacent to a given vertex a is called the *valence* of a and denoted by $\mathrm{val}(a)$. Note that many graph theorists use *degree* of a vertex a instead of $\mathrm{val}(a)$.

We say that an edge is *incident* to its endpoints. Since every edge is incident to two endpoints, summing over the vertex valences yields twice the number of edges. This simple but useful observation is called the *handshaking lemma*:

$$\sum_{a \in V} \mathrm{val}(a) = 2|E|.$$

If the vertices of graph G are ordered, then G can be conveniently encoded by its *adjacency matrix*, a $|V| \times |V|$ matrix of zeros and ones in which there is a 1 in position (i, j) if and only if the ith and jth vertices, in the given ordering, are

Fig. 2.1 The divisor graph on $\{2, 3, \ldots, 13\}$

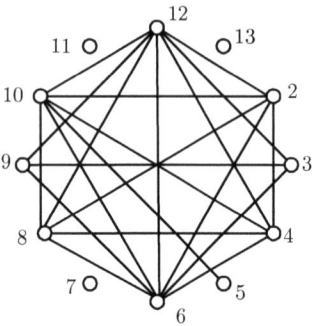

adjacent. Also of interest is the *incidence matrix*, which requires an ordering of both the vertices and the edges. It is an $|E| \times |V|$ matrix in which there is a 1 in position (i, j) if and only if the ith edge is incident to the jth vertex.

It is natural to represent a graph with a diagram by using a "dot" to signify each vertex and a (possibly curved) line segment connecting the two dots a and b if $a \sim b$. The larger the number of vertices and more complex the adjacency relation, the less will be our ability to visually gather useful information from such a figure. Even with only 12 vertices, the drawing of the divisor graph in Fig. 2.1 does not reveal to us very much of the structure. A graph is a purely abstract concept, and its representation as a diagram leaves a great deal of freedom, so perhaps a more judicious placement of the vertices would make apparent to us some property the present drawing obscures.

2.2 Examples of Graphs

The simplest examples of graphs need no figures. They are the *discrete graphs*, graphs in which no pair of vertices is adjacent. The discrete graph on n vertices is no more or less complicated than a set of n vertices. At the other extreme, we have the *complete graph*, K_n, on n vertices, in which every pair of vertices is adjacent:

$$K_n = (\{v_1, v_2, \ldots, v_n\}, \{(v_i, v_j) \mid i < j; i, j = 1, \ldots, n\}).$$

Although the complete graphs pictured in Fig. 2.2 have a certain charm, it is mainly derived from the symmetrical placement of the vertices and the pleasant pattern of the completely irrelevant edge crossings. Adjacency in the complete graph also gives no information beyond the set of vertices.

Nevertheless, the complete and discrete graphs become important as *subgraphs*. A subgraph of a graph $G = (V, E)$ is a graph $G' = (V', E')$ such that $V' \subseteq V$, $E' \subseteq E$. Every graph will have many subgraphs which are discrete, but finding subgraphs which are complete graphs is, quite literally, a hard problem; see, for example, [29].

2.2 Examples of Graphs

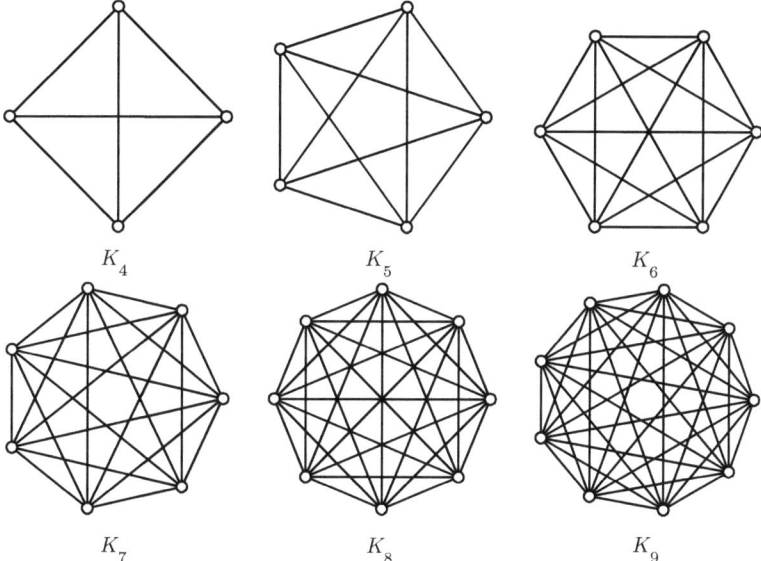

Fig. 2.2 Complete graphs $K_n, n = 4, 5, \ldots, 9$

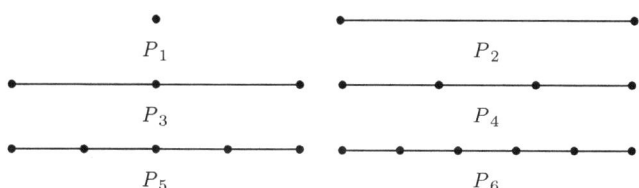

Fig. 2.3 Paths $P_n, n = 1, 2, \ldots, 6$

A subgraph G' of G is said to be *induced* if every pair of vertices in V' which is adjacent in G is also adjacent in G'. An independent set in a graph is an induced discrete subgraph, and finding independent sets in a graph is just as easy, and just as hard, as finding complete subgraphs.

2.2.1 Paths

The *path* on n vertices, $P_n = (V, E)$, is defined by

$$V = \{v_1, v_2, \ldots, v_n\} \qquad E = \{v_i v_{i+1} \mid i = 1, \ldots, n-1\}.$$

As with an edge, the vertices v_1 and v_n in the above example are called *endvertices* (Fig. 2.3) of P_n. The vertices which are not endvertices are called *internal*.

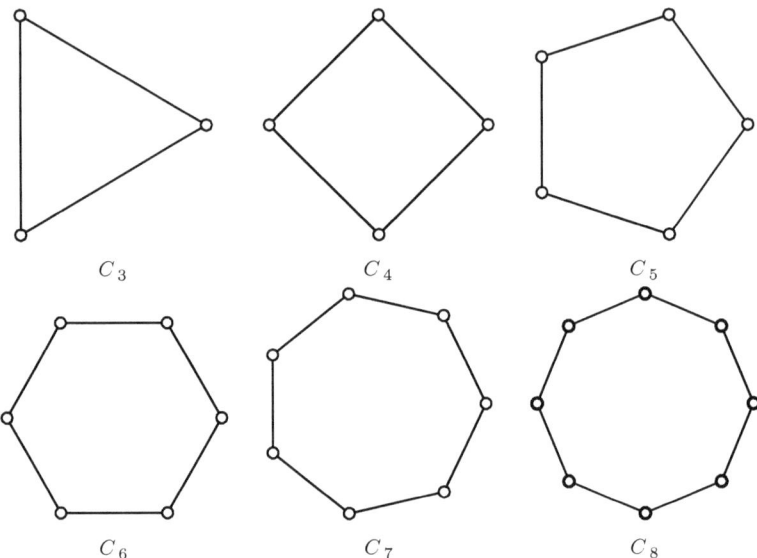

Fig. 2.4 Small cycles C_n, $n = 3, 4, \ldots, 8$

While not trivial, the adjacency relation is still very simple, just a simple ordering of the vertices, so as with the complete and discrete graphs, the main interest of P_n lies in its role as a subgraph. The most important concept in graph theory is defined via subgraphs which are paths: If each pair of vertices in a graph G is the endvertices of some path, then we say the graph G is *connected*. A graph which is not connected is said to be *disconnected*. A disconnected graph may be partitioned into maximal connected subgraphs, called *connected components*. The graph in Fig. 2.1 has three connected components which consist of a single vertex. Does it have any others? See Exercise 2.3.

Paths are also used to define the *distance* between the two vertices v and u as the fewest number of edges in any path in G connecting u and v.

2.2.2 Cycles

The cycle of length n, $C_n = (V, E)$, is defined by

$$V = \{v_1, v_2, \ldots, v_n\}, \quad E = \{v_1v_2, v_2v_3, \ldots, v_{n-1}v_n, v_nv_1\}.$$

See Fig. 2.4, and again, the adjacency relation is very simple, and these graphs derive their main interest as subgraphs. A graph with no subgraph which is a cycle is said to be an *acyclic graph*. In an acyclic graph, two vertices can be endvertices of at most one path. A connected acyclic graph is called a *tree*, and an acyclic graph is also called a *forest*, since it is the union of its connected components, each of which is a tree.

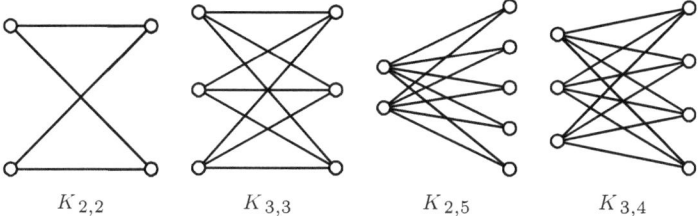

Fig. 2.5 Small complete bipartite graphs $K_{m,n}$

A subgraph $G' \leq G$ is said to be a *spanning* subgraph of G if $V(G') = V(G)$. It is not hard to show that a graph is connected if and only if it contains a spanning tree. Obviously, a graph with $|V|$ vertices and $|E|$ edges contains $2^{|V|}$- induced subgraphs and $2^{|E|}$ spanning subgraphs. These numbers indicate that it is sometimes difficult to find subgraphs of a particular kind in a large graph. Nevertheless, finding a spanning tree in a graph is computationally simple. By contrast, finding a spanning cycle, called a *Hamilton cycle*, is intractable. A graph is called *Hamiltonian graph* if it contains a Hamilton cycle as subgraph. To determine if a graph G is Hamiltonian, we have to provide a one-to-one correspondence f between the vertices of C_n and those of G which preserves adjacency.

2.2.3 Complete Bipartite Graphs and Multipartite Graphs

The *complete bipartite graph*, $K_{m,n} = (V, E)$, is defined by

$$V = \{a_1, a_2, \ldots, a_m, b_1, b_2, \ldots b_n\} \quad E = \{a_i b_j \mid i = 1, \ldots, m; j = 1, \ldots, n\}.$$

See Fig. 2.5. Alternatively, we can define a complete bipartite graph as a graph whose vertex set is partitioned into two sets (the as and bs in the example above), and two vertices are adjacent if they are in different sets of the bipartition. This naturally generalizes to any partition of the set V. The *complete multipartite graph*, $K_{n_1,n_2,\ldots,n_p} = (V, E)$, is defined by (Fig. 2.6)

$$V = \{a_{ij} \mid 1 \leq i \leq p; 1 \leq j \leq n_i\}$$
$$E = \{a_{ij} a_{kl} \mid i \neq k\}.$$

A complete multipartite graph has a vertex set partitioned into p sets, and two vertices are adjacent if they belong to different sets of the partition. For the graph $K_{n_1,n_2,\ldots,n_p}$ with $n_1 = n_2 = \cdots = n_p = n$, the more economical notation $K_{p(n)}$ is occasionally used.

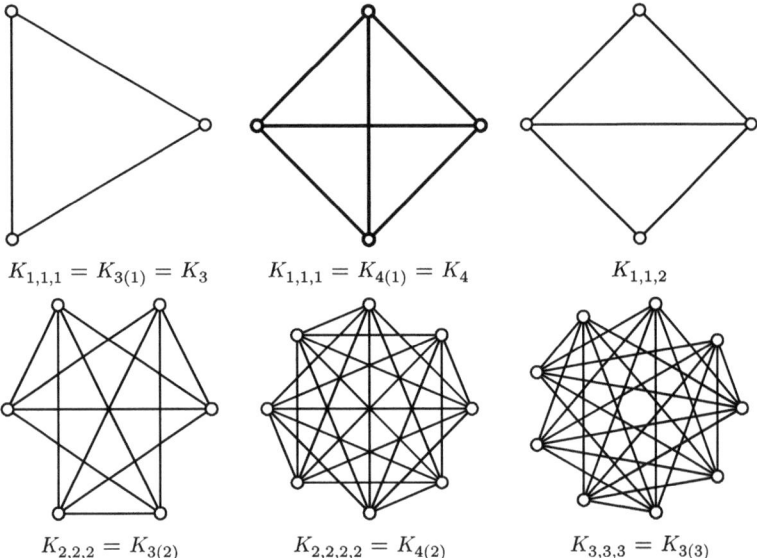

Fig. 2.6 Small complete multipartite graphs

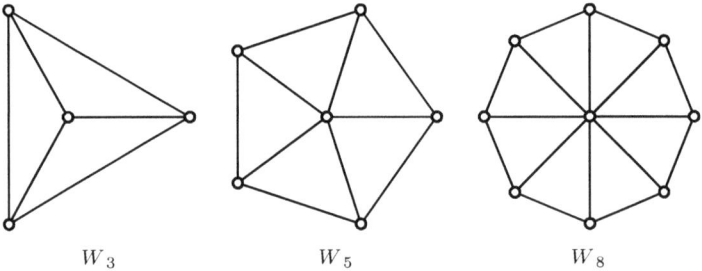

Fig. 2.7 Wheel graphs W_n, sometimes called *pyramid graphs*

2.2.4 Wheel Graphs

If you append to the vertex set of the cycle graph C_n a new vertex c and specify that the new vertex is adjacent to all the vertices u_i of the cycle, the graph created, $W_n = (V, E)$

$$V = \{c, u_1, \ldots, u_n\} \quad E = \{cu_i, u_i u_{i+1} \mid i = 1, \ldots, n\},$$

indices modulo n, is called the *wheel graph*; see Fig. 2.7. The adjacency relation on the wheel graph indicates that the bijection on the vertices defined by $f(c) = c$ and $f(u_i) = u_{i+1}$ takes adjacent pairs to adjacent pairs or, equivalently, defines a bijection on the edge set E. This is an example of a graph *automorphism*.

2.2 Examples of Graphs

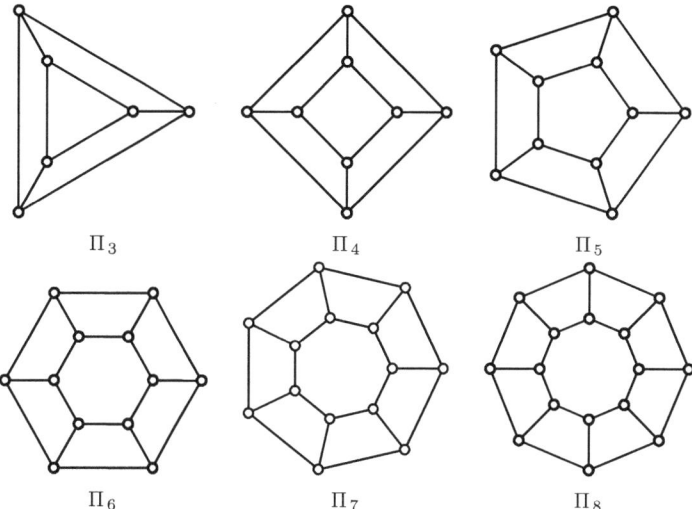

Fig. 2.8 Prisms Π_n, $n = 3, 4, ..., 8$

The symmetric drawing of the wheel graph W_n suggests a three-dimensional object, specifically a three-dimensional pyramid with an n-sided base, viewed from above. In the same way, any polyhedron made up of vertices, edges, and faces gives rise to a graph by simply retaining the adjacency information between the vertices and edges and forgetting the faces. This graph is called the *skeleton* or *1-skeleton* of the polyhedron. So the 1-skeleton of an n-sided pyramid is the wheel graph W_n. For this reason, the wheel graph is known also as the *pyramid graph*.

This brings us to several more interesting examples of graphs.

2.2.5 Prism Graphs

A regular prism is a polyhedron with two parallel opposite faces, called bases, that are congruent regular polygons. All the other faces, called lateral faces, are squares formed by the straight lines through corresponding vertices of the bases. $\Pi_n = (V, E)$, the n-sided *prism graph*, is the skeleton of a prism whose base is an n-gon:

$$V = \{u_1, \ldots, u_n, v_1, \ldots, v_n\} \quad E = \{u_i u_{i+1}, v_i v_{i+1}, u_i v_i \mid i = 1, \ldots, n\}.$$

with indices modulo n; see Fig. 2.8.

2.2.6 Antiprism Graphs

A regular antiprism of order n is a polyhedron with two regular n-gons as bases and $2n$ equilateral triangles as side faces. Its 1-skeleton is the graph $A_n = (V, E)$, the n–sided *antiprism graph*,

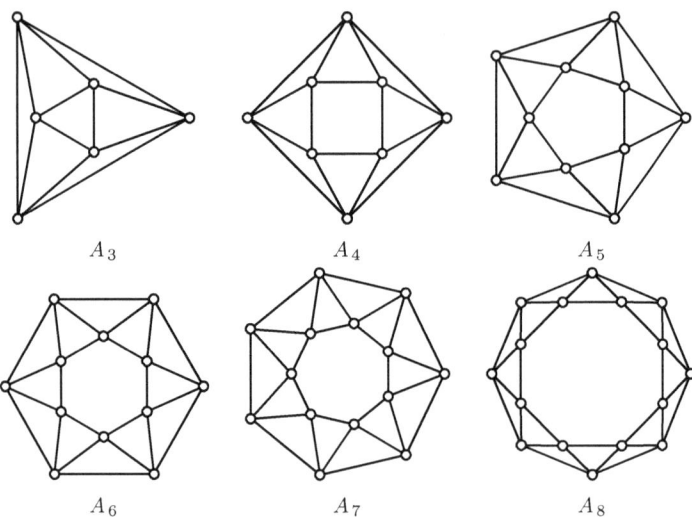

Fig. 2.9 Antiprism graphs $A_n, n = 3, 4, \ldots, 8$

$$V = \{u_1, \ldots, u_n, v_1, \ldots, v_n\} \quad E = \{u_i u_{i+1}, v_i v_{i+1}, u_i v_{i+1}, u_i v_i \mid i = 1, \ldots, n\},$$

with indices modulo n; see Fig. 2.9. Regular three-dimensional prisms and antiprisms may be constructed as follows. Take two identical regular n-gons in the plane. Translate one vertically out of the plane and, for the antiprism, rotate the other by π/n in the plane. The vertical translation is continued until the nearest neighbors between the two polygons have distance equal to the side length of the n-gons. The three-dimensional solid determined by the points of these two n-gons has symmetries generated by rotations and reflections. It may be expected that the graph, whose vertices are not held rigidly in place by the solid and in fact are not actually locations at all, will have other purely combinatorial automorphisms; however, we will see later that the automorphisms of the prism and antiprism graphs are exactly those which arise from the symmetries of the associated highly symmetric solid.

2.2.7 Platonic and Archimedean Graphs

The regular pyramids, prisms, and antiprisms each in general have two classes of faces, the bases on the one hand and the side faces on the other. For small cases, however, the side faces become indistinguishable from the bases, and new symmetries occur. The regular triangular pyramid, the regular quadrilateral prism, and the regular triangular antiprism become the regular tetrahedron, cube, and octahedron, respectively. These solids are completely regular in the sense that there is a symmetry between each pair of vertices, each pair of edges, and each

2.2 Examples of Graphs

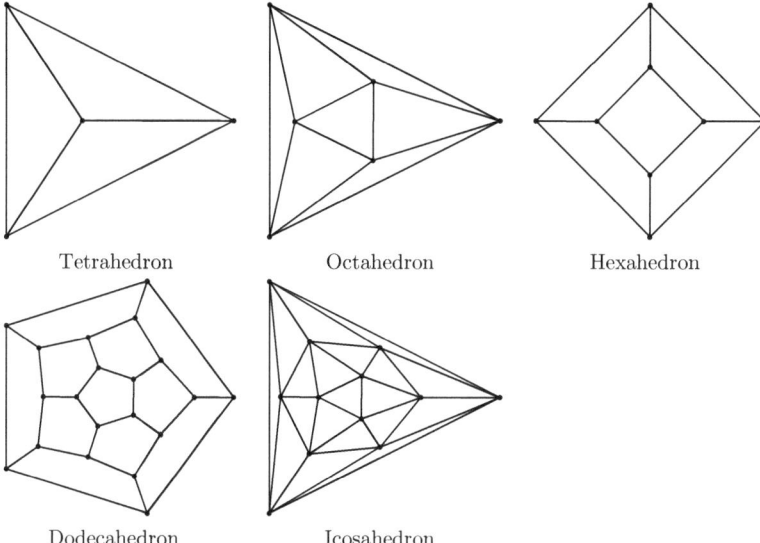

Fig. 2.10 Five platonic graphs: *tetrahedron, octahedron, hexahedron (cube), dodecahedron, and icosahedron*

pair of faces. There are two other solids with this degree of symmetry, namely, the dodecahedron and the icosahedron, and these five *Platonic solids* give us the five *Platonic graphs*; see Fig. 2.10. If we require symmetry between each pair of vertices but only require symmetry between pairs of regular polygonal faces of the same type, then the resulting solids are the Archimedean solids and give rise to the *Archimedean graphs*; see Fig. 2.11.

Platonic and Archimedean solids have a long and rich history and are well studied [22].

2.2.8 Polyhedral Graphs

Directly using the 1-skeleton is only one way of generating a graph from a polyhedron. More generally, we can consider the collection of all the vertices, edges, and faces in the solid and consider the incidences between them. The following example uses the cube graph to show how we can associate a graph to any convex polyhedron. The cube has 8 vertices,

$$1, 2, 3, 4, 5, 6, 7, 8,$$

12 edges,

$$a, b, c, d, e, f, g, h, i, j, k, l,$$

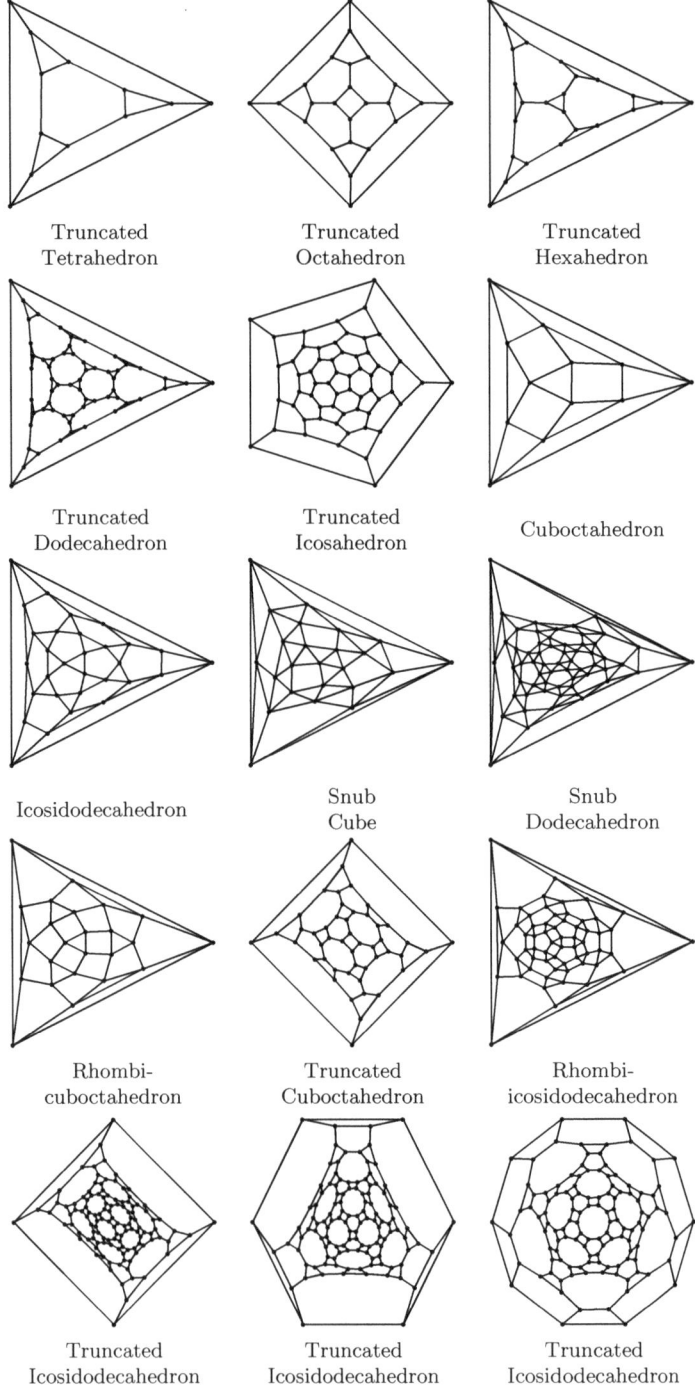

Fig. 2.11 Thirteen Archimedean graphs (The last one is shown in three different forms)

2.2 Examples of Graphs

Fig. 2.12 The incidence graph of the cube as drawn by VEGA

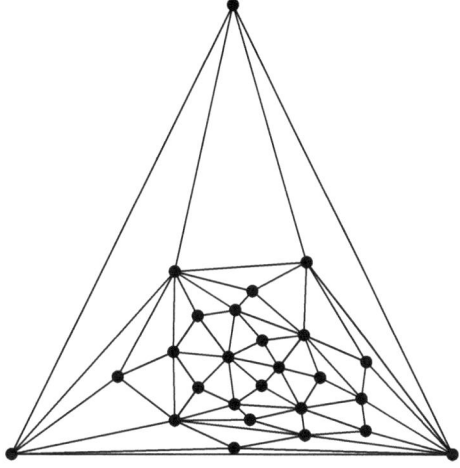

and 6 faces,

$$A, B, C, D, E, F.$$

Define a graph on $8+12+6 = 26$ vertices with the property that two elements x and y are adjacent in the graph if and only if they are incident on the cube. So each edge is incident to two faces as well as its two endpoints, and each face is incident to the vertices and edges on its boundary. In Fig. 2.12, the 12 vertices corresponding to the edges of the cube are easily distinguished from the 8 vertices corresponding to the vertices of the cube or the six corresponding to the faces of the cube by their valence: Edge vertices have valence 4, vertex vertices have valence 6, and face vertices have valence 8.

2.2.9 Generalized Petersen Graphs

All classes of graphs considered so far have arisen naturally from geometry or illustrate relations so regularly that they really do not require the full generality of graph theory. The next example, possibly the most celebrated graph, is a true native to the subject. Clearly related to the simple prism graphs, it regularly appears in statements of theorems in graph theory as an exceptional case. It is the *Petersen graph*, GP(5, 2); see Fig. 2.13 which gives the classic drawing of the Petersen graph, the drawing which inspired the following generalization.

For a positive integer $n \geq 3$ and $0 < r < n/2$, *the generalized Petersen graph* GP(n, r) has a vertex set $\{u_1, \ldots, u_n, v_1, \ldots, v_n\}$ and edges of the form $u_i v_i$, $u_i u_{i+1}$, $v_i v_{i+r}$ for $i \in \{1, \ldots, n\}$ with indices modulo n, ($i \in \mathbb{Z}_n$). In the diagrams, the vertices u_i form a cycle on the outside connected by the edges $u_i v_i$ to the vertices v_i arranged compatibly on the inside, where the n edges $v_i v_{i+r}$ form a pentagram in the case of the classic Petersen graph and form one or several cycles in the general case depending on whether r and n have a common divisor.

Fig. 2.13 The Petersen graph GP(5, 2)

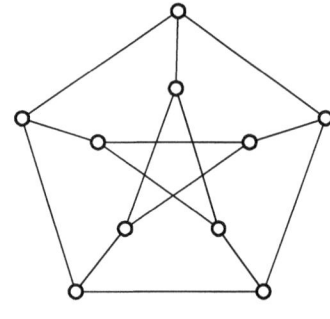

Fig. 2.14 The Dürer graph GP(6, 2) obtained as the skeleton of a truncated cube

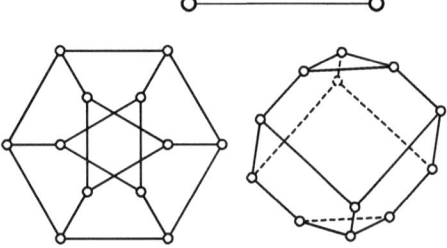

All vertices in GP(n, r) are of degree 3. The values of r are restricted to avoid, in the case $r \equiv 0 \pmod{n}$, a vertex adjacent to itself or, for even n, the case $r \equiv n/2 \pmod{n}$ which would have the vertices v_i having valence of only 2. The other values of r, $n/2 < r < n$ produce duplicates since $v_i v_{i+r} = v_{i+r} v_i = v_{i+r} v_{i+r-r} = v_{i'} v_{i'-r}$; thus, we have that GP($n, r$) = GP($n, n - r$).

If $r = 1$, then GP($n, 1$) is the same as the prism graph $\Pi(n)$, so $n = 5$ is the first interesting case, and GP(5, 2) is the only nonprism example. For $n = 6$, we also have a single nonprism example, the so-called *Dürer graph*; see Fig. 2.14 which gets its name from a solid, at first glance a truncated cube resting on one of the two triangular faces which are produced when two antipodal corners of the cube have been cut away. This mysterious solid appears in the famous medieval engraving "Melancholia I" by the Nürnberg artist Albrecht Dürer.

Up until now, each of our graph examples could be distinguished from one another by the numbers of vertices and edges. For the generalized Petersen graphs, this is no longer the case. For fixed n, each graph GP(n, r) has $2n$ vertices and $3n$ edges, and each vertex is of valence 3. So the question arises as to whether they have distinct graph structures. We say two graphs are *isomorphic* if there is a bijection between the vertex sets which preserves the property of adjacency.

For each of $n = 7$ and $n = 8$, there are two generalized Petersen graphs on our list; see Fig. 2.15. For GP(8, 2), the interior figure is two 4-cycles, while for GP(8, 3), it is a single 8-cycle. Moreover, since we can easily check that GP(8, 3) has no 4-cycles, GP(8, 2) cannot be isomorphic to GP(8, 3). For $n = 7$, both graphs have several 7-cycles, and the situation is less obvious. It is not hard to show that there is no isomorphism which sends the vertices of the outer 7-cycle of GP(7, 2) to the outer cycle of GP(7, 3), so let us consider an isomorphism which sends the vertices of the outer 7-cycle of GP(7, 2) to the inner 7-cycle of GP(7, 3) and vice versa.

2.2 Examples of Graphs

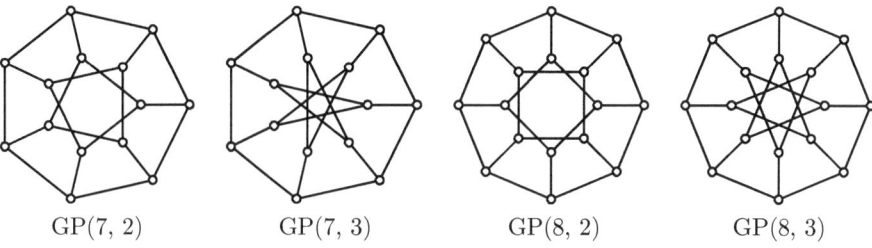

Fig. 2.15 Generalized Petersen graphs GP(n, r) for n = 7 and n = 8

We will consider the general case. Suppose that there is an isomorphism f between GP(n, r) and GP(n, s):

$$GP(n,r) = (\{a_i, b_i \mid i \in \mathbb{Z}_n\}, \{a_i a_{i+1}, a_i b_i, b_i b_{i+r} \mid i \in \mathbb{Z}_n\})$$

$$GP(n,s) = (\{c_i, d_i \mid i \in \mathbb{Z}_n\}, \{c_i c_{i+1}, c_i d_i, d_i d_{i+s} \mid i \in \mathbb{Z}_n\})$$

which interchanges the inner and outer n-cycles. Without loss of generality, we may say $f(a_0) = d_0$. So, using the ring edges, $f(a_1) = d_{\pm s}$, say $f(a_1) = d_s$, and then inductively $f(a_i) = d_{is}$ for all i, and using the spoke edges, $f(b_i) = c_{is}$ for all i, and in particular $f(b_{i+r}) = c_{(i+r)s}$ and, since f is an isomorphism, c_{is} and $c_{(i+r)s}$ must be adjacent, so $c_{(i+r)s} = c_{is \pm 1}$. Thus, adjacency is preserved if and only if $rs \equiv \pm 1 \pmod{n}$.

So, in particular, GP(7, 2) is isomorphic to GP(7, 3) by a ring-swapping isomorphism, and we write GP(7, 2) $\cong$ GP(7, 3).

For $n = 9$ and $n = 10$, there are three examples each of the generalized Petersen graphs; see Fig. 2.16, and since $2 \cdot 4 = 8 \equiv -1 \pmod 9$, we have GP(9, 2) $\cong$ GP(9, 4). In 2009, Staton and Steimle [93] proved the following result:

Theorem 2.1. *For $2 \leq r, s \leq n - 2$ with $\gcd(n, r) = \gcd(n, s) = 1$, the generalized Petersen graphs GP(n, r) and GP(n, s) are isomorphic if and only if either $r \equiv \pm s \pmod n$ or $r \cdot s \equiv \pm 1 \pmod n$.*

So, in other words, the only way two generalized Petersen graphs with connected inner rings can be isomorphic is either by an isomorphism which preserves the outer ring or one which exchanges the inner and outer rings. For more details, see [9, 75].

2.2.10 Cages

The next collection of examples may also be regarded as generalizations of the Petersen graph. The Petersen graph has many 5-cycles, not simply the outer and inner 5-cycles of the standard diagram. Moreover, it is easy to see that GP(5, 2) has no shorter cycles. There are smaller graphs than the Petersen graph with no 3-cycles

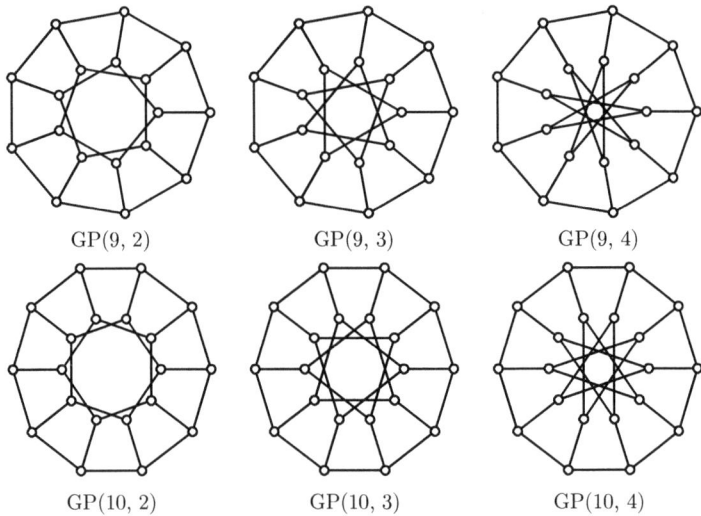

Fig. 2.16 More generalized Petersen graphs GP(n, r)

and no 4-cycles, but those also have few cycles of any kind since they are unions of simple cycles, trees, etc., and they avoid multiple cycles by having low valence. The Petersen graph, however, has many cycles since every vertex has valence 3. The cages generalize this property.

The length of the shortest cycle in a graph is called the *girth* of the graph. The girth of a graph without cycles, a tree or forest, is defined to be infinite, so the girth of a simple graph is at least 3.

The girth depends only on the isomorphism class of the graph, that is, it is a *graph invariant*. Computing the girth involves solving a nontrivial optimization problem. For the example graphs presented so far, the complete graphs have girth three, and the complete bipartite graphs have girth four (or infinity). It can be shown that the graphs arising as the 1-skeleta of the three-dimensional polyhedra illustrated all have girth equal to the number of edges in the smallest facial cycle, although this is not true in general. The girth of GP(5, 2), as remarked above, is 5.

A graph is said to be a *g-cage* if it is trivalent, has girth g, and there exists no trivalent graph with girth g having fewer vertices. Note that the definition does not preclude there being more than one cage of a particular size. The complete graph K_4 is the unique 3-cage, and the complete bipartite graph $K_{3,3}$ is the only 4-cage. The Petersen graph GP(5, 2) is the only 5-cage.

The previous examples were defined by a predetermined structure, so we could simply list examples. The g-cages are defined by the graph theoretic properties which they must satisfy, so it is neither clear which graphs belong on the list nor, given such a list, whether the graphs on the list do indeed belong on it. A complete structure theorem for g-cages is unknown, although many examples have been computed. At least we may establish a lower bound on the number of vertices a g-cage must have:

2.2 Examples of Graphs

Fig. 2.17 A level 4 binary tree with a root of valence 3

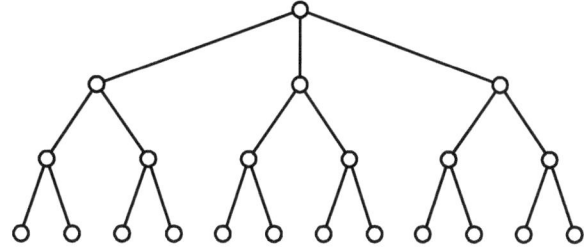

Theorem 2.2. *A $2k$ cage must have at least $2^{k+1} - 2$ vertices. A $2k + 1$ cage must have at least $3 \cdot 2^k - 2$ vertices.*

Proof. To establish a lower bound on the number of vertices of a g-cage, we start with a single vertex, list its three neighbors, each of those has two neighbors, etc. For the case $g = 2k + 1$ up to the k-level, each neighbor set gives rise to two new unrecorded vertices, creating a *binary tree* whose root vertex is of valence 3; see Fig. 2.17. Edges between vertices on the same level are only allowed on level k, so we have at least $1 + 3 + 3 \cdot 2 + 3 \cdot 4 + \cdots + 3 \cdot 2^{k-1}$ vertices, yielding the desired bound. For even girth, there is a similar construction; see Exercise 2.18. □

It turns out that the 6-cage is the Heawood graph; see Chap. 5, Fig. 5.32. The 7-cage has 24 vertices and is depicted in Chap. 3, Fig. 3.35. The 8-cage is known as the Cremona–Richmond graph; see Fig. 5.28; however, graph theorists prefer to call it the Tutte 8-cage. We will learn more about the Heawood graph and the Tutte 8-cage later and the relationship of the former to projective planes and the latter to the *hexagrammum mysticum* of Pascal.

It is interesting that the 9-cages were not found until quite recently. The search for 9-cages involved a lot of computer checking, and the result came as a surprise. There are 18 nonisomorphic 9-cages. All smaller cages have regular structure and are unique. However, the 9-cages do not show any apparent structure; they are computed in [13].

Balaban found one of the three 10-cages which is shown in Fig. 2.18. It is perhaps of interest to note that the 10-cages were known, see [76], before all the 9-cages were computed. The reason is simply that the gap between the easily proven lower bound and the actual size of the cage is larger for the 9-cage than for the 10-cage. By Theorem 2.2, there is no trivalent graph of girth 9 on fewer than 46 vertices and there is no such graph of girth 10 on fewer than 62 vertices. Since the 9-cage has 58 vertices [13] and the 10-cage has 70 vertices, the respective gaps are 12 for the 9-cage and only 8 for the 10-cage. For a survey on cages, see [108] where Wong states an interesting conjecture.

Conjecture 2.3. Every g cage with g even is bipartite.

Fig. 2.18 Balaban's 10-cage

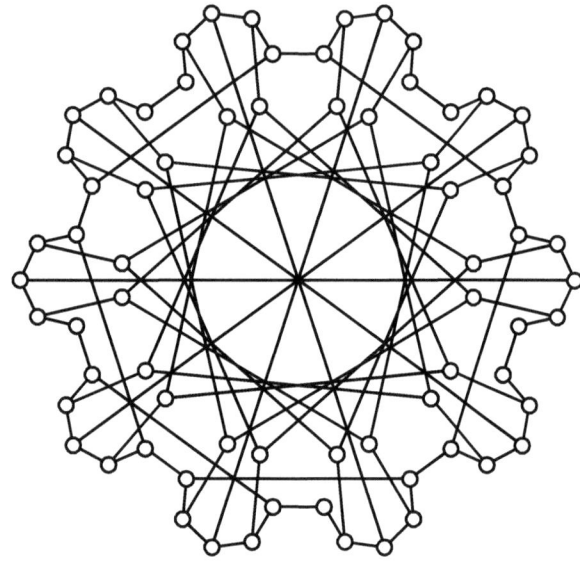

Fig. 2.19 A plane and a nonplane embedding of K_4

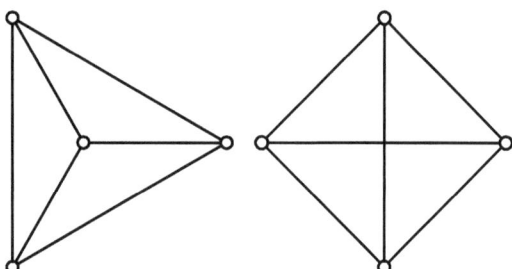

2.2.11 Planar Graphs

A graph that can be drawn in the plane so that edges intersect only at vertices is called planar. By a *drawing* of a graph G, we mean a representation of G in the plane such that vertices are represented by distinct points in the plane and edges by (curved) line segments connecting their endpoints. We will be more precise in Sect. 2.6.6. A drawing without edge crossings is called a *plane embedding* of the graph. Clearly, any tree can be drawn without edge crossings. Let G be a connected planar graph and consider a plane embedding of it. Such a drawing subdivides the plane into regions, one of which is unbounded. To avoid this special case, it is better to consider an embedding into the sphere, in which case we call the regions the *faces* of G. For example, in the plane embedding of K_4 in Fig. 2.19, we count four faces, namely, three triangles and the infinite outer face. Often, when we have a plane embedding of a graph and we count faces, we are implicitly regarding the plane as part of a large sphere and the exterior region then counts as one face.

Given a connected planar graph G on n vertices, together with a plane (sphere) embedding, choose a spanning tree T in G. T has $n-1$ edges. The plane drawing of T, induced by the plane drawing of G, has only one face. Inserting an additional edge of $E(G) - E(T)$ will divide this region into two. Inductively, we get one more face by inserting an additional edge. We started out with n vertices, $n-1$ edges, and 1 face. We add $e - (n-1)$ edges to get f faces, so $e - (n-1) = f - 1$ or $n - e + f = 2$. The alternating sum $n - e + f$ is called the *Euler characteristic*. It is a property of the surface in which the graph is embedded, and we say that the Euler characteristic of the sphere equals 2.

2.3 Regularity

2.3.1 Regular Graphs

Cycles, complete graphs, prisms, and antiprisms are all examples of *regular* graphs. A graph is *k-regular* or *k-valent* if all of its vertices have valence k. Cycles are 2-valent, and, conversely, 2-valent graphs are collections of disjoint cycles. Even graphs which are 1-valent graphs are not without interest. A 1-valent graph is a disconnected set of edges. One way of studying a large graph is to partition the edge set into spanning 1-valent graphs, called 1-*factors*. A spanning subgraph which is k-valent is called a k-factor. By contrast to the simple structure of 1-valent and 2-valent graphs, 3-valent graphs exist in much greater variety. Please note the important difference between the way the word "regular" is used in the contexts of geometry and graph theory. In geometry, a regular figure has a high degree of symmetry. The regular examples we have seen so far are irregular in the sense of being atypical. A regular graph typically has no symmetry at all; see Fig. 2.20 in which the outside 12-cycle has been rather haphazardly connected to interior cycles of various sizes.

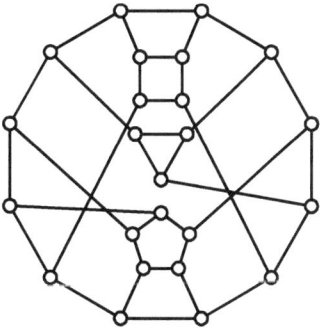

Fig. 2.20 A 3-regular but asymmetric graph

Fig. 2.21 LCF code:
$[-5, 2, 4, -2, -5, 4, -4, 5, 2, -4, -2, 5]$

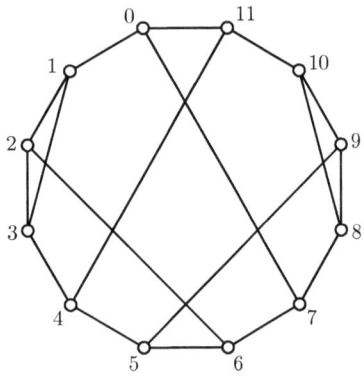

2.3.2 Cubic Graphs and LCF Notation

Three-valent graphs are also called *cubic*. Recall that a graph is Hamiltonian if it has a spanning cycle. If a cubic graph is Hamiltonian, we can draw it as a $|V|$-cycle with inserted chords, which leads to a convenient notation, the LCF notation, named for developers Lederberg, Coxeter, and Frucht. See [20, 33].

Given the Hamilton cycle, all we have to do to specify the graph is to list the lengths of chords measured in jumps when we traverse the vertices along the Hamilton cycle. Such a list is called the *LCF notation*. For instance, K_4 can be described by $[2, 2, 2, 2]$. $K_{3,3}$ is $[3, 3, 3, 3, 3, 3]$, and the cube Q_3 is $[3, 5, 3, 5, 3, 5, 3, 5]$ or $[3, -3, 3, -3, 3, -3, 3, -3]$ if we let a negative jump denote a chord measured in the opposite direction. We can also use exponent notation in order to shorten repeated subsequences. Here is an equivalent shorthand notation for the above examples: $\text{LCF}(K_4) = [2^4], \text{LCF}(K_{3,3}) = [3^6], \text{LCF}(Q_3) = [(3, -3)^4]$.

Example 2.4. The graph G with

$$\text{LCF}(G) = [-5, 2, 4, -2, -5, 4, -4, 5, 2, -4, -2, 5]$$

is depicted in Fig. 2.21.

2.3.3 Regularity and Bipartite Graphs

In Sect. 2.2.3, we examined the complete bipartite graphs. In general, a graph $G = (V, E)$ is *bipartite* if V can be partitioned into two nonempty sets V_1 and V_2 such that each edge has one of its endvertices in V_1, the other in V_2. Note that if G is connected and bipartite, the *bipartition* of the vertex set is uniquely determined, namely, two vertices are in the same set of the bipartition if and only if their distance in G is even. For disconnected graphs, bipartiteness clearly implies bipartiteness of

2.3 Regularity

Fig. 2.22 Bipartite graphs with *black* and *white* bipartition

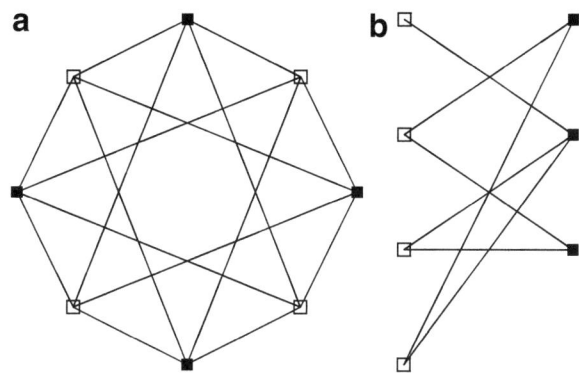

each of its connected components. There are several standard ways to indicate in a diagram that a graph is bipartite, for example, to arrange the vertex sets V_1 and V_2 on two different lines; see Fig. 2.22b. Except for very small graphs, a better method to indicate the bipartition is to color the vertices, say, black and white. It is easy to visually check if every edge has one black and one white endpoint; see, for example, Fig. 2.22a.

We have already seen that the incidences of a configuration can be encoded as a graph, see Fig. 1.12, and in this graph, the points and the lines form a vertex partition, so naturally bipartite graphs are of particular interest, and in particular regular bipartite graphs.

If every vertex of the bipartite graph $G = (V_1 \cup V_2, E)$ has valence k, then $k|V_1| = k|V_2|$, so unless $k = 0$, we have $|V_1| = |V_2|$, i.e. V must be partitioned into two sets of equal cardinality. In particular, $|V|$ must be even. For $k = 1$, we get a set of mutually nonincident edges. The following graph theoretic result was first formulated and proved in terms of configurations by Steinitz in his Ph.D. dissertation, [94].

Theorem 2.5. *Every k-valent bipartite graph G can be written as the edge disjoint union of k 1-factors.*

Proof. We use induction on k. For $k = 1$, there is nothing to show. We assume $k > 1$ and want to show that a k-valent bipartite graph G contains a 1-factor F. We then use the induction hypothesis on $G - F$ to obtain the desired decomposition of the edge set.

To construct a 1-factor, select mutually nonincident edges until every edge not yet selected is incident with at least one of the edges selected so far. Let us call this maximal set of mutually nonincident edges M. If M is not spanning, let v be a vertex not covered by M and consider the set A of all paths starting at v, then using an edge of M, an edge not in M, then an edge in M, etc. We can find a set of mutually nonincident edges which is of larger cardinality than M if there is at least one path in A that ends at another uncovered vertex u by removing from M all edges of M on this $u - v$-path and adding its edges not in M. If A does not contain such

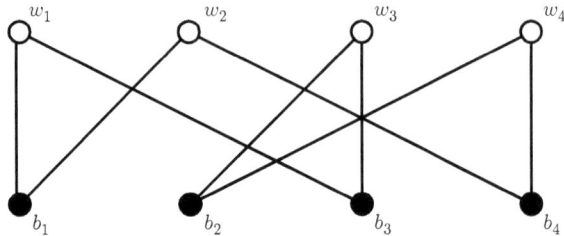

Fig. 2.23 The graph from Example 2.6

a path, then the subgraph of G induced by A, $I(A)$, has only one vertex, namely, u, which is not covered by M. Thus, the vertices of $I(A)$ in the same bipartition as u have valence k, and there is one vertex fewer in the other bipartition class, a contradiction. □

This theorem enables us to encode regular bipartite graphs on $2n$ vertices by k permutations of the set $\{1, 2, \ldots, n\}$. Given n black vertices $B = \{b_1, \ldots, b_n\}$ and n white vertices, $W = \{w_1, \ldots, w_n\}$ and a k-regular graph on these $2n$ vertices with a bipartition respecting the colors. Suppose the edges are decomposed into k 1-factors. For each 1-factor, let the black endpoints adjacent to $w_1, \ldots, w_n$ be $b_{i_1}, \ldots, b_{i_n}$ respectively, then set the permutation of $1, \ldots, n$ corresponding to the 1-factor to be $i_1, \ldots, i_n$.

Example 2.6. Suppose we have the simple 8-cycle of Fig. 2.23 with vertex set $\{w_1, w_2, w_3, w_4, b_1, b_2, b_3, b_4\}$ and edge set

$$\{(w_1, b_1), (b_1, w_2), (w_2, b_4), (b_4, w_4), (w_4, b_2), (b_2, w_3), (w_3, b_3), (b_3, w_1)\}.$$

There is a unique partition of the edges into two 1-factors:

$$\{(w_1, b_1), (w_2, b_4), (w_3, b_3), (w_4, b_2)\} \cup \{(b_3, w_1), (b_1, w_2), (b_2, w_3), (b_4, w_4)\}$$

and the two permutations

$$1\ 4\ 3\ 2$$
$$3\ 1\ 2\ 4$$

encode the graph. If we wish to augment the 8-cycle to a 3-valent bipartite graph, we need to add another 1-factor. Not any one factor will do, however, since many will correspond to sets of edges some of which we already have. The permutations we can allow must not have the same value in the ith position as either of the previous two; in other words, if we add the permutation as a row in the array above, there must be distinct elements in each column. So,

$$1\ 4\ 3\ 2$$
$$3\ 1\ 2\ 4$$
$$2\ 3\ 4\ 1$$
$$4\ 2\ 1\ 3$$

would be one way to complete the example to a complete bipartite graph.

In general, k permutations on n symbols give rise to a k-valent bipartite (simple) graph, provided that distinct permutations move a symbol to distinct symbols.

2.3.4 Semiregular Bipartite Graphs

If we want to relax the condition of regularity to allow unequal bipartition, we can at least require that all vertices of the same color have the same valence. In this case, we call the bipartite graph *semiregular*. We write $G = (V_1 \cup V_2; k_1, k_2)$ to indicate bipartition and vertex valences. Prescribing the size of the bipartition imposes restrictions on the values of k_1 and k_2. Certainly, if G is simple, the k_1 and k_2 are bounded by $|V_2|$ and $|V_1|$ respectively. Moreover, a semiregular bipartite graph $G = (V_1 \cup V_2; k_1, k_2)$ must satisfy

$$|V_1|k_1 = |V_2|k_2.$$

Given $|V_1|$ and $|V_2|$, we might ask for all possible values of k_1 and k_2 so that a semiregular bipartite $G = (V_1 \cup V_2; k_1, k_2)$ exists. $|V_1| = 5$ and $|V_2| = 3$, for example, allow the only possible solution $k_1 = 3$ and $k_2 = 5$, yielding the complete bipartite graph $K_{3,5}$ as the unique connected structure satisfying the requirements.

Given k_1 and k_2, we may ask for the smallest vertex set on which there is a semiregular bipartite graph with the prescribed regularity. Again, we get as a unique answer the complete bipartite graph.

It is not difficult to show, see Exercise 2.32, that the obvious necessary conditions on the parameters, namely, $|V_1|k_1 = |V_2|k_2$, $k_1 \leq |V_2|$ and $k_2 \leq |V_1|$, are also sufficient for the existence of a simple semiregular bipartite graph $G = (V_1 \cup V_2; k_1, k_2)$.

The situation for constructibility changes drastically if we add as extra requirement that G must have girth larger than 4. This is not an arbitrary condition. A quadrilateral in the incidence graph corresponds to two distinct lines having two distinct points in common.

To construct a graph $G = (V_1 \cup V_2; k_1, k_2)$ of girth larger than 4, we need to insure that all k_1 neighbors of a vertex in $|V_1|$ have disjoint sets of k_2-1 neighbors in $|V_1|$, and we get the necessary condition $|V_1| \geq 1+k_1(k_2-1)$. By symmetry, we have the corresponding requirement on the size of $|V_2|$, namely, $|V_2| \geq 1 + k_2(k_1 - 1)$. Unfortunately, these obvious necessary conditions are not sufficient to ensure the existence of G. According to Gropp [38], there does not exist any 5-valent bipartite graph on 44 vertices of girth larger than 6. The smallest parameter set satisfying the necessary conditions, but for which the existence of a bipartite semiregular graph is not known, is $|V_1| = 30$, $|V_2| = 20$, $k_1 = 4$, $k_2 = 6$. [38] gives several more examples.

2.3.5 Permutations

We have seen how permutations are useful to construct regular bipartite graphs, so we would like to recall a few facts about them. A *permutation* on the set V is a bijection **p** of V onto itself, $\mathbf{p} : V \to V$. The set of all permutations on V

is denoted by Sym(V). Usually, we consider permutations of the "standard set" $V = \{1, 2, \ldots, n\}$. In this case, we write Sym(V) = Sym(n) = S_n and call it the *symmetric group*.

Example 2.7. Each row of the configuration table of the Pappus configuration, see Fig. 1.9, is a permutation of the nine points:

$$\begin{array}{c} 1\,2\,3\,4\,5\,6\,7\,8\,9 \\ 8\,7\,2\,1\,6\,9\,3\,4\,5 \\ 6\,1\,8\,9\,7\,3\,4\,5\,2 \end{array}$$

The ith column gives the images of vertex i under the three permutations. Since the images are distinct, the three rows define three 1-factors of a regular bipartite graph which is, in fact, the incidence graph of the Pappus configuration.

Since a permutation maps V onto itself, it may be composed with itself, and listing the successive images of elements in cyclic order gives us the cycle notation for a permutation:

$$\mathbf{a}_1 = (1)(2)(3)(4)(5)(6)(7)(8)(9)$$
$$\mathbf{a}_2 = (184)(273)(569)$$
$$\mathbf{a}_3 = (163857492)$$

A permutation, such as $\mathbf{a}_3$, which consists of a single cycle is called *cyclic permutation*. Permutation $\mathbf{a}_2$ has all cycles of the same length, and such a permutation is called *polycyclic* or *semiregular*. Permutation $\mathbf{a}_1$, the *identity permutation*, is polycyclic in the trivial sense that all cycles are of length 1. Each of its cycles specifies element x for which $\mathbf{a}(x) = x$ called a *fixed point* of the permutation. Let Fix($\mathbf{p}$) denote Fix($\mathbf{p}$) = $\{x \in V | \mathbf{p}(x) = x\}$ and let fix($\mathbf{p}$) = |Fix($\mathbf{p}$)|. Hence, fix($\mathbf{a}_1$) = 9, fix($\mathbf{a}_2$) = fix($\mathbf{a}_3$) = 0. A fixed-point free permutation is also called a *derangement*. The set of derangements over V is denoted by Der(V). If seen as a subset of S_n, it is denoted by Der(n).

A permutation whose longest cycle has length 2 is called an *involution*. If, in addition, it has no fixed points, it is called a *fixed-point free involution*. The *order* of the permutation $\mathbf{p} \in$ Sym(n) is the least integer k, such that $\mathbf{p}^k$ is identity. The number n is called the *degree* of permutation $\mathbf{p}$.

Each permutation can be depicted in graphic form with each element $x \in V$ represented by a vertex and with vertices x and $\mathbf{p}(x)$ adjacent. Such a graph does not encode the directions of the cycles, so it is more common to draw an arrow from x to $\mathbf{p}(x)$. This is a *directed graph*. The graph so drawn is not in general a simple graph. Any fixed point will give an element adjacent to itself, and every cycle of length two will give two vertices joined by two arrows, one in each direction. See Fig. 2.24. The connected components of the graph of the permutation are called *orbits*.

2.3 Regularity

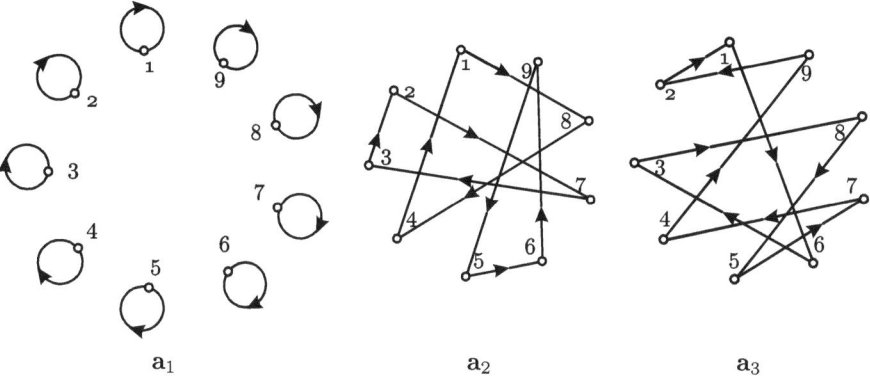

Fig. 2.24 Permutations a_1, a_2, a_3

2.3.6 Directed Graphs and Multigraphs

A graph which allows for loops and multiple edges is often called a *multigraph*. In order to accomplish a mathematical description of multigraphs, we consider two disjoint sets V and E, the vertex set, and the edge set, as well as a function that assigns each edge $e \in E$ a subset of V consisting of at least one and most two vertices which are, as before, called the *endvertices* of e. A *loop* is an edge whose image set has a single element, and two edges assigned to the same set are said to be *parallel*. If a direction is required, each edge is assigned an ordered pair of two vertices.

Let $S \subseteq V \times E$ denote the collection of vertex–edge incident pairs. We have $s \in S$ if and only if $s = (v, e)$ and e is an edge one of whose end-vertices is v. S is called the set of *semiedges* or *arcs*.

The most general definition that we will use defines a graph G as a quadruple (V, S, i, r) such that V and S are sets, i is a map $i : S \to V$ that assigns each arc $s \in S$ its end-vertex $i(s) \in V$, and $r : S \to S$ is an involution $r^2 = 1$ mapping each arc to its *opposite arc*. In this model, the set of edges E is given as the set of orbits of r. If r is allowed to have fixed points, the corresponding orbits have a single element and the corresponding edge is called a *half-edge*. Structures with half edges are sometimes called *pregraphs*.

If not clear from the context, for any graph X, we will use the sets $V(X)$, $E(X)$, $S(X)$, the adjacency relation $\sim_X$, the mapping i_X, and the involution r_X. Note that our definition of graph isomorphism was only given for simple graphs. In Exercise 2.50, we discuss this notion for general graphs.

2.4 Operations on Graphs

We shall now describe several operations on graphs that can be used to generate new, large graphs from old, simple ones.

For the following list of operations, the reader is encouraged to take pencil and paper and produce several drawings combining examples from Sect. 2.2.

2.4.1 Graph Complement

The *graph complement* $Y = X^c$ has $V(Y) = V(X)$ and $x \sim y$ in Y if an only if x is not adjacent to y in X.

2.4.2 Graph Union

We define the *graph union* $X \cup Y$ as the disjoint union of two graphs. So the vertex set is the disjoint union of $V(X)$ and $V(Y)$, and two vertices are adjacent if they are adjacent in X or adjacent in Y. Even if X and Y are connected, $X \cup Y$ is disconnected, having X and Y as connected components. On the other hand, a connected graph cannot be written as the graph union of any proper subgraphs.

If two graphs are isomorphic, we write $2X$ for $X \cup X$, $3X$ for $X \cup X \cup X$, etc.

2.4.3 Graph Join, Cone, and Suspension

The *graph join* $X * Y$ of graphs X and Y can be defined in terms of graph union and graph complement:
$$X * Y = (X^c \cup Y^c)^c.$$
Joining X to a single vertex, the *apex*, is called *coning*. We denote by $C(X)$ the *cone* over X. This operation generalizes to a k-fold cone $C^{(k)}(X)$ in which k new vertices are introduced. A twofold cone is known as *suspension*. Finally, $K_{m,n} = K_m^c * K_n^c$. (see Exercise 2.63.)

2.4.4 One-Point Union and Connectivity

Given a graph X with vertex u and graph Y with vertex v, the *one-point union* of X and Y with respect to u and v, $X \cup_{u,v} Y$, is obtained from the disjoint union by identifying the vertices u and v. So $G = X \cup_{u,v} Y$ then $|V(G)| = |V(X)| + |V(Y)| - 1$. Every path in $X \cup_{u,v} Y$ from a vertex in X to a vertex in Y must pass through the identified vertex. It is a *cut vertex*.

2.4 Operations on Graphs

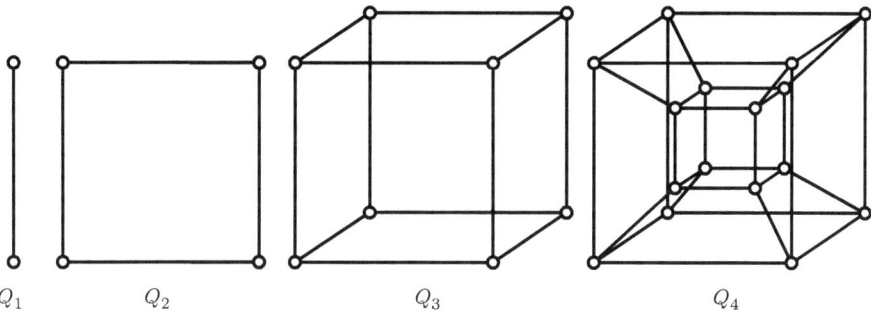

Fig. 2.25 Small hypercube graphs Q_n; $n = 1, 2, 3, 4$

If a graph G cannot be written as a disjoint union of subgraphs nor as a one-point union of subgraphs, it is said to be 2-connected, and between any pair of vertices, there must be two internally disjoint paths.

In general, a graph is called *n-connected* if it contains n internally disjoint paths between any pair of its vertices. The *connectivity* of a graph X is the largest k for which X is k-connected. Connectivity is a graph invariant. The connectivity of C_n, for example, is 2, P_n has connectivity 1, while K_n has connectivity $n - 1$. Note that in an n-connected graph, every vertex must have valence at least n.

2.4.5 Cartesian Product

Let X and Y be any two simple graphs. The *Cartesian product*, $X \square Y$, has vertex set

$$V(X \square Y) = V(X) \times V(Y).$$

Vertices (x, y) and (x', y') from $V(X \square Y)$ are adjacent if and only if either $x = x'$ and $y \sim y'$ or $x \sim x'$ and $y = y'$.

Clearly, $C_4 = K_2 \square K_2$ and the hexahedron or cube Q_3 is the threefold Cartesian product of K_2 with itself, i.e., $Q_3 = C_4 \square K_2$.

The prism Π_n, for example, can be expressed as the Cartesian product of a cycle of length n and the complete graph on 2 vertices, $\Pi_n = C_n \square K_2$.

Since $\square$ is associative (see Exercise 2.57), we can consider the Cartesian product of several factors. Taking n factors equal to K_2, we obtain the *hypercube graph* Q_n. $Q_n = K_2 \square K_2 \square \cdots \square K_2$. Small hypercube graphs are depicted in Fig. 2.25. Note that Q_2 is used as the symbol $\square$ to denote the Cartesian product.

In Exercise 2.58, an alternate definition of Q_n is given.

2.4.6 Tensor Product

Let X and Y be any two simple graphs. The *tensor product* $X \times Y$ has vertex set
$$V(X \times Y) = V(X) \times V(Y).$$
Vertices (x, y) and (x', y') from $V(X \times Y)$ are adjacent if and only if $x \sim x'$ and $y \sim y'$. The tensor product of two K_2's is the disjoint union of two edges and is used as symbol $\times$ for the tensor product.

2.4.7 Strong Product

Let X and Y be any two simple graphs. The *strong product* $X \boxtimes Y$ has the vertex set
$$V(X \boxtimes Y) = V(X) \times V(Y).$$
Vertices (x, y) and (x', y') from $V(X \boxtimes Y)$ are adjacent if and only if either $x \sim x'$ and $y \sim y'$ or $x = x'$ and $y \sim y'$ or $x \sim x'$ and $y = y'$. Again, the strong product of two K_2's is used as multiplication symbol.

2.4.8 Line Graph

For any simple graph X, let $L(X)$ denote the graph whose vertex set $V(L(X))$ is $E(X)$ and two vertices e and e' from $V(L(X))$ are adjacent if and only if e and e' are incident (as edges of X) with a common vertex of X. The line graph of K_4, for example, is the octahedron graph.

2.4.9 Subdivision Graph

The *subdivision graph* $S(X)$ has $V(S(X)) = V(X) \cup E(X)$, and two vertices x and e of $S(X)$ are adjacent if and only if $x \in V(X)$ and $e \in E(X)$ and x is incident to e in X. A drawing of the subdivision graph is obtained from a drawing of the graph by inserting one new vertex in the interior of each edge.

2.4.10 Graph Square

For a given graph X, its *square* X^2 is a graph on the same vertex set with two vertices adjacent if and only if they are at distance at most 2 in X. Each vertex in X^2 is contained in a clique of size $\deg_X(v) + 1$.

The *pure square* $X^2 - E(X)$ is a graph on the same vertex set as X, with two vertices adjacent if and only if they are at distance 2 in X.

2.5 Graph Colorings

2.5.1 Vertex Colorings

A mapping c from $V(X)$ to any finite set of colors C is called a *vertex coloring* if no two adjacent vertices are assigned the same color of C. The smallest number of colors needed for a (proper) vertex coloring of a graph G is called a *chromatic number* of a graph and is denoted by $\chi(G)$.

Example 2.8. It is not hard to see that the chromatic number of a cycle C_n is 2 if n is even and 3 if n is odd.

Example 2.9. Clearly, the chromatic number of the tetrahedron graph is 4. Since the octahedron graph is $K_{2,2,2}$, its chromatic number is 3. The cube is bipartite; therefore, it has a chromatic number equal to 2. Since the dodecahedron graph contains an odd cycle, its chromatic number is at least 3. It is not hard to find a proper 3-coloring of GP(10, 2). We leave the determination of the chromatic number of the icosahedron to the exercises.

The study of colorings of graphs constitutes an important branch of graph theory. The problem of determining the exact upper bound on the chromatic number of planar graphs was an outstanding open problem in graph theory for over a 100 years until it was solved by the aid of a computer in 1976 by Appel and Haken [1], yielding the *four color theorem*. Clearly, colorings of graphs have played an important role in the development of topological graph theory.

Theorem 2.10 (Brooks). *Let G be a connected graph. The chromatic number of G is less than or equal to the maximum valence of any vertex in G unless G is complete or an odd cycle.*

Proof. Let $\Delta(G)$ denote the maximum valence of any vertex in G. If $\Delta \leq 2$ then G is K_2, a path or a cycle, and we know that paths and even cycles are bipartite. We now assume $\Delta \geq 3$ and proceed by induction on the number of vertices of G. Clearly, a graph on 4 vertices which is not complete is 3-colorable. Let G be a graph on more than 4 vertices and $\Delta(G) \geq 3$. For any vertex v in G, note that $G - v$ has maximum valence at most Δ and cannot have $K_{\Delta+1}$ as a component; hence, $G - v$ is Δ-colorable by the induction hypothesis. If v is a vertex of valence less than Δ, then any Δ coloring of the components of $G - v$ easily extends to a Δ-coloring of G. So if G is not Δ-colorable, every vertex of G must have valence Δ. Moreover, in every Δ-coloring of $G-v$, all colors are used coloring the neighbors of v; otherwise, we could extend the coloring to G using the missing color. Fix a coloring of $G - v$ that does not extend to G and consider the subgraph $X_{i,j}$ induced by the vertices colored with color i or color j. The two neighbors of v colored i and j must be in the same component $X'_{i,j}$ of $X_{i,j}$ because, otherwise, we could interchange the roles of i and j in one of these components to obtain a Δ-coloring of $G - v$ where two neighbors of v have the same color. Consider now a path in $X'_{i,j}$ connecting two neighbors of v. Let u be the closest to v along this path so that u has valence larger than 2 in $X'_{i,j}$. Then u can be recolored, but this recoloring disconnects the $X_{i,j}$ so that the neighbors of v are in different parts, contradicting our earlier observation. Therefore, $X'_{i,j}$ is simply a path. In two such paths, $X'_{i,j}$ and $X'_{i,k}$, $i \neq j \neq k$ can only intersect in an endpoint, namely, the neighbor of v with color i, since any interior intersection point could be recolored leading as before to disconnect $X'_{i,j}$. If all neighbors of v are pairwise adjacent, G is complete and there is nothing to show, so assume without loss of generality that given a Δ-coloring of $G - v$, the neighbors of v colored i and j are not adjacent so that $X'_{i,j} = v_i u ... v_j$ is a path of length at least 3. We now change colors i and k on the path $X'_{i,k}$. In this new coloring, the vertex u is both on $X'_{i,j}$ and on $X'_{j,k}$ contradicting the established fact that these paths only intersect at their endpoint v_j. □

In Example 2.2.8, we obtained a graph together with a vertex coloring from a geometric object. Conversely, we will see in Chap. 5 that a graph together with a proper vertex coloring with k colors is sufficient to describe a geometric object, specifically a rank k incidence structure.

2.5.2 Edge Colorings

A mapping $c : E(G) \to C$ from the edge set $E(G)$ to some finite set C is called an (admissible or proper) edge coloring if for any two incident edges g and f we have $c(g) \neq c(f)$. The least number of colors needed to properly color the edges of g is called the *chromatic index* and is denoted by $\chi'(G)$.

Clearly, the maximum valence Δ is a lower bound for the chromatic index. Any edge coloring problem can be translated into a vertex coloring problem on $L(G)$; the line graph of G and Brooks' Theorem 2.10 provides an upper bound by observing that $\Delta(L(G)) \leq 2\Delta(G) - 2$. However, this bound is not tight; in fact, the difference between upper and lower bound in terms of Δ is surprisingly small.

Theorem 2.11 (Vizing). *The chromatic index of a simple graph G satisfies the following inequalities:*

$$\Delta(G) \leq \chi'(G) \leq \Delta(G) + 1.$$

Proof. The first inequality is trivial, so the second one is the only thing we have to prove. We use induction on the number of edges. Consider a graph G with maximum degree Δ. If G has fewer than Δ edges, there is nothing to show. Assume that, by induction hypothesis, $G - e$ has a $\Delta + 1$-coloring for every edge e, but none of these colorings can be extended to a coloring of G.

Consider an edge $e = vw_0$ and fix a $\Delta + 1$-coloring c_0 of $G_0 = G - vw_0$. Since the maximum degree of any vertex is Δ and c_0 is a $\Delta + 1$-coloring, there is at least one color missing at every vertex. If the same color is missing from both v and w_0, we are done. Let α be the color missing at v with respect to c_0 and β the color missing at w_0. So there is an edge of color β incident with v. We call a path whose edges are alternately colored α and β an $\alpha-\beta$-path and observe that any $\alpha-\beta$-path starting at w_0 must end in v; otherwise, we could exchange α and β along this path and extend c_0 to a coloring of G by coloring vw_0 with α after that switch.

Now choose a maximal sequence $w_0, w_1, \ldots, w_k$ of distinct neighbors of v such that the color of the edge xw_i is missing at w_{i-1}. For each of the graphs $G_i = G - vw_i$, we define a coloring c_i derived from c_{i-1} by coloring the edge v, w_{i-1}, which does not exist in G_{i-1}, by the color of vw_i in c_{i-1}. All these colorings differ only in edges incident to v, but the set of colors used for edges incident to v is the same for all these colorings.

Let β be a color missing at w_k with respect to the coloring c_0 (and subsequently with respect to c_k). By the maximality of k, there is an index $i \in \{1, \ldots, k\}$ such that $c_0(vw_i) = \beta$. An $\alpha-\beta$-path, P, from w_k with respect to c_k must end in v; in fact, it must end with the edge vw_{i-1}. With respect to the coloring, c_0 β is missing at w_{i-1}. Let P' be an $\alpha-\beta$-path with respect to c_{i-1} starting at w_{i-1} in G_{i-1}. P and P' are identical except for edges incident to v, so P' contains w_k. Since there is no β edge at w_k with respect to c_{i-1}, P' ends in w_k, contradicting the assumption that c_{i-1} cannot be extended to G. □

2.6 From Geometry to Graphs and Back

There are numerous paths leading from geometry to graphs and back. We have already met the skeleta of polyhedra as a rich source of interesting graphs. Here, we mention some more of such interesting connections. But first, let us recall the concept of *metric space*. This structure lies somewhere between geometry and topology. It captures those properties of usual Euclidean space that measure distance between any two points in space.

2.6.1 Metric Space and Distance Function

A set M together with a function $d : M \times M \to \mathbb{R}$ is called a *metric space* if the following are true:

1. $d(x, y) \geq 0$ for any two points $x, y \in M$, and $d(x, y) = 0$ if and only if $x = y$.
2. $d(x, y) = d(y, x)$ for any two points $x, y \in M$.
3. (Triangle inequality) $d(x, y) \leq d(x, z) + d(z, y)$ for any three points $x, y, z \in M$.

The function d is called the *distance function* of M.

Example 2.12. The Euclidean plane, $\mathbb{R}^2 = \{(x, y) | x, y \in \mathbb{R}\}$, is a metric space for $d((x, y), (x', y')) = \sqrt{(x - x')^2 + (y - y')^2}$

Given a metric space, we define a *closed ball* $B(x, r)$ with center $x \in M$ and radius $r > 0$ as follows:

$$B(x, r) = \{y \in M \,|\, d(x, y) \le r\}.$$

2.6.2 Distances in Graphs

In a connected graph G, we define the *distance* $d_G(u, v)$ between vertices $u, v \in V(G)$ to be the length of the shortest path between u and v. Clearly, d_G defines a metric space on the vertex set $V(G)$. This metric space is usually described by the *distance matrix* $D(G)$ with entry $D_{i,j} = d_G(v_i, v_j)$ for a given ordering $v_1, v_2, ..., v_n$ of the vertices of G. For an arbitrary vertex $v \in V(G)$, we define the *distance sequence* $d_{G,v} = (1, d_1, d_2, ...)$ where d_k denotes the number of vertices at distance k from v. Usually, we only consider $d_k > 0$.

Example 2.13. Prism graphs are but one example of graphs in which every vertex has the same distance sequence because for any pair of vertices u and v, there is an automorphism mapping u to v. For instance, for Π_3, we have $d_{(\Pi_3, v)} = (1, 3, 2)$. Similarly, we get $d_{(\Pi_4, v)} = (1, 3, 3, 1), d_{(\Pi_5, v)} = (1, 3, 4, 2)$.

2.6.3 Intersection Graphs

Given a family of sets $\mathcal{B} = \{B_1, B_2, \ldots, B_n\}$, we may define its *intersection graph*. The vertex set is $\mathcal{B}$, and two vertices are adjacent if and only if the corresponding sets have nonempty intersection. We note that there is a variation to this construction, namely, we may construct a general graph by putting $|B_i \cap B_j|$ edges between B_i and B_j.

Example 2.14. Consider the following seven sets in the plane: the three sides of a regular triangle, the three heights, and the inscribed circle. It is not hard to see that the corresponding intersection graph is K_7.

Intersection graphs are universal in the sense that any graph can be represented as an intersection graph. However, by selecting various geometric objects as sets, we get interesting families of graphs. For instance, the so-called *interval graphs* are intersection graphs of finite families of line segments in the $\mathbb{R}^1$ line.

2.6.4 Intersection Graphs of a Family of Balls

Given a set of n points $V = \{v_1, v_2, \ldots, v_n\}$ in some metric space and a positive number $r > 0$, we may draw n closed balls $B_i := B(v_i, r), i = 1, 2, \ldots, n$, each

2.6 From Geometry to Graphs and Back

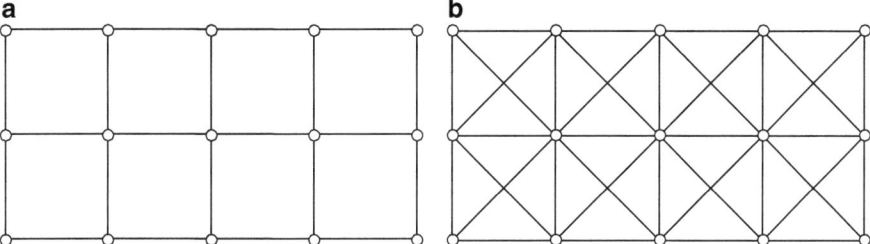

Fig. 2.26 The grid graph $Gr(3, 5) = P_3 \square P_5$ and $P_3 \boxtimes P_5$ as unit sphere graphs

ball B_i centered at v_i and having radius r. Define a graph $G(V, r)$ as follows: The vertices are the n selected points. Two vertices v_i and v_j are adjacent if and only if the corresponding balls intersect, i.e., if $B_i \cap B_j \neq \emptyset$. The radius r will be called the *unit* and the graph a *unit sphere graph*.

Here are some specific examples:

Example 2.15. Let us select the following points in the Euclidean plane: (x, y), $x \in \{1, 2, \ldots, a\}$, $y \in \{1, 2, \ldots, b\}$. Hence, $n = ab$. Let $r = 0.5$. The unit sphere graph is the well-known $a \times b$ grid graph $Gr(a, b)$, which we can simply describe as the Cartesian product of the paths P_a and P_b. Figure 2.26a shows the case for $a = 3$ and $b = 5$. If r is increased, there is no change in the structure of the graph until $r = \sqrt{2}/2$, when the diagonals of the 4-cycles appear; see Fig. 2.26b.

It would be interesting and useful to characterize the unit sphere graphs in $\mathbb{R}^2$ and $\mathbb{R}^3$. For instance, all platonic graphs arise as unit sphere graphs in $\mathbb{R}^3$. One has to take the vertices of the corresponding platonic solid and radius r to be one half of the edge length.

Example 2.16. In order to obtain the cube graph Q_3, one can take

$$V = \{(0,0,0), (0,0,1), (0,1,0), (0,1,1), (1,0,0), (1,0,1), (1,1,0), (1,1,1)\}$$

and $r = 1/2$.

This example shows that the cube graph can be described by a careful choice of 8 points in some metric space. There is another approach to this construction. It involves convex sets.

2.6.5 Convex Sets

A set of points $K \subseteq \mathbb{R}^3$ is *convex*, if for any two points $x, y \in K$, every point z on the line segment from x to y belongs to K. For any set $S \subseteq \mathbb{R}^3$, we can find the smallest convex set $S \subseteq \text{conv}(S) \subseteq \mathbb{R}^3$, called the *convex closure* or *convex hull* of S.

Example 2.17. The convex closure of the set

$$V = \{(0,0,0), (0,0,1), (0,1,0), (0,1,1), (1,0,0), (1,0,1), (1,1,0), (1,1,1)\}$$

is a cube.

This gives us another general mechanism for constructing graphs from simple geometric objects:

$$\text{Finite set } S \to \text{conv}(S) \to \text{skeleton}$$

Starting with a finite set of points in $\mathbb{R}^3$, its convex closure is a convex polyhedron whose 1-skeleton is a graph.

The intersection graph of the seven projective lines B_i of the Fano configuration, see Sect. 1.1.1 of Chap. 1, is K_7 and does not capture the whole combinatorial structure of the configuration. Taking in addition to the sets B_i also all the sets $C_{i,j} = B_i \cap B_j$ that are not empty, the resulting intersection graph captures all the combinatorial structure of the Fano plane. Deleting all edges $B_i B_j$ yields a cubic bipartite graph on 14 vertices still containing all combinatorial information about the Fano configuration. The vertices labeled C may be considered the vertices of the configuration, while the B's are the lines. Edges of the graph indicate what point is on what line or which line goes through which point.

2.6.6 Representations and Drawings of Graphs

Let G be a graph and let S be a set, and let $\mathcal{P}(S)$ denote the *power set* of S, that is, the set of all subsets of S. A pair of mappings

$$\rho_V : V(G) \to S, \qquad \rho_E : E(G) \to \mathcal{P}(S)$$

is called a *graph representation* or an S-representation of the graph G if $\rho_V(v) \in \rho_E(e)$ provided v is incident with e. If there is no fear for confusion, we omit the subscripts of ρ since the argument determines which mapping is considered. We only consider representations for which no pair of vertices is mapped to the same element of S.

Sometimes, we only specify ρ_V and have no need for ρ_E. In such a case, we may tacitly assume that for each $e = uv \in E(G)$, we have $\rho_E(uv) := \{\rho_V(u), \rho_V(v)\}$.

If S is a vector space, the representation is called a *vector representation*. If S is a metric space, the representation is called a *metric representation*. In a metric representation, we define the *length* of each edge $e = uv$ relative to representation ρ as $||e||_\rho = d(\rho(u), \rho(v)) > 0$. In a simple graph G, the length of each edge is strictly positive.

2.6 From Geometry to Graphs and Back

Fig. 2.27 Each generalized Petersen graph is a unit distance graph. In particular, this is true for the Dürer graph GP(6, 2)

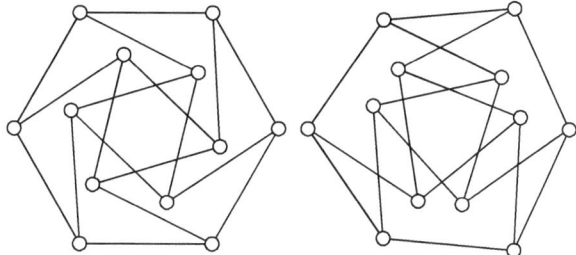

$S = \mathbb{R}^n$ is of particular importance to us because real n-space is both a metric and a vector space. An $\mathbb{R}^2$ representation is called *planar* and an $\mathbb{R}^3$ representation is called *spacial representation*. In both cases, we define $\rho_E(uv) := conv(\rho(u), \rho(v))$. Each edge is therefore represented as the line segment connecting the two represented vertices. Such a representation is called *graph drawing*.

Each figure depicting a graph in this book has now a formal description as a graph drawing defined above. We have to define when two drawings are equal (or equivalent). Obviously, we may consider two drawings that differ by an isometry equivalent. But we may also neglect the difference in scale. This means, for instance, we can always set the barycenter to be the origin and set the shortest edge length to be 1 to obtain a "standard" drawing. We define the *energy* of a drawing to be the sum of the lengths of all line segments representing the edges.

The *dilation coefficient* is the quotient between the longest and shortest edge of the drawing.

Graph drawings with dilation coefficient 1 are known as *unit distance graphs*.

2.6.7 Generalized Petersen Graphs as Unit Distance Graphs

All generalized Petersen graphs GP(n, k) can be drawn in the plane as unit distance graphs. We embed the outer rim as a regular polygon with side length 1. We also embed the inner rim as a collection of star polygons of side length 1. If $k = 1$, the inner polygon is congruent with the outer polygon. Translating one polygon by a unit vector yields the appropriate coordinates for the representation. Note that every prism graph Π_n can be drawn in the plane as a unit distance graph. If $k \neq 1$, the radius of the inner rim is different from the radius of the outer rim. This means that if the radii differ by less than 1, we can rotate the inner rim so that the distance between the two adjacent vertices along a spoke becomes 1. The vertices of the outer rim are given the coordinates $\rho(v_i) = (R\cos(i\pi/n), R\sin(i\pi/n))$, and the vertices in the inner rim are given the coordinates $\rho(u_i) = (r\cos(\phi + i\pi/n), r\sin(\phi + i\pi/n)$, where $R = 1/(2\sin(\pi/n)), r = 1/(2\sin(r\pi/n))$, and $\phi = \arccos((R^2 + r^2 - 1)/(2rR))$. This method works if $R - r < 1$. In particular, the case of the Dürer graph GP(6, 2) is shown in Fig. 2.27.

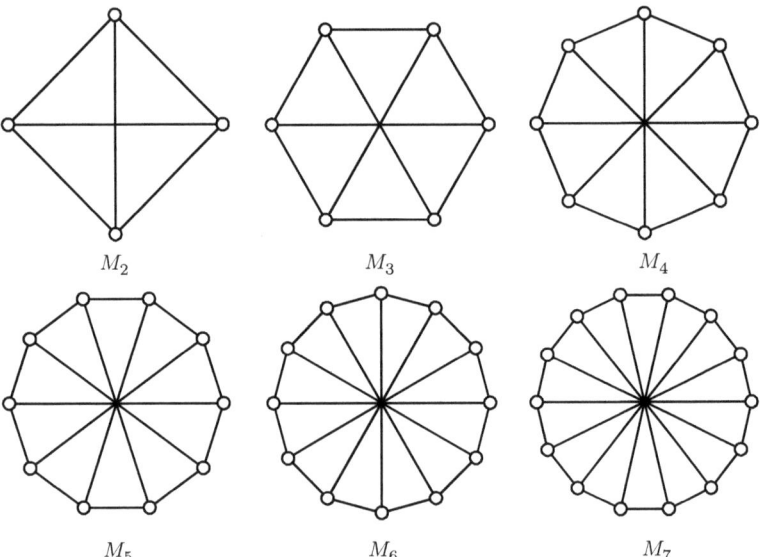

Fig. 2.28 Small Möbius ladders M_n

2.7 Exercises

Exercise 2.1. Consider a regular polygon with n sides in the plane. Use it to define a graph X_n whose vertex set consists of vertices of the polygon and two vertices are adjacent if and only if they belong to the same edge of your polygon. Prove that X_n is isomorphic to the cycle C_n.

Exercise 2.2. Find all independent sets of size greater than 3 in the divisor graph on $\{2, 3, \ldots, 13\}$.

Exercise 2.3. How many connected components does the common divisor graph of Fig. 2.1 have?

Exercise 2.4. The *Möbius ladder* M_n is obtained from the cycle C_{2n} by adding n main diagonals:

$$V = \{v_1, v_2, \ldots, v_{2n}\}$$
$$E = \{v_1 v_2, v_2 v_3, \ldots, v_{2n-1} v_{2n}, v_{2n} v_1, v_1 v_{n+1}, v_2 v_{n+2}, \ldots, v_n v_{2n}\}.$$

Prove that the Möbius ladder M_n can be obtained from the prism graph Π_n by deleting and reattaching only two edges (Fig. 2.28).

Exercise 2.5. Draw all nonisomorphic trees on n vertices for $n = 1, 2, 3, 4, 5$.

Exercise 2.6. Show that a graph is a forest if and only if each of its connected components is a tree.

2.7 Exercises

Exercise 2.7. Prove the following "two out of three" theorem for trees.

Let G be a graph on n vertices. Then, any two of the following conditions imply the third:

- G is acyclic.
- G is connected.
- G has $n - 1$ edges.

Exercise 2.8. A *fullerene* is a trivalent convex polyhedron whose faces are only pentagons and hexagons. We also call its skeleton by the same name. Prove that the smallest fullerene has 20 vertices.

Exercise 2.9. Prove that any fullerene (see Exercise 2.8) has exactly 12 pentagons.

Exercise 2.10. Prove that there are no fullerenes (see Exercise 2.8) on 22 vertices.

Exercise 2.11. Find all fullerenes (see Exercise 2.8) among the generalized Petersen graphs.

Exercise 2.12. Find all fullerenes (see Exercise 2.8) among the Platonic and Archimedean graphs.

Exercise 2.13. Consider the vertices and edges of the tetrahedron graph. Say that a vertex v is *across* from any edge which is part of a 3-cycle not containing v. Further, say that two edges are *across* from one another if they do not belong to any common 3-cycle. Define a graph on the ten vertices and edges of the tetrahedron where adjacency is determined by across. Show that this graph is isomorphic to the Petersen graph.

Exercise 2.14. Prove directly that the generalized Petersen graphs $GP(7, 2)$ and $GP(7, 3)$ are isomorphic.

Exercise 2.15. Prove that there is an isomorphism between $GP(n, r)$ and $GP(n, s)$ preserving the outer n-gon if and only if $r \equiv \pm s \pmod{n}$.

Exercise 2.16. Prove that the generalized Petersen graphs $GP(8, 2)$ and $GP(8, 3)$ are not isomorphic.

Exercise 2.17. Decide whether each of the generalized Petersen graphs pictured in Figs. 2.15 and 2.16 is planar.

Exercise 2.18. Complete the argument in Theorem 2.2 in the case of even girth. Note that the base of the construction does not have to be a single vertex.

Exercise 2.19. What is the girth of the graph in Exercise 2.34?

Exercise 2.20. Determine the size of the smallest cubic bipartite graph of girth larger than 4 and construct an example.

Exercise 2.21. Prove that the girth of the Petersen graph $GP(5, 2)$ is 5.

Exercise 2.22. Determine all generalized Petersen graphs $GP(n, r)$ of girth 5.

Exercise 2.23. Prove the following result. A graph is bipartite if and only if it contains no cycles of odd length.

Exercise 2.24. Show that if G has $c+1$ nonempty bipartite connected components, there are 2^c bipartitions of the vertex set.

Exercise 2.25. Write an LCF code for the Dürer graph.

Exercise 2.26. Write an LCF code for K_4.

Exercise 2.27. Show that M_n (defined in Exercise 2.4) admits a description via LCF notation. Show that it is isomorphic to the graph $[(n)^{2n}]$.

Exercise 2.28. Show that the Heawood graph admits a description via LCF notation. Show that it is isomorphic to the graph $[(5, -5)^7]$.

Exercise 2.29. Prove that the Heawood graph has no cycles of length less than 6.

Exercise 2.30. Show that the Tutte 8-cage is isomorphic to the graph
$$[(-7, 9, 13, -13, -9, 7)^5].$$

Exercise 2.31. Show that the Balaban 10-cage is Hamiltonian and find an LCF notation for it.

Exercise 2.32. Given the parameters $|V_1|, |V_2|, k_1, k_2$, satisfying $|V_1|k_1 = |V_2|k_2$, $k_1 \leq |V_2|$, and $k_2 \leq |V_1|$, construct a semiregular bipartite graph $G = (V_1 \cup V_2; k_1, k_2)$. Hint: Let the ith vertex of V_1 be adjacent to vertices $\{i, i+1, \ldots, i+k_1 - 1 \pmod{|V_2|}\}$.

Exercise 2.33. Formulate and prove a structure theorem analogous to Theorem 2.5 for semiregular bipartite graphs.

Exercise 2.34. Here is a table for the Fano configuration:

$$\begin{matrix} 1 & 1 & 1 & 2 & 2 & 3 & 3 \\ 2 & 4 & 6 & 4 & 5 & 4 & 5 \\ 3 & 5 & 7 & 6 & 7 & 7 & 6 \end{matrix}$$

Draw the corresponding regular bipartite graph and rewrite the table to reflect the partition into 1-factors.

Exercise 2.35. Prove that the Möbius ladder M_n is bipartite if and only if n is odd.

Exercise 2.36. Redraw the polyhedral graph of the cube, Fig. 2.12, coloring the vertices to indicate the tripartition. Make your drawing as symmetrical as possible.

Exercise 2.37. Write down all permutations in Sym(3).

Exercise 2.38. Write down all involutions in Sym(4).

Exercise 2.39. Write down all fixed-point free permutations in Sym(5). In other words, determine the set Der(5).

2.7 Exercises

Exercise 2.40. Write down all fixed-point free involutions in Sym(4) and in Sym(5).

Exercise 2.41. Determine the number of semiregular permutations in Sym(2006).

Exercise 2.42. Let p be a prime. Show that the number of semiregular permutations in Sym(p) equals $(p-1)!$

Exercise* 2.43. Determine the number of semiregular permutations in Sym(n).

Exercise* 2.44. Determine the number of derangements in Der(n).

Exercise* 2.45. Determine the number of involutions in Sym(n).

Exercise* 2.46. Determine the number of fixed-point free involutions in Sym(n).

Exercise 2.47. Express the order of a permutation in terms of its cycle structure.

Exercise 2.48. The definition of a cycle C_n in Sect. 2.2.2 applies to cycles with $n \geq 3$. Define C_1 (the loop) and C_2 as general graphs (see 2.3.6).

Exercise 2.49. Prove that there is—up to isomorphism—only one cubic graph on 4 vertices with 3 loops.

Exercise 2.50. Define the notion of isomorphism for general graphs and pregraphs.

Exercise 2.51. Show that the wheel graph W_n is isomorphic to the cone over C_n.

Exercise 2.52. Show that $K_{2,2,2}$ is isomorphic to the suspension over C_4.

Exercise 2.53. Show that K_{n+1} is isomorphic to the cone over K_n.

Exercise 2.54. Show that the prism graph Π_n is isomorphic to the Cartesian product $K_2 \square C_n$.

Exercise 2.55. Show that C_4 can be expressed using only single vertex graphs and the operations of graph union $\cup$ and graph join $*$.

Exercise 2.56. Show that a graph G can be expressed using only single vertex graphs and the operations of $\cup$ and $*$ if and only if G has no induced subgraph isomorphic to P_4.

Exercise 2.57. Show that the Cartesian product $\square$ is associative.

Exercise 2.58. We defined Q_n, the hypercube in dimension n, as Cartesian product of n factors equal to K_2. Show that Q_n may also be defined as follows: The vertex set of Q_n consists of n-tuples of 0's and 1's. Two vertices are adjacent if they differ in exactly one coordinate.

Exercise 2.59. Show that $G_1 \square G_2$ is connected if and only if both G_1 and G_2 are connected.

Exercise 2.60. Show that if G_1 and G_2 are both connected, then $G_1 \square G_2$ is 2-connected.

Fig. 2.29 $(GP(6, 2)^2)^c$

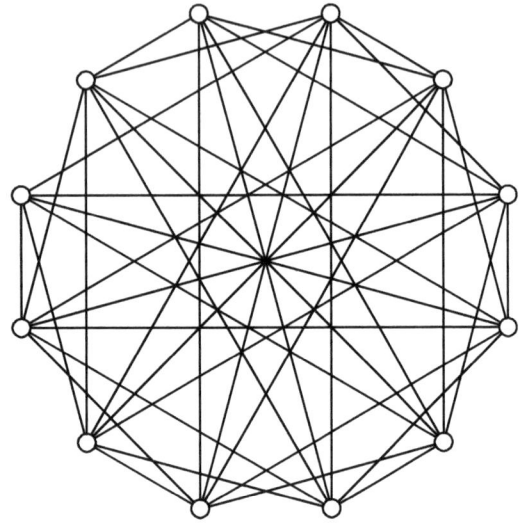

Exercise 2.61. Given the valence of a vertex $v_1 \in V(G_1)$ and the valence of $v_2 \in V(G_2)$, what can you say about the valence of $(v_1, v_2) \in V(G_1 \square G_2)$?

Exercise 2.62. Prove that the strong product of any two paths is a unit sphere graph.

Exercise 2.63. Show that $K_{m,n} = K_n^c * K_m^c$.

Exercise 2.64. Show that K_4, $K_{2,2,2}$, Q_3, and $GP(10, 2)$ are four out of the five Platonic graphs. Design a graph theoretical method for constructing the missing icosahedron graph.

Exercise 2.65. Show that the graph in Fig. 2.29 is indeed the complement of the square of the Dürer graph.

Exercise 2.66. Show that all hypercube graphs are Hamiltonian.

Exercise 2.67. A graph is called a *benzenoid graph* if it can be obtained by selecting a connected subset of hexagons in an infinite planar hexagonal lattice (representing graphene). Show that all benzenoid graphs can be described as unit sphere graphs in the plane.

Note that in theoretical chemistry, a benzenoid graph is sometimes defined in various slightly different ways. Benzenoid graphs represent molecules of polyhexes, i.e., polycyclic aromatic hydrocarbons. Vertices correspond to carbon atoms, edges to the carbon–carbon bonds, while hydrogen atoms are not shown.

Exercise 2.68. Let $K(X)$ denote the number of 1-factors in a graph G. Show that benzene has two 1-factors: $K(B_1) = 2$.

Exercise 2.69. Prove that a bipartite graph with bipartition sets of unequal size has no 1-factor. Use this result to show that $K(B_5) = 0$. Note that in chemistry, a 1-

2.7 Exercises

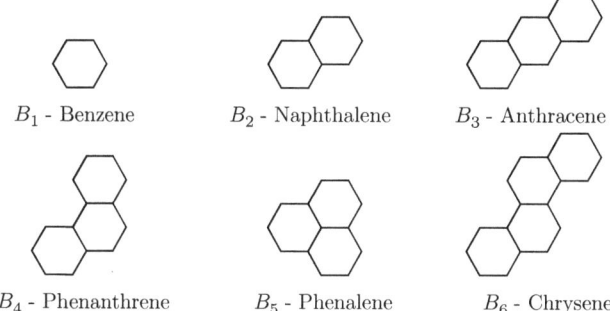

Fig. 2.30 Small benzenoid graphs

Fig. 2.31 The Moser graph

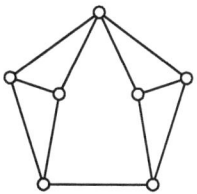

factor is called a *Kekulé structure*. It is known that polyhex hydrocarbons without Kekulé structures are extremely unstable, for instance, phenalene.

Exercise 2.70. Determine the number of Kekulé structures $K(B_n)$ for the benzenoid graphs in Fig. 2.30.

Exercise 2.71. Prove that every graph is an intersection graph of some family of sets. In particular, determine a family of sets whose intersection graph is isomorphic to GP(5, 2).

Exercise 2.72. Determine the coordinates for the vertices of a 4-valent convex polyhedron whose skeleton is isomorphic to the cuboctahedron graph.

Exercise 2.73. Determine the chromatic number for each Archimedean graph.

Exercise 2.74. Find the minimal dilation coefficient of any planar representation of the Moser graph in Fig. 2.31.

Exercise 2.75. It is easy to verify that K_4 is not a unit distance graph in the plane. Consider a drawing of K_4 in the plane with only two distinct edge lengths. How many such nonisomorphic drawings are there? (Hint: there are six). Compute the dilation coefficient for all such drawings.

Exercise 2.76. Show that there exists a spatial drawing of Q_3 with dilation coefficient 1. Is it unique?

Exercise* 2.47. Find a planar drawing of a generalized Petersen graph GP(n, r) with dilation coefficient 1.

Chapter 3
Groups, Actions, and Symmetry

3.1 Groups

3.1.1 Graph Automorphisms

An isomorphism of a graph into itself is called a *graph automorphism*. Every graph has at least one automorphism, namely, the identity, and for totally asymmetric graphs, that is the only one. The path of length $n > 1$, P_n, has exactly one nonidentity automorphism, sending vertex $k, k \in \{1, \ldots, n\}$ to vertex $n - k + 1$. The graph in Fig. 3.1 which can be described by the notation developed in Sect. 2.4 as the cone $C(K_2 \bigcup 2K_1)$ has four automorphisms in all: one that interchanges the two vertices of valence 1, one that interchanges the two vertices of valence 2, one that switches both pairs, and one that fixes everything. The graph $C(K_2 \bigcup 2K_2)$ is more symmetric than $C(K_2 \bigcup 2K_1)$ so we would expect $C(K_2 \bigcup 2K_2)$ to have more automorphisms than $C(K_2 \bigcup 2K_1)$. For simple graphs, every automorphism is determined by its action on the vertex set and can therefore be considered a permutation of the vertices. It is important not to confuse the automorphisms of a graph with the symmetries of its drawing. The graph on the left in Fig. 3.2 has four automorphisms, corresponding to the four symmetries of its drawing, while its seemingly less symmetric cousin on the right has 120 automorphisms.

3.1.2 Definition of Groups

The set of all automorphisms Aut (X) of a graph X together with composition of mappings form an algebraic structure, called a *group*. As an abstract structure, a group **G** of *order* n consists of n elements, $g_1, g_2, \ldots, g_n$, together with a law of composition which assigns to each ordered pair (g_i, g_j) a third element g_k,

Fig. 3.1 $C(K_2 \cup 2K_1)$ and $C(K_2 \cup K_2)$

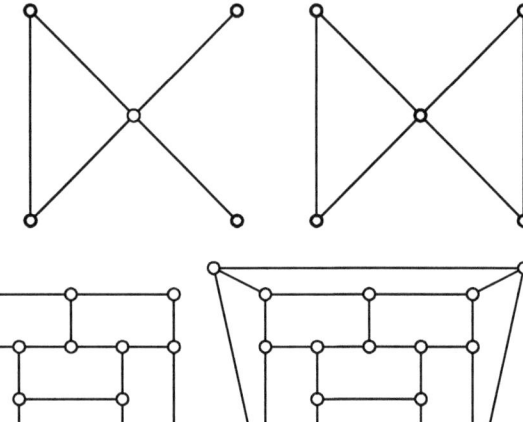

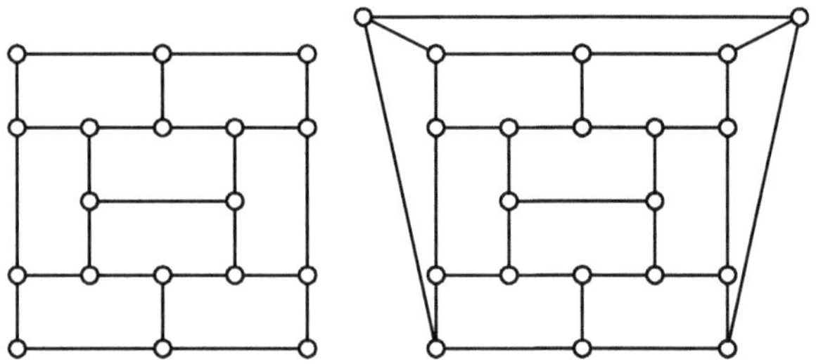

Fig. 3.2 Two symmetric planar graphs

which we usually call the product and write $g_i \cdot g_j = g_k$ provided that the law of composition satisfies the following two axioms:

Axiom 1 Replacing any two of the symbols in the equation $a \cdot b = c$ by group elements uniquely determines a third group element.

Axiom 2 The associative law holds, $(g_i \cdot g_j) \cdot g_k = g_i \cdot (g_j \cdot g_k)$.

Any composition closed set of invertible transformations forms a group; in particular, the automorphisms of a graph form a group in this sense. In fact, the converse is also true; given an abstract group satisfying the axioms above, there exists a realization of the elements of the group as permutations of a set.

Example 3.1. The real numbers, $\mathbb{R}$, form a group with respect to addition. If we want to consider multiplication as group operation on a set of real numbers, we have to exclude 0, because the equation $a \cdot 0 = 0$ does not uniquely define a real number a, in violation of Axiom 1. Also, if we have a set of reals and multiplication is the operation, we may safely discard all the negative reals from the set and still satisfy the axioms. The positive reals cannot be so cavalierly discarded.

Example 3.2. The integers, $\mathbb{Z}$, form a group with respect to addition as well as $\mathbb{Z}[\sqrt{3}] = \{n + m\sqrt{3} \mid n, m \in \mathbb{Z}\}$. The rational numbers $\mathbb{Q}$ also form an additive group, see Exercise 3.7.

From the group axioms, it is easy to deduce, see Exercise 3.1, the existence of a unique *unit element*, e, for which $g \cdot e = e \cdot g = g$ for all $g \in \mathbf{G}$. We say that two group elements g_i and g_j *commute* if $g_i \cdot g_j = g_j \cdot g_i$. Thus, every group element commutes with the unit element e. If every pair of elements in a group commutes, then the group is said to be a *commutative group*, or *abelian*. Most interesting groups are not abelian.

3.1 Groups

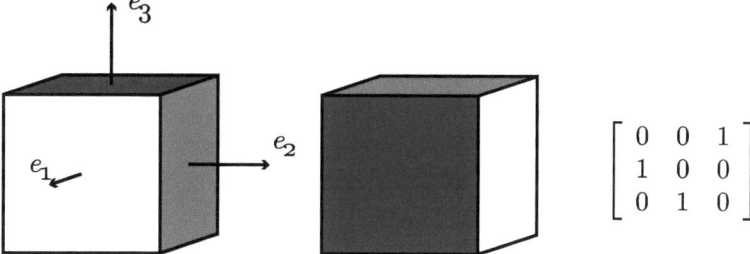

Fig. 3.3 120° rotation on the axis (1, 1, 1)

Example 3.3. The set $GL_n(\mathbb{R})$ of $n \times n$ real matrices M with $\det(M) \neq 0$ form a noncommutative group under matrix multiplication for $n > 1$. We may further restrict $\det(M)$ to yield other matrix groups, e.g., $O_n(\mathbb{R})$ with $|\det(M)| = 1$ and $SO_n(\mathbb{R})$ with $\det(M) = 1$.

The k-fold composition of a group element g with itself is written g^k, with $g^0 = e$ and $g^{-1} \cdot g = e$. The element g^{-1} is called the *inverse* of g and is uniquely defined.

In a group **G** of finite order n, there must be a natural number k for which $g^k = e$. The smallest such positive k is called the *order* of the element g. The order of an element divides the cardinality of the group (see Exercise 3.5). Consequently, in a group whose cardinality is a prime number p, each nonunit element has order p. For finite groups, we also have $g^{-1} = g^{k-1}$, where k is the order of g.

Example 3.4. The rotations of the plane about a fixed origin through multiples of $2\pi/n$ form a group with composition of rotations. The rotations by $2k\pi/n$ are determined by $k \in \{0, \ldots, n-1\}$, so this group can also be described by the set $\{0, \ldots, n-1\}$ with addition modulo n as the group operation and denoted by $\mathbb{Z}_n$.

Example 3.5. The set of numbers $\{1, -1, i, -i\} \subseteq \mathbb{C}$, where i is the imaginary unit, forms a group. Multiplication by these complex values rotates the complex plane by multiples of $\pi/2$, so this group is again a manifestation of $\mathbb{Z}_4$.

Example 3.6. Consider the specific cube in 3-space whose vertex coordinates are $(\pm 1, \pm 1, \pm 1)$ (Fig. 3.3). Any isometric transformation of the cube onto itself preserves the center of the cube, $(0, 0, 0)$, and is in fact given by a linear transformation, represented by a matrix. We say that this transformation group is *representable*. The specific matrices for the transformation can be readily determined by considering the effect of the transformation on the front, right, and top faces of the cube, which correspond the three standard basis elements e_1, e_2, and e_3, so the matrix of the transformation T is the matrix with columns $T(e_1)$, $T(e_2)$, and $T(e_3)$. Candidates for transformations would be matrices whose three columns contain one selection from each of $\{\pm e_1\}$, $\{\pm e_2\}$, and $\{\pm e_3\}$, giving a total of 48.

Fig. 3.4 Is the i in the pyramid really imaginary?

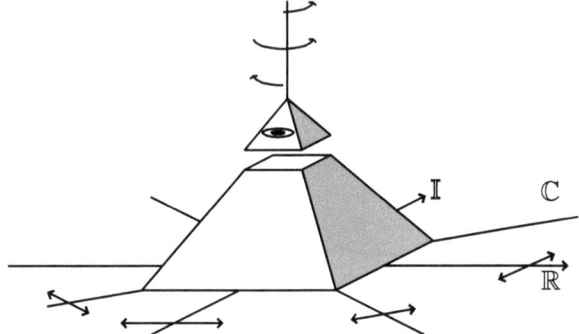

Example 3.7. Consider the pyramid of Cheops in Egypt as an object in 3-space. Orientation-preserving symmetries of this object form a group which is isomorphic to $\mathbb{Z}_4$. Groups of symmetries of geometric objects are important examples for us. Of course, Pharaoh was not concerned with physically manipulating the pyramid. Symmetry was included in the design for aesthetic reasons, so the reflective symmetries are also essential. Consider the pyramid placed on the complex plane $\mathbb{C}$ centered at the origin, see Fig. 3.4. The transformations of the pyramid are encoded by the complex functions $\{z, -z, iz, -iz, \bar{z}, -\bar{z}, i\bar{z}, -i\bar{z}\}$.

3.1.3 Subgroups

A *subgroup* **H** of a group **G** consists of a subset of the elements of **G** together with the same law of composition as for **G**. So **H** is itself a group. To check if a finite subset **H** of **G** forms a subgroup, it is enough to check if **H** is closed under composition, i.e. if the product of any two elements of **H** is again an element of **H**. This is not true if **H** is infinite, see Exercise 3.4.

For $\mathbb{Z}_n$, we can easily determine that all subgroups are isomorphic to $\mathbb{Z}_k$ where k divides n. There are no other subgroups. In particular, $\mathbb{Z}_p$, with p prime, has only trivial subgroups, namely, $\mathbb{Z}_1$ and $\mathbb{Z}_p$. In fact, any group of prime cardinality has only trivial subgroups.

Example 3.8. A common method of constructing a subgroup is to start with a group of transformations and consider the subset of those transformations which preserve some set of elements or some property. Consider the group **C** of symmetries of a cube described in Example 3.6, and suppose that the opposite faces of the cube have been colored: top and bottom red, left and right white, and front and back blue. Every transformation of the cube permutes these three colors. A 120° counterclockwise rotation about the top-right-front corner permutes the colors in a 3-cycle (red, blue, white). The identity fixes all three colors but so does a 180° rotation about an axis through the front face. Altogether there are eight

Fig. 3.5 A face coloring of a truncated hexagonal prism

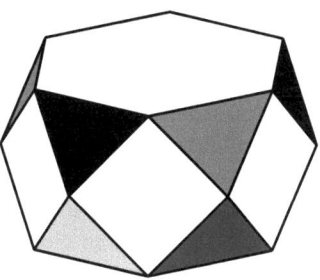

transformations which fix all three colors, see Exercise 3.32, but without listing what they are, we know a priori they form a group, $\mathbf{C}_{rwb}$, since they are closed under composition.

Consider instead those symmetries which are just required to assign red faces to red faces. As before, this set is closed under composition so these transformations form a subgroup $\mathbf{C}_r$. Similarly we can define the subgroups $\mathbf{C}_w$ and $\mathbf{C}_b$ of transformations which preserve white and blue, respectively. Clearly the three subgroups $\mathbf{C}_r$, $\mathbf{C}_w$, and $\mathbf{C}_b$ are isomorphic; however, there is a closer relationship between them. Let h be any symmetry which transforms white faces to red, and let $g \in \mathbf{C}_r$. Then clearly $h^{-1}gh \in \mathbf{C}_w$, so $h^{-1}\mathbf{C}_r h = \{h^{-1}gh \mid g \in \mathbf{C}_r\} \subseteq \mathbf{C}_w$. On the other hand, if $f \in \mathbf{C}_w$, then $f = h^{-1}(hfh^{-1})h$ and $hfh^{-1} \in \mathbf{C}_r$ so $h^{-1}\mathbf{C}_r h = \mathbf{C}_w$. We say the subgroups $\mathbf{C}_r$ and $\mathbf{C}_w$ are *conjugate by h* or simply *conjugate*. We also say that the pair of elements $g \in \mathbf{C}_r$ and $h^{-1}gh \in \mathbf{C}_w$ are *conjugate*.

Example 3.9. Consider the group of symmetries of the truncated hexagonal prism where the triangular faces are colored alternatingly red and black incident to the top face and alternatingly green and blue for those incident to the bottom face (Fig. 3.5). The 24 symmetries **P** of the prism permute the colors just as in Example 3.8, and the four 6-element subgroups $\mathbf{P}_r$, $\mathbf{P}_k$, $\mathbf{P}_b$, and $\mathbf{P}_g$, preserving, respectively, red, black, blue, and green, are conjugate, as before. Now, however, $\mathbf{P}_r = \mathbf{P}_k = \mathbf{P}_b = \mathbf{P}_g$, and we say the subgroup $\mathbf{P}_r$ is *normal*.

What other colorings of the 12 triangles have normal subgroups fixing particular colors?

A subgroup $\mathbf{H} \subseteq \mathbf{G}$ is *normal* if $g^{-1}\mathbf{H}g = \mathbf{H}$ for all $g \in \mathbf{G}$.

3.1.4 Cosets

Consider again the Example 3.8. There are 16 elements in $\mathbf{C}_r$, all of which preserve red. The other 32 elements of $\mathbf{C}$ are divided into two classes, 16 of which map red to blue, $\mathbf{C}_{rb}$, and 16 of which map red to white $\mathbf{C}_{rw}$. Given any element $t_{rb} \in \mathbf{C}_{rb}$, the remaining elements of $\mathbf{C}_{rb}$ can conveniently be listed as $\mathbf{C}_{rb} = \{t_{rb}t_r \mid t_r \in \mathbf{C}_r\}$, and we write $\mathbf{C}_{rb} = t_{rb}\mathbf{C}_r$. Similarly $\mathbf{C}_{rw} = t_{rw}\mathbf{C}_r$, for some fixed element t_{rw} which sends red to white. So the subsets $\mathbf{C}_r$, $t_{rb}\mathbf{C}_r$, and $t_{rw}\mathbf{C}_r$ partition $\mathbf{C}$ by the effect on red. Notice that, of these three subsets, only $\mathbf{C}_r$ is a subgroup. On the other hand, considering again two fixed transformations t_{wr} and t_{br} which transform white and blue, respectively, to red, $\mathbf{C}_r t_{wr}$, $\mathbf{C}_r t_{br}$, and $\mathbf{C}_r$ partition $\mathbf{C}$ by which color is transformed *to* red. Clearly these are distinct partitions.

Let $\mathbf{G}$ be a group of cardinality n and $\mathbf{H} = \{h_1, \ldots, h_k\}$ a subgroup of $\mathbf{G}$. For an element $g \in \mathbf{G}$, we call the set $\mathbf{H}g = \{h_1 \cdot g, h_2 \cdot g, \ldots, h_k \cdot g\}$ a *right coset* of $\mathbf{H}$. Cosets corresponding to the same subgroup are either identical or disjoint, since if $\mathbf{H}g_1$ and $\mathbf{H}g_2$ have an element $h_1 \cdot g_1 = h_2 \cdot g_2$ in common, then $g_2 \cdot g_1^{-1}$ and its inverse $g_1 \cdot g_2^{-1}$ are elements of $\mathbf{H}$, so $\mathbf{H}g_1 = \mathbf{H}g_2$. Moreover, all cosets have the same number, k, of elements; therefore, k divides n. The quotient $i = n/k$ is called the *index* of $\mathbf{H}$ in $\mathbf{G}$, written $i = [\mathbf{G} : \mathbf{H}]$. We formulate this as a theorem.

Theorem 3.10 (Lagrange). *The cardinality k of a subgroup $\mathbf{H}$ of a group $\mathbf{G}$ divides the cardinality n of $\mathbf{G}$ and $n = ik$, where $i = [\mathbf{G} : \mathbf{H}]$ is the index of $\mathbf{H}$ in $\mathbf{G}$.*

Right cosets of $\mathbf{H}$ therefore partition the group $\mathbf{G}$ into equivalence classes of the same size. In a similar way, one can define *left cosets*.

Consider again Example 3.9. Just as in Example 3.8, the partition of $\mathbf{P}$ by right cosets, $\{\mathbf{P}_r t_{gr}, \mathbf{P}_r t_{kr}, \mathbf{P}_r t_{br}, \mathbf{P}_r\}$, partitions $\mathbf{P}$ according to which colored faces are transformed to red, and the partition by left cosets, $\{t_{rg}\mathbf{P}_r, t_{rk}\mathbf{P}_r, t_{rb}\mathbf{P}_r, \mathbf{P}_r\}$, is according to which faces are the image of red faces. Unlike Example 3.8, in this case, the two partitions are the same. That is, all the permutations in each of the cosets induce the same permutation of the colors: Each element of $t_{rk}\mathbf{P}_r$ gives (red black)(green blue); each element of $t_{rb}\mathbf{P}_r$ gives (red blue)(black green); each element of $t_{rg}\mathbf{P}_r$ gives (red green)(blue black); and each element of $\mathbf{P}_r$ fixes all the colors. In fact, these cosets act as a group permuting the colors.

If $\mathbf{H}$ is a normal subgroup of $\mathbf{G}$, then it follows immediately that corresponding left and right cosets of $\mathbf{H}$ are identical, $g\mathbf{H} = \mathbf{H}g$, and in this case, we can define a natural multiplication of the cosets by $(\mathbf{H}g_1)(\mathbf{H}g_2) = \mathbf{H}g_1g_2$, which satisfies the axioms for a group, called the *quotient group* of $\mathbf{G}$ by $\mathbf{H}$, written $\mathbf{G}/\mathbf{H}$. The cardinality of $\mathbf{G}/\mathbf{H}$ is just the number of cosets, that is, the index of $\mathbf{H}$ in $\mathbf{G}$, so we have by Lagrange's theorem:

$$|\mathbf{G}| = |\mathbf{H}|[\mathbf{G} : \mathbf{H}] = |\mathbf{H}||\mathbf{G}/\mathbf{H}|.$$

Example 3.11. The quaternions $\mathbb{H}$ are a four-dimensional noncommutative generalization of the complex numbers $\mathbb{C}$ which were introduced by Hamilton in the nineteenth century. Just as a complex number can be written as $z = a + bi$, a quaternion is of the form $q = a + bi + cj + dk$ with $a, b, c,$ and d real. Addition in $\mathbb{H}$ is componentwise, and the multiplication distributes over addition, so in order to compute the product in $\mathbb{H}$, we need only note the products of the 8 elements forming the quaternion group, $\mathbf{H}$ given in Table 3.1.

In the table, the elements of the cosets with respect to the subgroup $\{1, -1\}$ have been paired to show as well the multiplication of the four elements in $\mathbf{H}/\{1, -1\}$.

3.1 Groups

Table 3.1 The quaternion group

⊙	1	−1	i	$-i$	j	$-j$	k	$-k$
1	1	−1	i	$-i$	j	$-j$	k	$-k$
−1	−1	1	$-i$	i	$-j$	j	$-k$	k
i	i	$-i$	−1	1	$-k$	k	j	$-j$
$-i$	$-i$	i	1	−1	k	$-k$	$-j$	j
j	j	$-j$	k	$-k$	−1	1	$-i$	i
$-j$	$-j$	j	$-k$	k	1	−1	i	$-i$
k	k	$-k$	$-j$	j	i	$-i$	−1	1
$-k$	$-k$	k	j	$-j$	$-i$	i	1	−1

3.1.5 Group Homomorphisms and Isomorphisms

Given two groups $(\mathbf{G}, \cdot)$ and $(\mathbf{H}, \cdot)$, a *homomorphism* $\phi : \mathbf{G} \to \mathbf{H}$ is a map with the property that $\phi(g_i) \cdot \phi(g_j) = \phi(g_i \cdot g_j)$ for all pairs (g_i, g_j) of elements of $\mathbf{G}$. Every homomorphism maps the identity of $\mathbf{G}$ to the identity of $\mathbf{H}$ because $\phi(g) = \phi(1 \cdot g) = \phi(1) \cdot \phi(g)$. Moreover, $\phi(g^{-1}) \cdot \phi(g) = \phi(1) = 1$ implies that $\phi(g^{-1}) = (\phi(g))^{-1}$.

A homomorphism $f : \mathbf{G} \longrightarrow \mathbf{H}$ which is both injective and surjective is called *isomorphism*, and we write $\mathbf{G} \cong \mathbf{H}$

There are three essential subgroups associated with any group homomorphism $f : \mathbf{G} \longrightarrow \mathbf{H}$.

$$\ker f = \{g \in \mathbf{G} | f(g) = 1\}$$
$$f(\mathbf{G}) = \{h \in \mathbf{H} | f(g) = h \text{ for some } g \in \mathbf{G}\}$$
$$\mathbf{G}/\ker f \cong f(\mathbf{G})$$

The fact that the kernel, $\ker(f)$, and the image, $f(\mathbf{G})$, are in fact subgroups; that $\ker(f)$ is always a normal subgroup so that $\mathbf{G}/\ker f$ is defined; and that the groups $\mathbf{G}/\ker f$ and $f(\mathbf{G})$ are isomorphic is a standard exercise. Conversely, every normal subgroup $\mathbf{N}$ of a group $\mathbf{G}$ is the kernel of a homomorphism, namely, the map $\mathbf{G} \longrightarrow \mathbf{G}/\mathbf{N}$ which sends each $g \in \mathbf{G}$ to its coset $\mathbf{N}g$, so it is common to establish the normality of a subgroup by exhibiting a homomorphism of which it is the kernel.

Example 3.12. If $\mathbf{G}$ is a group of invertible matrices such as in Example 3.6, then the determinant function, since it satisfies $\det(MN) = \det(M)\det(N)$, defines a homomorphism $\mathbf{G} \longrightarrow (\mathbb{R} - \{0\})$ into the multiplicative group of the reals. For this homomorphism on the symmetries of the cube, the image group is $\det(\mathbf{G}) = \{-1, 1\}$, and the kernel, $\ker(\det)$, is the group of rotational symmetries, which therefore has index 2. The nontrivial coset is the set of symmetries with determinant -1 and consisting of reflections, rotary reflections, and the antipodal map.

By Lagrange's theorem, a homomorphism is an isomorphism onto its image if and only if the kernel is trivial.

Example 3.13. The multiplicative group on $\{1, -1, i, -i\}$ is isomorphic to the additive group $\mathbb{Z}_4$ on the elements $\{0, 1, 2, 3\}$ with $\phi(1) = 0, \phi(-1) = 2, \phi(i) = 1, \phi(-i) = 3$ as an isomorphism.

For a prime p, all groups of cardinality p are isomorphic, so up to isomorphism, there is only one group of cardinality p, see Exercise 3.16.

3.2 Cayley Graphs

3.2.1 Definition of the Cayley Graph

The *Cayley graph* is a very fruitful way of using graphs to study groups. Let **G** be a group with identity 1, and let S be a set of *generators*, that is, every element of **G** can be written as a product of elements in S. We define a graph $G(\mathbf{G}, S)$ by taking the vertices of $G(\mathbf{G}, S)$ to be the elements of **G** and setting g and h adjacent if either $g = hs$ or $h = gs$ for some $s \in S$. The graph $G(\mathbf{G}, S)$ is called the (right) Cayley graph of **G** with respect to S.

The edges are naturally partitioned by the set S, and it is common to color the edges of the Cayley graph with colors corresponding to generators, see Fig. 3.6.

The Cayley graph depends on the choice of generating set. In Fig. 3.6, we have two regular graphs which are both Cayley graphs for $\mathbf{C}_6$. If there are too many generators, the uncolored Cayley graph contains little information beyond the number of elements in the group. In fact, every group has a Cayley graph which is complete, e.g., for $\mathbf{C}_6$ take $S = \{\alpha, \alpha^2, \alpha^3\}$.

Note that nonisomorphic groups may have isomorphic Cayley graphs, see Fig. 3.7 in which the two graphs on the left are isomorphic. The first corresponds to a cyclic (see Exercise 3.14 for the definition) group of order 4, and the second corresponds to the Klein 4-group, as defined in Example 3.17. More intriguing examples of this nature were examined in [57]. The graph on the right of Fig. 3.7 corresponds to the quaternion group. The reader is invited to find an abelian group with the same Cayley graph.

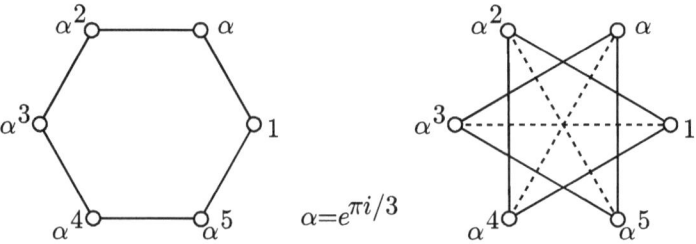

Fig. 3.6 Cayley graphs for $\mathbf{C}_6$ with generators $\{\alpha\}$ and $\{\alpha^2, \alpha^3\}$

3.2 Cayley Graphs

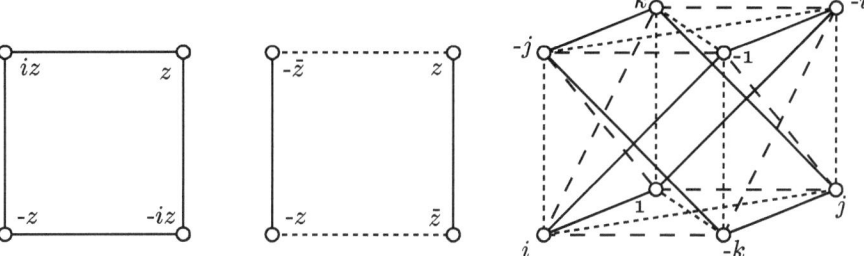

Fig. 3.7 Cayley graphs for C_4, **K**, and **H**

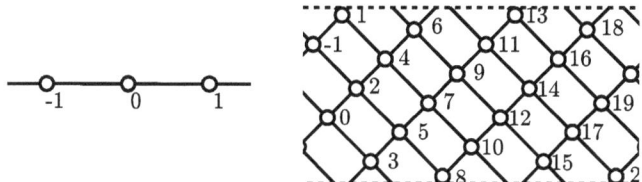

Fig. 3.8 Cayley graphs for the integers

Fig. 3.9 $C(\mathbb{Z}_{10}, \{1\})$ and $C(\mathbb{Z}_{10}, \{2, 5\})$

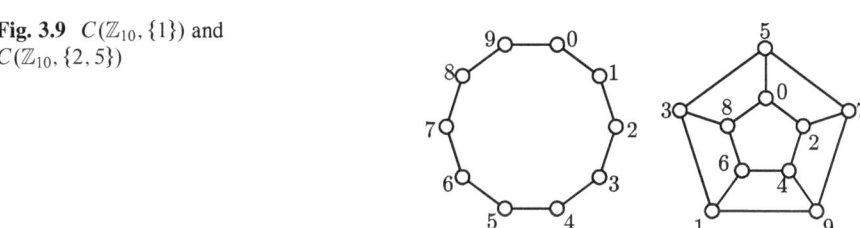

3.2.2 Examples of Cayley Graphs

Example 3.14. The cyclic group, Cyc(n). A group is *cyclic* if it has a single generator. All elements of a multiplicative cyclic group with generator α are of the form α^k. If the powers are all distinct, then the group is infinite and isomorphic to the integers, $\mathbf{G} = \text{Cyc}(\infty) \cong \mathbb{Z}$. The Cayley graph of Cyc(∞) with $S = \{\alpha\}$ is an infinite path; however, with other generating sets, the graph can be more complicated, see Fig. 3.8 in which $G(\mathbb{Z}, \{2, 3\})$ is drawn on an infinite cylinder. If α has finite order n, then the group is $\text{Cyc}(n) = \{\alpha^0, \alpha^1, \ldots, \alpha^{n-1}\}$ and is isomorphic to $\mathbb{Z}_n$, see Fig. 3.9. The infinite cyclic group Cyc(∞) generated by α has two cyclic generators, α and α^{-1}. The element α^k generates a subgroup of Cyc(∞) of index k. The element α^k generates a subgroup of Cyc(n) of index $\gcd(n, k)$, so Cyc(n) has $\phi(n)$ cyclic generators, where ϕ, the *Euler totient ϕ function*, is defined to be the number of positive integers less than n and coprime to n.

Graphs which are Cayley graphs of cyclic groups are often called *circulant graphs*, or simply circulants.

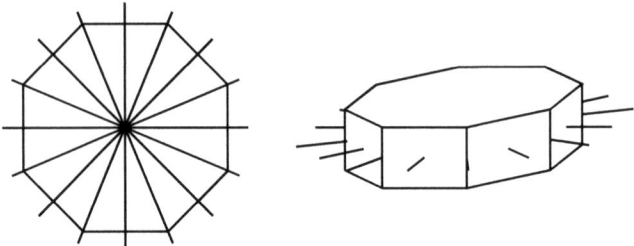

Fig. 3.10 Cardinality 2 subgroups of Dih(8)

Example 3.15. The direct product of cyclic groups. The *direct product* of **G** and **H** is denoted by **G** × **H** and consists of all |**G**||**H**| elements of the form (g, h) with multiplication coordinatewise. The group $\mathbb{Z}_n \times \mathbb{Z}_2$, $n \geq 3$ is generated by $S = \{(1, 0), (0, 1)\}$, and its corresponding Cayley graph is isomorphic to the prism graph Π_n. If n is odd, then $\mathbb{Z}_n \times \mathbb{Z}_2$ is also generated by $(1, 1)$, and so $\mathbb{Z}_n \times \mathbb{Z}_2 \cong \mathbb{Z}_{2n}$ for n odd. In general, if $\gcd(n, m) = 1$, then $\mathbb{Z}_n \times \mathbb{Z}_m \cong \mathbb{Z}_{nm}$. The fundamental theorem of abelian groups details how to decompose every finitely generated abelian group into the direct product of cyclic groups, see [83].

For the remainder of the examples in this section, if no Cayley graph is given, then the reader is expected to produce a diagram.

Example 3.16. The dihedral group, Dih(n), $n > 3$. The dihedral group is the group of symmetries of a regular n-gon in the plane. It contains, together with n rotations about its center through integer multiples of $2\pi/n$, the n reflections about the lines of symmetry. The group is therefore of cardinality $2n$ and is denoted by Dih(n). The subgroup of all rotations is of index 2 in Dih(n) and is hence a normal subgroup. The n subgroups of cardinality 2 generated by the reflections through the center of the n-gon are not normal: For odd n, any two are conjugate to one another by a reflection whose mirror line passes through one vertex and bisects one side, while for even n, they fall into two conjugacy classes, one of which uses mirrors bisecting two sides and one of which uses mirrors passing through two vertices.

The dihedral group gets its name not from this manifestation but from its related role as the group of rotational symmetries of the n-fold prism, in which the n subgroups of cardinality 2 are 180° rotations about the horizontal symmetry axes, see Fig. 3.10. The full symmetry group of the n-fold prism is Dih$(n) \times \mathbb{Z}_2$, with the nontrivial element of $\mathbb{Z}_2$ being the reflection in the horizontal plane.

Example 3.17. The Klein 4-Group. If we regard a regular 2-gon as two vertices joined by two semicircular arcs, the dihedral group Dih(2) would consist of 4 elements: two reflections, m_1 and m_2, in perpendicular mirror lines, one 180° rotation, r, and the identity. This group of four elements is not cyclic and is usually denoted by **K**. The matrices of these transformations are

3.2 Cayley Graphs

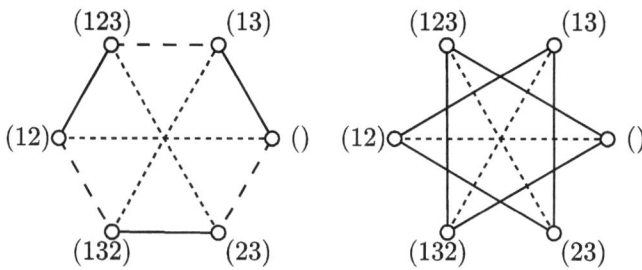

Fig. 3.11 $C(\text{Sym}(3), \{(12), (23), (13)\})$ and $C(\text{Sym}(3), \{(12), (123)\})$

$$m_1 = \begin{bmatrix} -1 & 0 \\ 0 & 1 \end{bmatrix}, m_2 = \begin{bmatrix} 1 & 0 \\ 0 & -1 \end{bmatrix}, r = \begin{bmatrix} -1 & 0 \\ 0 & -1 \end{bmatrix}, \text{ and } 1 = \begin{bmatrix} 1 & 0 \\ 0 & 1 \end{bmatrix}.$$

Geometrically we can distinguish the two orientation-reversing reflections from the orientation-preserving rotation. Algebraically, however, they are not distinguishable; in fact, any permutation of the nonidentity elements gives an automorphism of **K**. We will see **K** recur in several forms.

Example 3.18. We have already considered in Sect. 2.3.5 the symmetric group Sym(n). Sym(n) is non-abelian for $n > 2$. The $n!$ elements of the symmetric group are generated by the $n(n-1)/2$ *transpositions*, those bijections which fix all but two elements. Generating Sym(n) with fewer transpositions is possible, see Exercise 3.36. For viewing purposes, one prefers a Cayley graph with few generators, see Fig. 3.11 in which we have two- and three-generator Cayley graphs of Sym(3). One of these graphs is planar. Which one? The other arises from a generating set containing a generator whose removal leaves a connected Cayley graph. This means that the generating set is redundant. The 24 elements of Sym(4) also have a generating set inducing a planar Cayley graph, see Exercise 3.9.

Example 3.19. The alternating group, Alt(n). There is an homomorphism from the symmetric group to $n \times n$ matrices which sends each permutation π to the matrix with columns $[\mathbf{e}_{\pi(1)}, \mathbf{e}_{\pi(2)}, \mathbf{e}_{\pi(3)}, \ldots, \mathbf{e}_{\pi(n)}]$, where $\mathbf{e}_i$ is the ith column of the identity matrix. The determinant of the matrix corresponding to a transposition is -1 so, since Sym(n) is generated by transpositions, all determinants are ± 1. The homomorphism det : Sym(n) $\longrightarrow \{-1, 1\}$ has kernel Alt(n), so Alt(n) is a normal subgroup consisting of all those permutations whose associated matrices have determinant 1. Alt(n) is called the *alternating group*. We have shown the nonobvious fact that each element of Sym(n) can either only be written as the product of an even number of transpositions, i.e. those in Alt(n), or can only be written as product of an odd number of transpositions.

The alternating group Alt(2) is trivial. Alt(3) is isomorphic to $\mathbb{Z}_3$ and so has no nontrivial subgroups. Alt(4) has twelve elements and corresponds to the rotational group of a regular tetrahedron. To verify this fact, notice first that the rotation group **T** of the tetrahedron must permute the four vertices, so there is a homomorphism

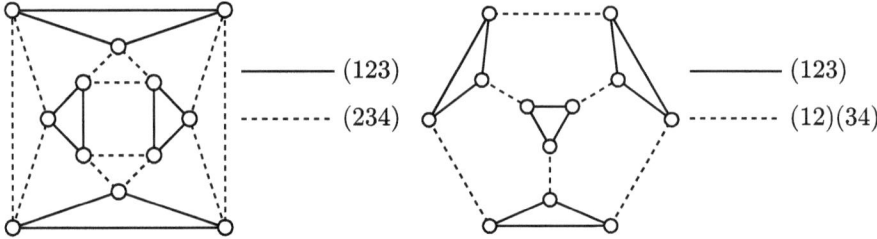

Fig. 3.12 $C(\text{Alt}(4), \{(123), (234)\})$ and $C(\text{Alt}(4), \{(123), (12)(34)\})$

T ⟶ Sym(4). Next, note that the axes of rotation are lines through the centroid of the tetrahedron which intersect its surface at a vertex, the center of an edge, or the center of a face. Those that pass through a vertex pass through the center of the opposite face and correspond to one of the eight 3-cycles (123), (132), (124), (142), (134), (143), (234), and (243). If the axis passes through the center of an edge, it passes as well though the opposite edge, and the three pairs of opposite edges correspond to the three products of disjoint transpositions (12)(34), (13)(24), and (14)(23). These eleven transformations, plus the identity, comprise the whole rotational group, and all correspond to the even elements of Sym(n) and Alt(n) ≅ **T**.

Alt(4) has some nice Cayley graphs, including the graphs of a truncated cube and a truncated tetrahedron, see Fig. 3.12. Another interesting fact about Alt(4) is that it is the only alternating group which has a nontrivial normal subgroup, see Exercise 3.12.

Example 3.20. The free group, $\mathbf{F}_n$. The *free group* on a set $\{x_1, \ldots, x_n\}$ has the property that, given any group **H** and any sequence of n elements $\{h_1, \ldots, h_n\}$ in **H**, there is a unique homomorphism $f : \mathbf{F}_n \longrightarrow \mathbf{h}$ such that $f(x_i) = h_i$. This property is restrictive enough to uniquely define $\mathbf{F}_n$ up to isomorphism and specifies that $\mathbf{F}_n$ is the most general group with generators $\{x_1, \ldots, x_n\}$. If we want to specify the generating set, we write $\mathbf{F}_n = \langle x_1, x_2, \ldots, x_n \rangle$. It is not hard to show that $\mathbf{F}_1 = \langle a \rangle \cong \mathbb{Z}$. In general, the free group $\mathbf{F}_n = \langle x_1, \ldots, x_n \rangle$ may be constructed from all *words* on the generators, i.e. sequences of (positive and negative) powers of the generators. The multiplication is by concatenation of words, and two words are equivalent if they are related by a finite sequence of insertions and deletions of the trivial words, $x_i x_i^{-1}$ and $x_i^{-1} x_i$. In a free group, there are only trivial equations between words in the generators. The details of the construction are rather delicate, see [61]. The Cayley graph of **F** with respect to $\{x_1, \ldots, x_n\}$ is an infinite $2n$-valent tree.

Example 3.21. Finitely presented groups. In Example 3.19, we determined that the rotational group of the tetrahedron **T** was generated by a vertex rotation of order 3, $r = (123)$, and an involution $s = (12)(34)$. There is therefore a homomorphism from $\mathbf{F} = \langle r, s \rangle$, the free group on r and s to **T**, $p : \langle r, s \rangle \longrightarrow \mathbf{T}$. Since the rotations r and s satisfy $r^3 = s^2 = (rs)^3 = 1$, we have $r^3, s^2, (rs)^3 \in \ker(p)$. In fact, see Exercise 3.17, the kernel is generated by the words r^3, s^2, $(rs)^3$, and their conjugates; in other words, the equation $r^3 = s^2 = (rs)^3 = 1$ are sufficient to describe **T** in terms of the generators r and s. We have $\mathbf{T} \cong \langle r, s \rangle / \ker(p)$, and we write $\mathbf{T} = \langle r, s \mid r^3 = s^2 = (rs)^3 = 1 \rangle$ and say that **T** is *presented* by generators r and s with *relations* $r^3 = 1$, $s^2 = 1$, and $(rs)^3 = 1$.

3.3 Group Actions

3.3.1 Permutation Groups

We have so far considered groups in the context of various transformations: the automorphism groups of graphs, symmetry groups of geometric objects, groups of matrices, and groups of linear transformations. The concept of *group action* unifies all these ideas and can be regarded as a generalization of the permutation group. A *permutation group* **G** is simply a subgroup of Sym(X). For example, Alt(X) is a permutation group. More generally, if there is a homomorphism $\phi : \mathbf{G} \longrightarrow$ Sym(X), we say that **G** *acts* on the set X and that ϕ gives an *action* of **G** on X. If we have an action, each element of **G** corresponds to a bijection $\phi(g)$ on X; however, in most cases, the notation for the homomorphism is suppressed, and we write simply $g(x)$.

Given any group action of **G** on X and given any $x \in X$, the *orbit*, Orbit(x), is defined to be the set of images of x under the action of **G**:

$$\text{Orbit}(x) = \{gx \mid g \in \mathbf{G}\}.$$

The fewer the number of orbits, the more symmetric the object. The action of **G** on X is called *transitive* if there is only one orbit. Clearly, the action of Sym(X) itself is transitive on X. We also define the *stabilizer* of x, Stab(x), to be the set of all group elements which fix $x \in X$. Since the composite of two bijections which fix x also fixes x, we have that Stab(x) is a subgroup of **G**.

Theorem 3.22. *If* **G** *acts on* X, *and* $x \in X$, *then*

$$|\mathbf{G}| = |\text{Orbit}(x)||\text{Stab}(x)|$$

so the cardinality of the orbit of x is the index of its stabilizer,

$$|\text{Orbit}(x)| = [\mathbf{G} : \text{Stab}(x)].$$

Proof. Associate to each element $g(x) \in \text{Orbit}(x)$ the set $X_{g(x)}$ of all elements of **G** which map x to $g(x)$. The sets $X_{g(x)}$ partition **G**. Moreover, if $g' \in \text{Stab}(x)$, then $gg'(x) = g(x)$, so $g\text{Stab}(x) \subseteq X_{g(x)}$. On the other hand, if $g'(x) = g(x)$, then

$g^{-1}g'(x) = x$ and $g^{-1}g' \in \text{Stab}(x)$, so $g' = gg^{-1}g' \in g\text{Stab}(x)$ and $g\text{Stab}(x) = X_{g(x)}$. So the partition of **G** by the sets $X_{g(x)}$ is in fact the partition by left cosets, $|X_{g(x)}| = |\text{Stab}(x)|$ and the result follows. □

Given any group action of **G** on X and given any $g \in \mathbf{G}$, the subset of all fixed points of g is denoted by $\text{Fix}(g)$, the *fixed point set* of g.

Example 3.23. The rotational group of the prism P_n, with 2 n-gonal and n rectangular faces, acts on the set of vertices, edges, and faces. (If the prism has square faces, then P_4 is a cube, which has additional symmetry.) The only element in the stabilizer of a vertex v is the identity, $|\text{Stab} v| = 1$. There is one vertex orbit so the rotational group acts transitively on the vertices, $|\text{Orbit} v| = |V| = 2n$, so $\text{Dih}(n) = 2n$, by Theorem 3.22. The rotational group does not act transitively on the set of faces. The two n-gonal faces belong to one orbit, while the rectangular faces belong to the other. Each n-gonal face f is stabilized by the n rotations about the vertical axis, and so $\text{Stab}(f)$ is normal of cardinality n. Each rectangular face f' belongs to an orbit of size n, and each is stabilized by the 180° rotation about the center of the face, $\text{Stab} f' \cong \mathbb{Z}_2$. These n copies of $\mathbb{Z}_2$ are all conjugate, i.e., $\text{Stab} f'$ is not normal. The reader may similarly analyze the action on the edges of the prism.

Example 3.24. Quaternions were mentioned in Example 3.11. The unit quaternions,

$$\mathbb{H}_1 = \{a + bi + cj + dk \mid a^2 + b^2 + c^2 + d^2 = 1; a, b, c, d \in \mathbb{R}\}$$

can be expressed as $\sin(\theta/2) + \cos(\theta/2)\mathbf{u}$, with $\mathbf{u}$ a unit vector in $\mathbb{R}^3$ expressed in the standard basis i, j, and k. These unit quaternions form a multiplicative group and act on $\mathbb{R}^3$ by rotating $\mathbb{R}^3$ about the axis in direction $\mathbf{u}$ through the origin by an angle θ. That is, there is a homomorphism $\phi : \mathbb{H}_1 \longrightarrow O(3)$. In this case, the homomorphism is not one to one, since the unit quaternions 1 and -1 both map to the identity in $O(3)$, expressing the fact that a rotation by θ about axis $\mathbf{u}$ is the same as rotating by $-\theta$ about the opposite vector $-\mathbf{u}$.

The eight-element quaternion group $\mathbf{H} = \{\pm 1, \pm i, \pm j, \pm k\}$ also acts on $\mathbb{R}^3$ with ± 1 acting as the identity, $\pm i$ a 180° rotation about the x-axis, $\pm j$ a 180° rotation about the y-axis, and $\pm k$ a 180° rotation about the z-axis. So the index 4 normal subgroup $\ker(\phi) = \{\pm 1\}$ consists of those elements which act as the identity, and the action is carried by the quotient, which is isomorphic to **K**, the Klein four-group.

Example 3.25. Given any group **G**, and any fixed $g \in \mathbf{G}$, left multiplication by g defines a permutation π_g on **G** by $\pi_g(x) = gx$, since $gx_1 = gx_2$ implies $x_1 = g^{-1}gx_1 = g^{-1}gx_2 = x_2$. The map $\pi : \mathbf{G} \longrightarrow \text{Sym}(\mathbf{G})$ defines a homomorphism since $\pi_{gh}(x) = ghx = \pi_g(hx) = \pi_g(\pi_h(x))$. This action is called the *left regular action* of **G**. A related but algebraically distinct action is the *right regular action* defined by right multiplication; $\rho_g(x) = xg^{-1}$.

3.3 Group Actions

Fig. 3.13 Cayley graphs for S_4 with generators a transposition and four-cycle

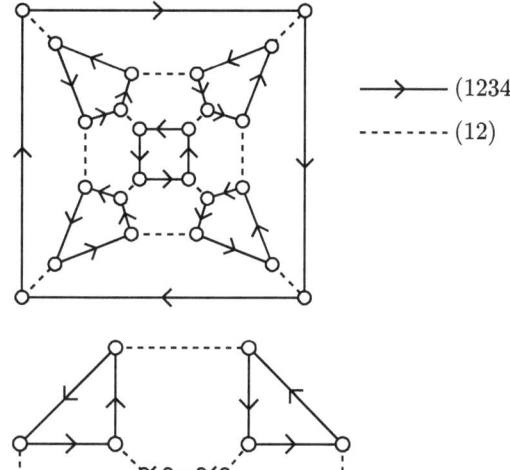

→ (1234)
------ (12)

Fig. 3.14 Cayley graphs for S_4 with generators a transposition and three-cycle

→ (234)
------ (12)

3.3.2 Actions on Cayley Graphs

The seemingly trivial action of Example 3.25 is very useful in the study of **G**. It is transitive, since $\pi_{gh^{-1}}$ maps any h to any g. Moreover, every stabilizer is trivial, since $gh = h$ implies $g = 1$. So we have

Theorem 3.26. **G** *is isomorphic to a subgroup of* Sym(**G**).

The elements of **G** have already been considered as the vertices of the Cayley graph with generating set S. If (g, sg) is an edge in the left Cayley graph, then the right regular action defines an automorphism of $C(\mathbf{G}, S)$. Similarly, the left regular action defines an automorphism of the right Cayley graph, with these two actions coinciding if and only if the group is abelian. We have the following:

Theorem 3.27. **G** *is isomorphic to a subgroup of* Aut $(C(\mathbf{G}, S))$.

G $\not\cong$ Aut $(C(\mathbf{G}, S))$ in general essentially because there may be underlying symmetry in the generating set itself. If, however, we color the edges of the Cayley graph according to which generator they correspond and direct the edges, say for the left Cayley graph, from g to sg, we get the directed colored Cayley graph, see Figs. 3.13 and 3.14, in which the involution (12) in Sym(4) can either correspond to two oppositely directed parallel edges or, equivalently, a single undirected edge.

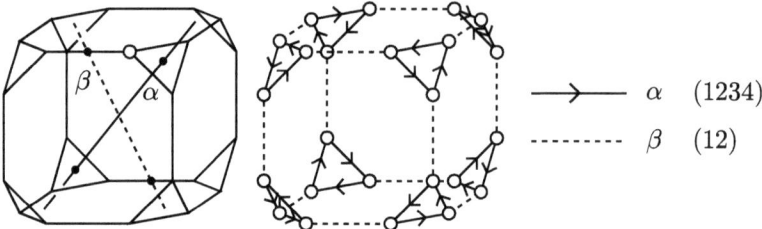

Fig. 3.15 The rotation group of the truncated cube

We have the following:

Theorem 3.28. **G** *is isomorphic to the group of color-preserving automorphisms of the directed colored Cayley graph* $C(\mathbf{G}, S)$.

Proof. The right regular action on the vertices preserves the color and direction of the edges. The only thing that needs to be checked is that there are no other color- and direction-preserving automorphisms. We first note that among color- and direction-preserving automorphisms, the only automorphism that fixes an element must fix as well all the vertices adjacent to that element. Since $C(\mathbf{G}, S)$ is connected, it follows that the colored and directed stabilizer of every vertex is trivial. Now, if there is a colored direction-preserving automorphism f which maps vertex v to u, then $\rho_{v^{-1}u}$ also maps v to u, and so $f(\rho_{v^{-1}u})^{-1}$ fixes the vertex u and is hence the identity, so $f = \rho_{v^{-1}u}$, see Example 3.25. □

So, for the colored directed Cayley graph, one need not label the vertices with the group elements. We can always recover the labels up to isomorphism by choosing one vertex and labeling it with the identity and labeling any other vertex with the sequence of generators corresponding to any path in the Cayley graph from 1 to that vertex. This method can be applied to realize a graph as a Cayley graph if its automorphism group is sufficiently rich. If the action of the group **G** on the graph G is transitive on the vertices and has only trivial vertex stabilizers, so $|\mathbf{G}| = |G|$, then the action is said to be *regular*. Sabidussi [85] characterized Cayley graphs in terms of regular actions.

Theorem 3.29 (Sabidussi). *A graph* $G = (V, E)$ *is a Cayley graph if and only if there is a subgroup* **G** *of* $Aut(G)$ *acting regularly on* V. *The generating set S for the Cayley graph may be taken to be those elements of* **G** *which map some fixed vertex of G to its neighbors.*

Example 3.30. It is easy to verify that the rotation group of the truncated cube acts regularly on the vertices, so the graph of the truncated cube is the Cayley graph of the rotation group with generating set $S = \{\alpha, \beta\}$, a set of rotations selected among those rotations α, α^{-1}, and β, which transform a selected vertex to its neighbors, see Fig. 3.15.

3.3 Group Actions

In the undirected, uncolored case, it can happen that the right action on the left Cayley graph yields all the automorphisms of the Cayley graph,

$$\mathbf{G} \cong \mathrm{Aut}\,(C(\mathbf{G}, S)).$$

In such a case, we say that the graph is a *GRR (graphical regular representation)* for **G**. Often finding a GRR for a group involves simply choosing an asymmetric generating set, and for almost all groups, there will exist many generating sets giving a GRR, see [5]. On the other hand, the Klein 4-group has no GRR.

Probably the most famous highly symmetric graph which is not a Cayley graph is the Petersen graph. Among generalized Petersen graphs, there are other examples which are not Cayley graphs, see [72].

It is interesting to consider the symmetry of generalized Petersen graphs. Here is a theorem of Frucht, Graver, and Watkins [34].

Theorem 3.31. *A generalized Petersen graph* $GP(n, k)$ *is vertex transitive if and only if* $(n, k) = (10, 2)$ *or* $k^2 \equiv \pm 1 \pmod{n}$.

Theorem 3.31 was used in [72] to prove the following result:

Theorem 3.32. $GP(n, k)$ *is a Cayley graph if and only if* $k^2 \equiv 1 \pmod{n}$.

$GP(n, k)$ *is a vertex-transitive graph that is not a Cayley graph if and only if* $k^2 \equiv -1 \pmod{n}$ *or* $(n, k) = (10, 2)$, *the exceptional graph being isomorphic to the 1-skeleton of the dodecahedron.*

3.3.3 Primitive Versus Imprimitive Actions

A transitive action of **G** on X is called *imprimitive* if the elements of X can be partitioned into k sets, $X = \bigcup_{i=1}^{k} X_i$, with $1 < k < |X|$ and each $g \in \mathbf{G}$ induces a setwise permutation of the X_i's, i.e., for each i, there exists a j such that $g(X_i) = X_j$.

Example 3.33. Consider the automorphism group of the graph of the truncated cube, see Fig. 3.15. The action on the vertices permutes the eight triangular faces which partition the vertex set, so the action on the vertices is imprimitive. The action on the faces is also imprimitive since no automorphism maps a triangular face to an octagonal face. The action on the triangular faces alone is also imprimitive, since every automorphism preserves pairs of antipodal faces.

If an action of a group on a set is not transitive, then the orbits are trivially permuted setwise, and the action is imprimitive. Looking for a finer partition, we consider the transitive case.

Suppose an action of **G** on a set X is imprimitive, so we have a nontrivial partition of X into sets X_i which are setwise permuted by **G**. Suppose also that the action is transitive. In this case the action on the sets X_i is also transitive, and all the sets X_i

are of the same cardinality, in particular $|X_i| \geq 2$ for all i. Let **H** be the subgroup of **G** of all elements which fix X_1 setwise. Let $x \in X_1$. Then we have that every element of Stab(x), since it fixes x, must fix all of X_1 setwise, so

$$\text{Stab}(x) \subseteq \mathbf{H} \subseteq \mathbf{G}.$$

If Stab(x) = **H**, so the elements of **H** fix X_1 pointwise; no element of **G** can map x to another element of X_1, violating transitivity. On the other hand, if **H** = **G**, then every element of **G** maps x to some element of X_1, so transitivity again contradicts the fact that the partition is nontrivial. Thus, if a transitive action is imprimitive, the point stabilizers cannot be maximal subgroups of **G**. The converse of this is also true.

Theorem 3.34. *Suppose* **G** *acts transitively on a set* X. *Then the action is imprimitive if and only if the point stabilizers are not maximal subgroups of* **G**.

Proof. Let $x \in X$. We have Stab(x) $\neq$ **G** by transitivity. Suppose that there is a group **H** strictly between Stab(x) and **G**. Associate to each element y of X the set of elements of **G** which map x to y; in other words, associate to each y in X that left coset $g(\text{Stab}(x))$, where $g(x) = y$. Now, the action of **G** on X is determined by the action of **G** on the associated left cosets, i.e. if $g(y) = z$ and $g_y(x) = y$, $g_z(x) = z$, then $g(g_y\text{Stab}(x))$ contains the element gg_y and $gg_y(x) = g(y) = z$, so $g(g_y\text{Stab}(x)) = g_z(\text{Stab}(x))$. Now, the cosets of Stab(x) give a subpartition of the cosets of **H**, so we partition elements of X corresponding to which cosets of Stab(x) belong to which cosets of **H**. The group **G** permutes the cosets of **H** setwise, so the action is imprimitive. □

3.3.4 Burnside's Theorem

Burnside's theorem, sometimes referred to as the Cauchy–Frobenius theorem in the literature, enables us to find the number of orbits induced on X by the action of **G** from the sizes of the sets Fix(g). Some history of the theorem is contained in [73, 109], and it still has its place in advanced texts, see [92]. The proof is an example of a standard counting technique in combinatorics, that is, to count the same set in two different ways to obtain an interesting result.

Theorem 3.35 (Burnside's Theorem). *The number, t, of orbits of the action of a group* **G** *on the set* X *is given by*

$$t = \frac{1}{|\mathbf{G}|} \sum_{g \in \mathbf{G}} |\text{Fix}(g)|$$

3.3 Group Actions

Proof. The set of pairs

$$\{(g, x) \mid g \in \mathbf{G}, x \in X, g(x) = x\}$$

can be counted by choosing an element $g \in \mathbf{G}$, calculating $|\text{Fix}(g)|$, and then summing over $g \in \mathbf{G}$ or, on the other hand, by choosing $x \in X$, calculating $|\text{Stab}(x)|$, and then summing over $x \in X$, so we have

$$\sum_{g \in \mathbf{G}} |\text{Fix}(g)| = \sum_{x \in X} |\text{Stab}(x)|.$$

If $\text{Orbit}(x_1) = \text{Orbit}(x_2)$, then the stabilizers of x_1 and x_2 have the same size as they are conjugates. That means that for each element in the same orbit, we get the same contribution on the right-hand side, so each orbit contributes $|\text{Orbit}(x)||\text{Stab}(x)| = |\mathbf{G}|$ and $\sum_{g \in \mathbf{G}} |\text{Fix}(g)| = t|\mathbf{G}|$. □

Example 3.36. How many ways are there to color, up to rotational symmetry, the faces of the cube from the set of colors $\{r, w, b\}$?

The 24-element rotational group of the cube acts on the colorings, and in terms of the problem, two colorings are regarded as the same if and only if they lie in the same orbit. The problem thus requires us to count the number of orbits. We consider each conjugacy class of rotation separately and compute the number of colorings fixed by it. The identity fixes all 3^6 colorings. The three 180° face rotations have 4 orbits, so 3^4 colorings. The six 90° counterclockwise face rotations have three face orbits, so 3^3 fixed colorings. The six 180° edge rotations have 3 orbits, so 3^3 colorings. The eight 120° vertex rotations have 2 orbits, so 3^2 colorings.

Summing, there are

$$(1/24)[1 \cdot 3^6 + 3 \cdot 3^4 + 6 \cdot 3^3 + 6 \cdot 3^3 + 8 \cdot 3^2] = 57$$

orbits by Burnside's theorem, giving the number of colorings up to rotational symmetry.

3.3.5 The Escher Problem

As an application of Burnside's theorem, we consider a problem due to the Dutch graphic artist M. C. Escher about patterns in the plane. It is well known that there are 17 types of repeating patterns in the plane. One of Escher's experiments was to start with an asymmetric motif and place four copies of it in a 2×2 stamp, subject to some restrictions on placement. Then, using the 2 × 2 stamp, he could generate a doubly periodic pattern, moving his stamp with only horizontal and vertical translations, see Fig. 3.16. Escher's problem was to determine the number of patterns one could generate by this method when the restrictions are removed.

Fig. 3.16 Generating a wallpaper pattern

Fig. 3.17 Making an Escher stamp pattern

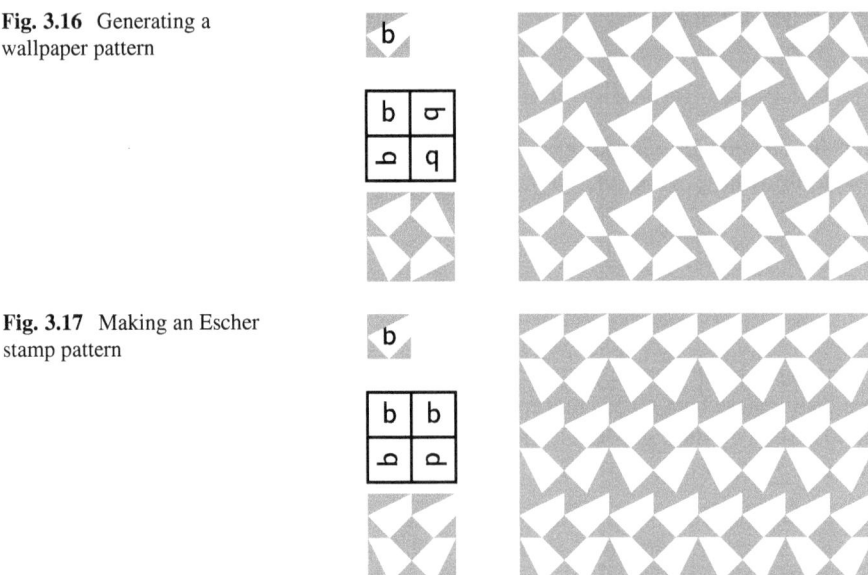

Example 3.37. We wish to apply Burnside's theorem to investigate the number of 2×2 Escher stamp patterns. The pattern is made by starting with an asymmetric motif, represented by b, carved into a square. The stamp is made by a 2×2 array of the motif in any of its four possible orientations, and the (infinite) pattern is obtained by using the stamp B and horizontal and vertical translations, see Fig. 3.17.

Since there are four possible orientations of b and we can choose any one of them for the four spots on B, there are 4^4 different stamps B; however, several of these stamps will produce the same pattern on the plane. We call two of the resulting patterns identical if we can rotate or translate the plane so that the two patterns are identical. In the language of Burnside's theorem, C_4 acts on the set of 4^4 stamps, and we wish to count the number of orbits, except that these are not the only transformations of the stamp that yield equivalent patterns.

If we tile the plane with a tile derived from B by switching the top and bottom row, we also will get an equivalent infinite pattern. Similarly, we can switch the left and right columns or switch pairs of opposite corners, so the Klein 4-group also acts on the set of tiles, see Fig. 3.18.

It is not hard to see that the group **G** acting on the set of stamps has 16 elements and consists of all products of the form

$$\{a^{\epsilon_1} b^{\epsilon_2} r^{\epsilon_3} \mid \epsilon_1, \epsilon_2 \in \{0, 1\}, \epsilon_3 \in \{0, 1, 2, 3\}\}$$

with r the 90° counterclockwise rotation, a the interchange of the two columns, and b the interchange of the two rows of the stamp, see Fig. 3.19 in which the action of each element is seen by comparing the shading of the stamp shown with the shading of the identity.

3.3 Group Actions

Fig. 3.18 Column and row switching yields equivalent patterns

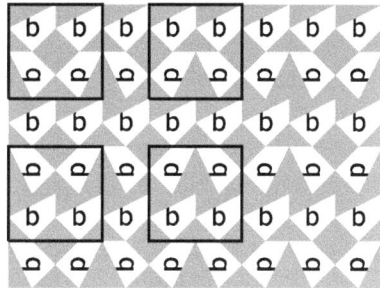

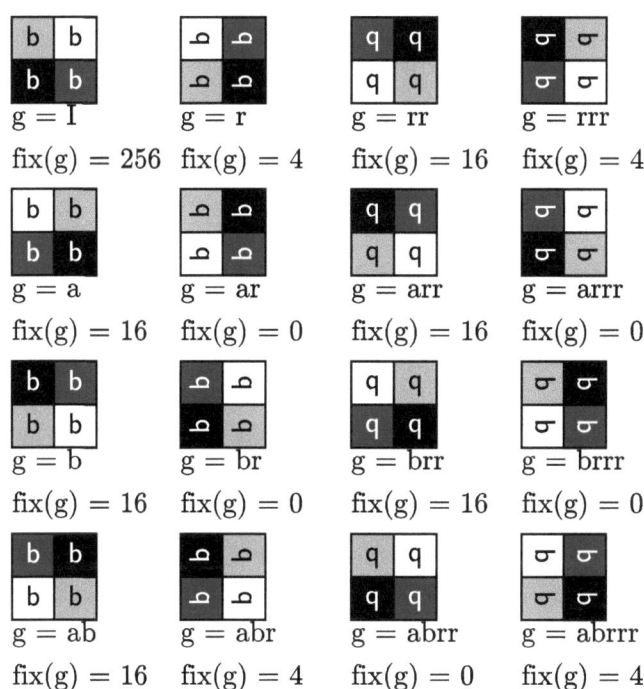

Fig. 3.19 The elements of the Escher stamp group

For Burnside's theorem, we are required to compute $|\text{fix}(h)|$ for each element h of $K \times C_4$. For example, $|\text{fix}(ar)| = 0$, because the stamped symbol in the right-upper corner of a tile in $\text{fix}(ar)$ should be identical to its 90-degree rotated version, which is impossible if the stamp pattern is asymmetric. Also $|\text{fix}(ar^2)| = 16$ since the tile symbols get pairwise exchanged under ar^2. $|\text{fix}(abr)| = 4$, since the abr permutes the symbols cyclically. Proceeding similarly for all elements of H, we get 6 fixed point sets of size 16, 4 of size 4, 5 empty ones, and of course $\text{fix}(1) = H$, see Fig. 3.19. Burnside's theorem gives us $1/16(256 + 6 \cdot 16 + 4 \cdot 4) = 23$ for the number of orbits and hence the number of geometrically distinguishable patterns.

Fig. 3.20 Different stamp patterns with the same wallpaper group

Note that we found 23 distinguishable stamp patterns, but we claimed that there are only 17 wallpaper groups. Therefore, some of the stamp patterns must have the same symmetry group, see Fig. 3.20. In Exercise 3.24 you are asked to determine the wallpaper group for each of the stamp patterns. For more information on tiling problems and combinatorics, see, for example, [87] and [81].

3.4 Symmetry and Transitivity

3.4.1 Vertex- and Edge-Transitive Graphs

If all vertices of G belong to a single orbit under the action of the automorphism group, we say that G is *vertex transitive*. We sometimes say that the vertices are *indistinguishable*. The graphs of the Platonic and Archimedean solids are all vertex transitive. The regular representation implies that all Cayley graphs are vertex transitive.

We may also view automorphisms of G acting on the edges of G and say that a group is *edge transitive* if there is a single edge orbit.

The graph of a regular prism Π_n is obviously vertex transitive but in general is not edge transitive. The automorphism group does not distinguish among the vertices, but one can tell apart lateral edges from the base edges except in the case Π_4, whose graph is the same as that of the cube. For $n \neq 4$, the action of Aut Π_n on the vertex set $V(\Pi_n)$ is imprimitive.

3.4.2 Semisymmetric Graphs

Edge transitivity and vertex transitivity are independent qualities. It is possible for a graph to be vertex transitive and not edge transitive, e.g., the graphs of the Archimedean solids, or to be edge transitive and not vertex transitive, e.g., the complete bipartite graph $K_{2,3}$. A graph which is vertex transitive but not edge transitive may have many kinds of edges, that is, there may be many edge orbits,

3.4 Symmetry and Transitivity

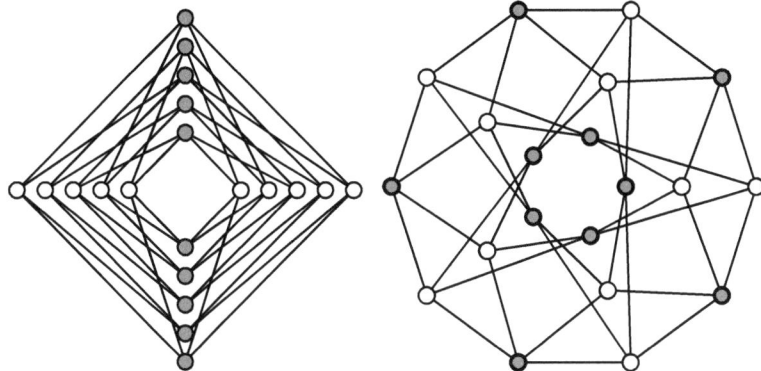

Fig. 3.21 The standard drawing of the Folkman graph and another drawing of the same graph

strictly bounded by the valence of a vertex. A graph which is edge transitive but not vertex transitive must have exactly two vertex orbits, since every automorphism which fixes an edge must fix both its endpoints.

The simplest family of graphs which are edge but not vertex transitive is the family of complete bipartite graphs $K_{n,m}$ with $n \neq m$. It is much more difficult to find regular graphs with this property. A d-valent edge-transitive but not vertex-transitive graph is called a *semisymmetric* graph.

Every semisymmetric graph is necessarily bipartite, with one example being the Folkman graph [31] on 20 vertices which is 4-valent and is edge transitive but not vertex transitive, see Fig. 3.21.

The smallest known cubic semisymmetric graph has 54 vertices and is known as the *Gray graph*, denoted hereafter by $\mathcal{G}$. The first published account on the Gray graph is due to Bouwer [11] who mentioned that this graph had in fact been discovered by Gray, in 1932. Brower gave two ways of constructing $\mathcal{G}$. First, three copies of the complete bipartite graph $K_{3,3}$ were taken, and each edge in each copy was subdivided by a vertex. Finally, each triple of vertices of valence 2 subdividing corresponding edges in the three copies of $K_{3,3}$ was joined to a new vertex. So the original 18 vertices were augmented with 27 valence -2 vertices, in turn joined by 9 attaching vertices, giving 54. The second construction identifies a particular Hamilton cycle in $\mathcal{G}$ and the corresponding 27 chords (see Fig. 3.22). The Gray graph $\mathcal{G}$ is cubic and bipartite. Its automorphism group Aut $(\mathcal{G})$ acts transitively on each of the bipartition sets; however, the vertices of the two bipartition sets have different distance sequences, so the graph is not vertex transitive.

Indeed, much activity has centered on deducing the surprising properties of the Gray graph, see [62, 65, 66, 70, 79], work that would have been greatly simplified by using the simple description of the Gray graph as the Levi graph of the geometric configuration which we will see described in Chap. 6.

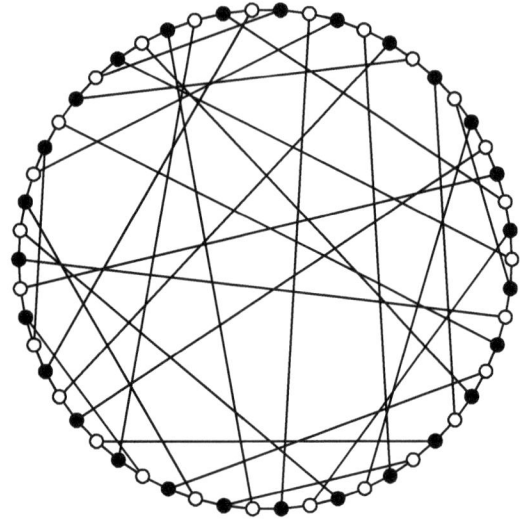

Fig. 3.22 The Gray graph with an identified Hamilton cycle as in [11]

3.4.3 Arc-Transitive Graphs

A graph G is called *arc transitive* if the action of its automorphism group is transitive on the ordered pairs of adjacent vertices. Thus, if G is arc transitive, then for any pair of edges $(u, v), (u', v') \in E$, one can find two automorphisms π_1 and π_2 such that $\pi_1(u) = u', \pi_1(v) = v'$ and $\pi_2(u) = v'$ and $\pi_2(v) = u'$. In particular, arc transitivity is stronger than edge transitivity. It means that we can map any edge to any other edge in an arbitrary direction. Clearly arc transitivity implies both vertex and edge transitivity.

For example, the Petersen graph GP(5, 2) is arc transitive, and among generalized Petersen graphs GP(n, k), the only arc-transitive examples are GP(4, 1), GP(8, 3), GP(10, 2), GP(10, 3), GP(12, 5), and GP(24, 5), of which only GP(4, 1), GP(8, 3), GP(12, 5), and GP(24, 5) are arc-transitive Cayley graphs by Theorem 3.31.

3.4.4 s-Arc Transitivity

Just as arc transitivity is a refinement of edge transitivity, by considering the action of the automorphism group on more complex substructures of the graph, we obtain further refinements. We follow the notation of Biggs [8] and define a *t-arc* $[\alpha]$ in a graph Γ to be a sequence $(\alpha_0, \alpha_1, \ldots, \alpha_t)$ of $t + 1$ vertices of Γ, such that α_{i-1} is adjacent to α_i, for $1 \leq i \leq t$ and $\alpha_{i-1} \neq \alpha_{i+1}$, for $1 \leq i < t$. We may identify 0-arcs with vertices and 1-arcs with arcs.

3.4 Symmetry and Transitivity

Clearly Aut(G) acts on the set of t-arcs. A graph G is t-*arc transitive* if its automorphism group acts transitively on the set of t-arcs and is *not* transitive on the set of $t+1$-arcs.

For cubic graphs it is quite easy to determine their arc transitivity since there are only two ways in which an s-arc may be extended to an $s+1$ arc. The tetrahedron, for example, is 2-arc transitive because, adding an edge to a path of length 2, we can either complete a cycle or get a path of length 3.

The cube is also 2-arc transitive. Here a path of length 2 may be extended to a path of length 3 such that its endvertices are either adjacent or not adjacent. The Petersen graph is 3-arc transitive.

The preceding examples suggest that arc transitivity and girth of the graph are related.

Theorem 3.38 ([98]). *Let G be a k-regular graph, $k \geq 3$ which is also s-arc transitive and of girth g. Then $s \leq g/2 + 1$.*

Proof. Consider a cycle C of length g. Since the degree of each vertex is at least 3, there is a vertex not on C but adjacent to a vertex of C, so there is a path of length g which is not closed, so thus, G cannot be g-transitive, i.e. $s < g$.

Since G is s-arc transitive, and C contains s-arcs, every s-arc must be in a cycle of length g. Moreover, each $(s-1)$-arc may be extended to an s-arc in at least two distinct ways, yielding two s-arcs intersecting in an $(s-1)$-arc, so there are two circuits of length g intersecting in $(s-1)$ edges. Their symmetric difference contains another circuit of length at least g, so

$$2g - 2(s-1) \geq g \text{ or } s \leq g/2 + 1. \qquad \square$$

3.4.5 1/2-Arc Transitivity

We have seen that 1-arc transitivity implies vertex and edge transitivity. The converse is not true, and we say graph G is $1/2$-*arc-transitive* if it is vertex transitive and edge transitive but not 1-arc-transitive.

Theorem 3.39. *A $1/2$-arc-transitive graph G must be regular of even valence.*

Proof. Regularity of G follows from vertex transitivity. Let $e = (v, u)$ be an edge. If there was an automorphism mapping the vertex–edge pair (v, e) to (u, e), necessarily mapping (u, e) to (v, e) as well, edge transitivity would imply 1-arc transitivity. So (v, e) and (u, e) must be representatives of the two 1/2-edge orbits. Color one red and the other blue. By the handshaking lemma, there are equal numbers of red and blue 1/2 edges, $|E|$ of each. Since G is vertex transitive, there are half edges of both colors incident to v and so to every vertex, with the same number of each type. The red valence and blue valence at each vertex are $|E|/|V|$, and the result follows. $\square$

A graph is *1-regular* if the automorphism group acts regularly on the set of arcs. By the well-known result of Tutte [98], a cubic arc-transitive graph is at most 5-arc transitive, with the degree of transitivity related to the size of the corresponding vertex stabilizers. For example if a cubic graph is 1-arc transitive, that is, its automorphism group acts transitively on the 1-arcs but not the 2-arcs, then the corresponding vertex stabilizers are isomorphic to a cyclic group of cardinality 3. In this case the automorphism group is regular on the set of 1-arcs, and the graph is 1-regular. Such graphs are of particular interest to us, for their line graphs are tetravalent 1/2-arc-transitive graphs as is seen by the following result, which is Proposition 1.1 in [67]. Recall that the vertices of the line graph $L(X)$ of a graph X are the edges of X, with adjacency corresponding to the incidence of edges in X.

Proposition 3.40 (Marušič and Xu). *A cubic graph is 1-regular if and only if its line graph is a tetravalent 1/2-arc-transitive graph.*

3.4.6 Automorphisms of Generalized Petersen Graphs

To study the automorphisms of the generalized Petersen graph $GP(n, k)$, it helps to first consider the subgroup of those automorphisms which preserve the outer ring $\{u_0, \ldots, n_{n-1}\}$, all of which necessarily also preserve the inner ring $\{u_0, \ldots, n_{n-1}\}$. Any automorphism of this outer ring extends uniquely to the inner ring if $r \equiv -r \pmod{n}$, so this subgroup of the automorphism group is the group of the n-cycle, the dihedral group $\text{Dih}(n)$ generated by the two involutions σ and τ,

$$\sigma(u_i) = u_{1-i}, \quad \tau(u_i) = u_{-i}, \quad \rho(u_i) = (\sigma\tau)(u_i) = u_{i+1}$$
$$\sigma(v_i) = v_{1-i}, \quad \tau(v_i) = v_{-i}, \quad \rho(v_i) = (\sigma\tau)(v_i) = v_{i+1}$$

Otherwise only the rotational subgroup generated by ρ extends to the inner ring. If we expand the set of transformations under consideration to those which either preserve the outer ring or swap the inner and outer ring, the orbit/stabilizer theorem, Theorem 3.22, implies that, if there are any new transformations, this will be a group of cardinality $2|\text{Dih}(n)| = 4n$ automorphisms generated by σ, τ, and any transformation which swaps the inner and outer rings.

We saw in Sect. 2.2.9 that there will be transformations swapping the inner and outer rings if and only if $k^2 \equiv \pm 1 \pmod{n}$, in which case a ring swapping transformation can be taken to be ϕ; defined by $\phi(u_i) = v_{ki}$ and, necessarily because of the spoke edges, $\phi(v_i) = u_{ki}$.

If $k^2 \equiv 1 \pmod{n}$, then ϕ is an involution, while if $k^2 \equiv -1 \pmod{n}$, then $\phi^2 = \tau$, so ϕ is of order 4 and τ is redundant in the generating set.

So now we have the subgroup of the automorphism group which preserves or swaps the rings or, which is the same, the subgroup of automorphisms which stabilize the perfect matching made up of the spoke edges. Again by the orbit/stabilizer theorem, we will know the cardinality of the full automorphism group as soon as we compute the orbit of this perfect matching. For $GP(4, 1)$, the stabilizer has 16

3.4 Symmetry and Transitivity

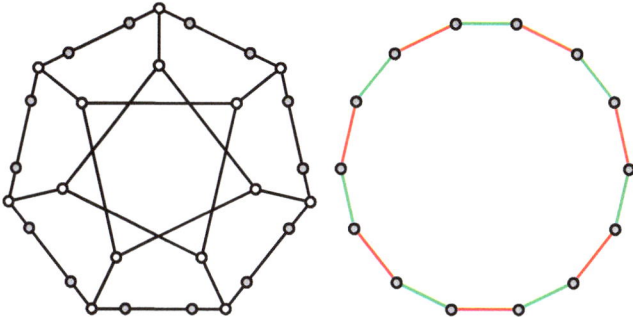

Fig. 3.23 The Cayley graph of Aut (GP(7, 3))

elements, since $1^2 \equiv 1 \pmod{4}$. There are nine perfect matchings, but only three separate the graph into two quadrilaterals. The orbit of the perfect matching of the spoke edges has cardinality 3, and $|\text{Aut}(GP(4,2))| = 48$, which is consistent with the fact that GP(4, 2) is the cube graph.

For GP(5, 2), the classic Petersen graph, there are six perfect matchings, each of which separates the graph into two 5-cycles. The stabilizer has 20 elements since $2^2 = 4 \equiv -1 \pmod{5}$ and $|\text{Aut}(GP(5,2))| = 120$.

Altogether there are only 5 other generalized Petersen graphs for which the orbit of the spoke edges is nontrivial: GP(8, 3), GP(10, 2), GP(10, 3), GP(12, 5), and GP(24, 5), see [34, 72]. For all other cases, the automorphism group is generated by σ, τ, and possibly ϕ. Each of these special groups has a story to tell. The graph GP(10, 2) stands out because it is the only one for which $k^2 \neq \pm 1 \pmod{n}$, so it is the only one for which the vertex transitivity is not correlated with an automorphism interchanging the inner and outer rings. Indeed, GP(10, 2) has no inner ring. Its spokes connect the outer 10-cycle to two inner 5-cycles. It is an interesting exercise to study the orbit of the spoke edges while blocking out from one's mind the revealing fact that GP(10, 3) is the graph of the dodecahedron.

Putting aside the automorphisms exhibited by the seven exceptional generalized Petersen graphs, let us consider only the automorphisms which stabilize the spokes. We have already computed the cardinality and provided generating vertex permutations. Let use compute the Cayley graphs in each of the three cases. By Theorem 3.29, we need to define a graph on which the group acts regularly on the vertices, so transitively with trivial stabilizer. The actual vertices of the generalized Petersen graph are not candidates since they have nontrivial stabilizers. For GP(7, 4), the action on the outer ring does act regularly on the semiedges, which we have decorated with gray dots, see Fig. 3.23. The automorphism group does act regularly on the gray dots. Now consider any gray dot, say u_0, and take the orbits of the pairs $(u_0, \sigma(u_0))$ and $(u_0, \tau(u_0))$ as two classes of edges, which we have colored. The spoke-stabilizing automorphism group of the generalized Petersen graph, in this case the whole automorphism group, acts regularly on the gray vertices of the connected colored graph, so it is the Cayley graph, the familiar Cayley graph of the dihedral group generated by two involutions.

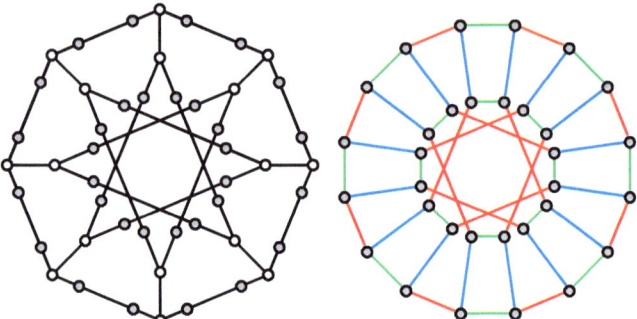

Fig. 3.24 The Cayley graph of the spoke-stabilizing automorphisms of Aut $(GP(8,3))$

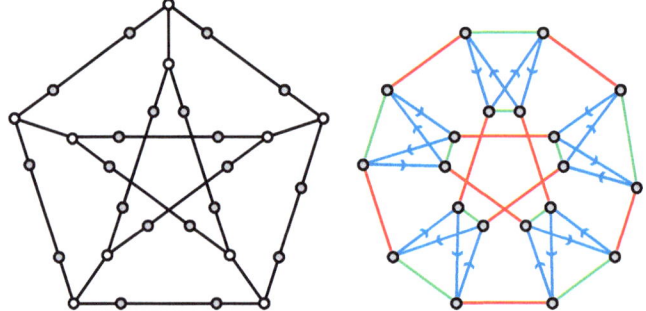

Fig. 3.25 Constructing a Cayley graph from the Petersen graph

For GP(8, 3), which is also an exceptional case, the orbit of the semiedges of the outer ring under the spoke-stabilizing automorphisms includes the semiedges of the inner ring, see Fig. 3.24. In this case, we must include ϕ, an involution since $3^2 = 9 \equiv 1 \pmod{8}$, among the generating transformation; thus, we include the orbits of the pair $(u_0, \phi(u_0))$ among the edges of the Cayley graph.

Lastly, for the case where $k \equiv -1 \pmod{n}$, we can consider the spoke-preserving automorphisms of the classical Petersen graph, see Fig. 3.25 where it is easy to observe that the generator τ is redundant since deleting the green edges leaves a connected Cayley graph.

3.5 Voltage Graphs and Covering Graphs

3.5.1 Quotient Graphs

Suppose we have the graph of a hexagonal prism H. The 180° rotation of the prism about its main axis generates an action of $\mathbf{C}_2$ on H. $\mathbf{C}_2$ acts on the set of vertices of H as well as the set of edges and, combining these, the set of pairs $\{(v, e) \mid v \in$

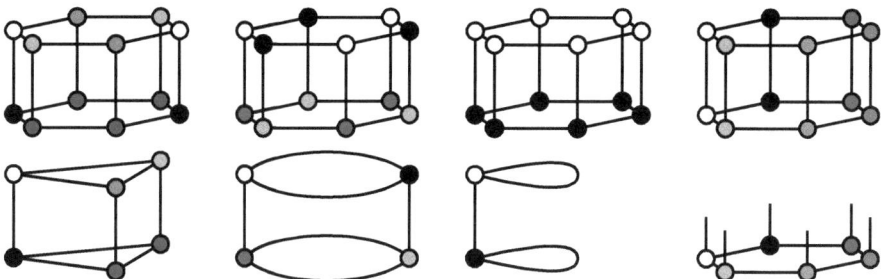

Fig. 3.26 Actions on the prism H with quotients

$V, e \in E\}$. It is natural to define a graph on the set of orbits and set Orbit(v) incident to Orbit(e) if there is $v' \in$ Orbit(v) and $e' \in$ Orbit(e) such that v' is incident to e'. Let us compute this *orbit graph*. There will be 6 vertex orbits, three from the top hexagon and 3 from the bottom, and 9 edge orbits. With the incidence relation among the orbits inherited from that in H, we obtain the triangular prism as the orbit graph in a natural way, see Fig. 3.26.

The situation is a little more complicated if we take a rotation by 120°, so that there are only 4 vertex orbits. In this case, there are 6 edge orbits; however, the two edge orbits corresponding to the upper hexagon are both incident to the same two vertex orbits. Similarly for the bottom hexagon, if we examine the vertex–edge pairs, the arcs, we see there are four orbits corresponding to the upper hexagon and four corresponding to the lower. It seems natural in this case to say that there are two parallel edges in each case, see Fig. 3.26.

If we consider the generating automorphism to be the rotation by 60°, then the 12 arcs of the upper hexagon are partitioned into just two orbits, both incident to the same vertex orbit, and we would say that the orbit graph has a loop there, see Fig. 3.26.

Lastly, if the generating automorphism is the horizontal mirror reflection of the prism, the vertex–edge pairs corresponding to an edge connecting the upper and lower hexagon are not only incident to the same vertex orbit, but they belong to the same arc orbit. This must correspond to a half edge at that vertex orbit, see Fig. 3.26.

Considering these cases, the most natural definition of the *orbit graph*, or *quotient graph*, by a group of automorphisms will in general be a pregraph (introduced in Sect. 2.3.6).

In the next section we want to reverse this process, that is, start with a pregraph and a group, and construct a graph which contains the prescribed group as a subgroup of its automorphisms, with the given pregraph as the quotient.

3.5.2 Pregraphs Revisited

A *pregraph* G is a quadruple $G = (V, S, i, r)$ where V is the set of vertices; S is the set of *arcs* (otherwise known as *semiedges*, darts, ...); i is a mapping $i : S \to V$,

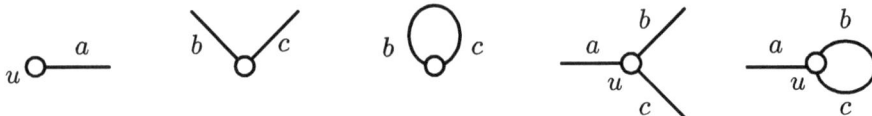

Fig. 3.27 $B(0; 1)$, $B(0; 2)$, $B(1; 0)$, $B(0; 3)$, and $B(1; 1)$

specifying the origin or *initial vertex* for each arc; and r is the *reversal involution*: $r : S \to S$, with $r^2 = 1$ whose orbits mark which half edges are associated. So defining $t : S \to V$ by $t(s) := i(r(s))$ specifies the terminal vertex for each arc. An arc s is part of an *edge* $e = \{s, r(s)\}$ which is called *proper* if $|e| = 2$ and is called a *half edge* if $|e| = 1$. Define $\partial(e) = \{i(s), t(s)\}$. Note that an edge is a pair of arcs, while $\partial(e)$ is a pair of vertices. A pregraph without half edges is called a (*general*) *graph*. G is a graph if and only if the involution r has no fixed points. A proper edge e with $|\partial(e)| = 1$ is called a *loop* and two edges e, e' are *parallel* if $\partial(e) = \partial(e')$. A graph without loops and parallel edges is called *simple*. The *valence* of a vertex v is defined as $\mathrm{val}(v) = |\{s \in S \mid i(s) = v\}|$. A pregraph of valence 3 is called *cubic*.

An automorphism of a pregraph is a bijection ϕ on the vertices, together with a bijection η on the arcs such that $\phi(i(s)) = i(\eta(s))$ and $r(\eta(s)) = \phi(r(s))$.

3.5.3 Pregraphs on a Single Vertex

Pregraphs on a single vertex, u, must have $i(e) = u$ for all e and so are completely determined by the involution r. So there is a unique pregraph on a single vertex of valence 1, $B(0; 1)$, with $r(a) = a$. There are two single-vertex pregraphs of valence 2, one with r the identity, $B(0; 2)$, and the other with r a transposition on the two half edges, $B(1; 0)$. Pregraphs with a single vertex underlie several different maps and maplike structures, see [74]. There are also two cubic pregraphs on a single vertex depending on whether r is the identity, $B(0; 3)$, or r contains a transposition, $B(1; 1)$, see Fig. 3.27.

We may define a family of *generalized bouquets of circles*: $B(k; p)$ consisting of a single vertex of valence $2k + p$ with k loops and p half edges; in other words, r has k transpositions. Clearly there are $1 + \lfloor d/2 \rfloor$ single-vertex d-valent pregraphs.

3.5.4 Voltage Graphs and Regular Coverings

Voltage graphs are obtained from pregraphs by assigning group elements to arcs. More precisely, a *voltage graph* X is a 6-tuple $X = (V, S, i, r, \Gamma, \alpha)$ where (V, S, i, r) is the underlying pregraph, Γ is a group, and α is a mapping $\alpha : S \to \Gamma$ such that the group elements assigned to arcs associated with the same edge are inverses of one another: So for each $s \in S$, we have $\alpha(r(s)) = \alpha^{-1}(s)$.

3.5 Voltage Graphs and Covering Graphs

Any voltage graph X defines the so-called *derived graph* or *regular covering graph* Y as follows:

$$V(Y) := V \times \Gamma$$
$$S(Y) := S \times \Gamma$$
$$i(s, \gamma) := (i(s), \gamma),$$
$$r(s, \gamma) := (r(s), \alpha(s) \cdot \gamma),$$

Let $\beta \in \Gamma$. Then β induces a bijection on Y by $(v, \gamma) \to (v, \gamma\beta)$ and $(s, \gamma) \to (s, \gamma\beta)$. Since the initial point of the image arc $i(s, \gamma\beta) = (i(s), \gamma\beta)$ is the image of the initial point of the arc $(i(s), \gamma)$, and the arc associated with the image arc $r(s, \gamma\beta) = (r(s), \gamma\beta)$ is the image of the associated arc, $(r(s), \gamma)$, the bijections induced by β are automorphisms of the regular cover, and we say the group Γ acts by right multiplication on Y. Clearly the quotient graph is X.

If $s \in S$ is a half edge, i.e. if $r(s) = s$, its voltage $\alpha(s)$ is of order at most two: $\alpha(s) = \alpha^{-1}(s)$. Hence, $\alpha = \alpha^{-1}$, or equivalently, $\alpha^2 = 1$. In the case when the group Γ is cyclic, $\Gamma = \mathbb{Z}_n$, the only voltages that can be assigned to a half edge are 0 and $n/2$. A voltage 0 on a half edge implies that the derived graph remains a pregraph having half edges. If we are interested in simple covering graphs, then the only admissible voltage for a half edge remains $n/2$; furthermore, n must be even.

Note that voltage graphs enable us to develop a combinatorial analog of the well-known theory of covering spaces in algebraic topology. The reader may find more on the theory of voltage graphs and covering graphs in [42, 78, 80]. We note in passing that there is a convention in drawing voltage graphs. Usually the voltage on only one of the pair of reversing arcs is specified. The choice of the arc is denoted by placing an appropriate arrow. When the voltage is an involution (in our case, a voltage from $\{0, n/2\}$), no arrow is drawn.

3.5.5 Voltage Assignments on $B(1; 1)$

There are four possible $\mathbb{Z}_2$-voltage assignments over $B(1; 1)$ which we will now describe in detail.

The trivial voltage assignment associates 0 to each half edge and yields $2B(1; 1)$, two disjoint copies of $B(1; 1)$, see Fig. 3.28.

Assigning voltage 1 to each arc gives rise to the so-called canonical double cover or Kronecker cover. In this case, the canonical double cover is the theta graph θ_3 on two vertices and three edges, known also as the cubic *dipole*, see Fig. 3.29.

Assigning voltage 0 to the half edge and 1 to the loop gives rise to a two-valent theta graph θ_2, augmented to a cubic pregraph by attaching a half edge at each vertex, see Fig. 3.30.

Fig. 3.28 $B(1;1)$ with voltage assignment deriving $Y = 2B(1;1)$

Fig. 3.29 $B(1;1)$ with voltage assignment deriving $Y = \theta_3$

Fig. 3.30 $B(1;1)$ with voltage assignment and derived graph

Fig. 3.31 Voltage on $B(1;1)$ with handcuff graph as cover

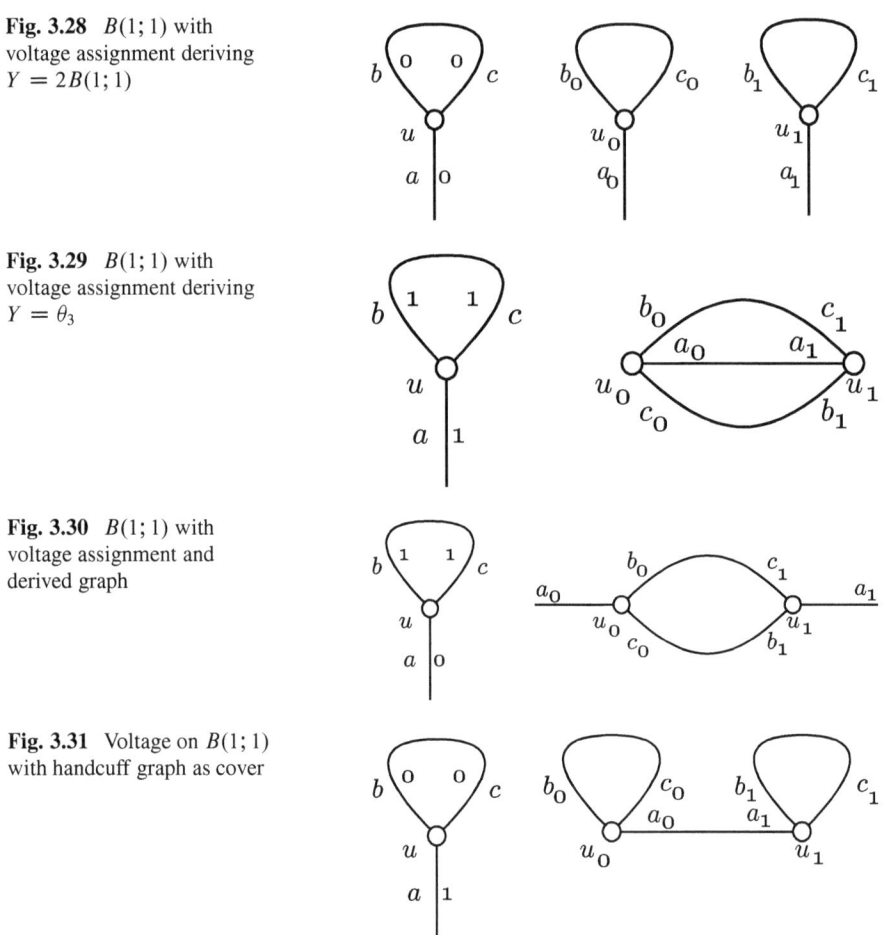

Finally, assigning voltage 1 to the half edge and 0 to the loop gives rise to the so-called *handcuff graph*, see Fig. 3.31.

In each of these cases, adding 1 to the subscripts modulo 2 defines an automorphism of the graph. This is the action of Z_2 on the cover. Notice particularly how this automorphism acts on the edges of the theta graph, interchanging two and reflecting one of its edges.

3.5.6 Generalized Petersen Graphs as Coverings

The *handcuff graph*, H, has two loops attached to the endpoints of a single edge. If we relax the requirement that a generalized Petersen graph be simple, one may describe the handcuff graph as $GP(1, 1)$. We have just seen in Fig. 3.31 that the handcuff graph $H = GP(1, 1)$ is a twofold regular covering over $B(1;1)$.

3.5 Voltage Graphs and Covering Graphs

Fig. 3.32 The voltage assignment for GP(n, r)

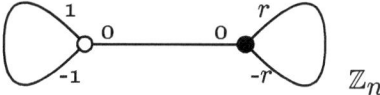

By assigning voltages from $\mathbb{Z}_n$ to H as in Fig. 3.32, we obtain a derived graph which is isomorphic to the generalized Petersen graph GP(n, r). The construction defines a graph mapping from GP(n, r) to GP(1, 1). The mapping is called the *covering projection* and is a local isomorphism.

The outer circle is projected to the loop at the white vertex, and the inner star polygon is projected to the black loop, with the spokes projected to the edge connecting the black and white vertices.

There is a general rule: In any voltage graph, one may change the direction of any arrow and replace the corresponding voltage by its inverse without changing the covering graph. In particular, this means one can always choose p^{-1} instead of p on any loop. This gives a short argument why GP(n, r) and GP(n, n − r) are isomorphic.

3.5.7 {0, 1} Voltage

Given a graph X with voltages in $\mathbb{Z}_2$, we can imagine the cover, which has twice as many vertices, as being drawn in two layers over X, one above the other. An edge with voltage 0 is covered by two edges which stay in the same layer, corridors, and an edge with voltage 1 is covered by two edges which switch layers, staircases. The action of $\mathbb{Z}_2$ on the cover is to simply switch layers. If all the voltages are 0, then the cover is the disjoint union of two copies of X. If all the voltages are 1, then the cover is the Kronecker double cover introduced above. It is easy to show that the double cover is always bipartite as is, in fact, the tensor product of X with K_2; $X \times K_2$, which we met in Sect. 2.4. The most compelling example is the double cover of K_4, the graph of the regular tetrahedron, which yields the graph of the cube. Once again, we are reminded of the fact that the cube contains two disjoint inscribed tetrahedra. Another classic example is GP(5, 2), the Petersen graph, whose Kronecker double cover is GP(10, 3), the Desargues graph.

One can also consider voltage assignments of 0 and 1 where the group is not $\mathbb{Z}_2$ but $\mathbb{Z}_n$. Then, instead of two layers, we have n layers with corridors covering the edges of voltage 0 and staircases covering those of voltage 1, each staircase raising one level until the top where, with an elevator presumably, one returns to the ground floor. Such covers are called *rotagraphs*, see [77].

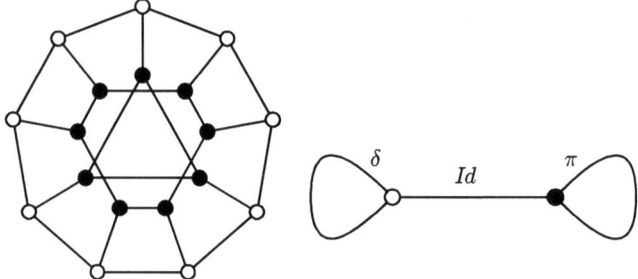

Fig. 3.33 Doubly generalized Petersen graph GP(9, π) and voltage graph

3.5.8 Permutation Voltage Assignments and Ordinary Coverings

In the definition of the covering graph, we can get more general but less symmetric covers by using voltage assignments not from a group but merely from a set on which a group acts. Topologically speaking, the theory of covering graphs is extended from regular coverings to ordinary coverings.

A *permutation voltage graph* $(X, \Gamma, A, h, \alpha)$ is a pregraph X, together with a group action (Γ, A, h), so $h : \Gamma \to \text{Sym}(A)$, and a mapping $\alpha : S(X) \to \Gamma$ such that $\alpha(r(s)) = \alpha^{-1}(s)$ for any $s \in S(X)$. Any permutation voltage graph defines the so-called *permutation derived graph* or *ordinary covering graph* Y as follows:

$$V(Y) := V \times A$$
$$S(Y) := S \times A$$
$$i(s, a) := (i(s), a)$$
$$r(s, a) := (r(s), h(\alpha(s))(a))$$

For simplicity, choose A to be the set $\{1, 2, \ldots, n\}$ and $\Gamma = \text{Sym}(n)$.

Example 3.41. Let $id = ()$, $\delta = (123456789)$, and $\pi = (147)(235689)$ be permutations of nine elements. The permutation voltage assignment on the handcuff graph given here has the permutation derived graph isomorphic to doubly generalized Petersen graph GP(9, π) depicted in Fig. 3.33.

With voltage assignments from Sym(9), the regular covering construction would have given a graph with $2 \cdot 9!$ vertices. Notice that not only does Sym(9) fail to act on the graph in a natural way, neither does the subgroup $\langle \pi, \delta \rangle$ generated by the assigned voltages nor even the cycle subgroups generated by the individual assigned voltages.

3.5 Voltage Graphs and Covering Graphs

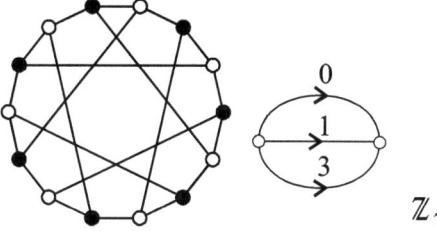

Fig. 3.34 The theta graph, equipped with voltages from $\mathbb{Z}_7$ for the 6-cage, the Heawood graph

With the *permutation voltage assignment* the doubly generalized Petersen graph does cover the handcuff graph in some sense but not with the handcuff graph playing the role of an orbit graph.

The *voltage graph* is also called the *base graph* and the derived graph is called the *covering graph*. The terminology follows the one of covering spaces in algebraic topology. See [42] for further information about graph coverings.

3.5.9 Cages as Covering Graphs

The definition of the cage is strictly in terms of its vertex valence and its girth. There is no symmetry requirement, yet many cages do exhibit a high degree of symmetry and so may profitably be described as covering graphs.

The 2-cage is only defined for multigraphs and is the cubic *theta graph*. As we know, see Fig. 3.29, we can describe it as a 2-cover over $B(1; 1)$.

The 3-cage, K_4, is a $\mathbb{Z}_4$-covering graph over $B(1; 1)$. The loop gets voltage 1, and the half edge gets the voltage 2.

The 4-cage, $K_{3,3}$, is the threefold covering graph over the cubic theta graph with the voltages 0, 1, and 2 in $\mathbb{Z}_3$.

The 5-cage, $GP(5, 2)$, is a covering of the handcuff graph.

To construct the 6-cage we again use the cubic theta graph with the voltages 0, 1, and 3 in $\mathbb{Z}_7$. This gives us the Heawood graph, see Fig. 3.34.

Note that if we use the same voltage assignments but regard them as being from $\mathbb{Z}_4$, we get the cube graph Q_3.

The 7-cage is an eightfold covering graph over the voltage graph on 3 vertices, see Fig. 3.35.

The 8-cage can be represented as a fivefold covering graph over a graph on 6 vertices, see Fig. 3.36.

The 4- and 6-cages have just been described as covers of the theta graph with voltages in an abelian group. Such a graph has been called a *Haar graph*, [52].

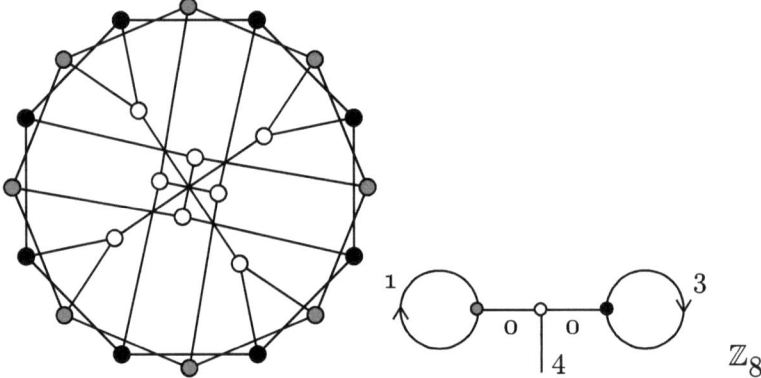

Fig. 3.35 The 7-cage with its voltage graph. This is also known as the McGee graph

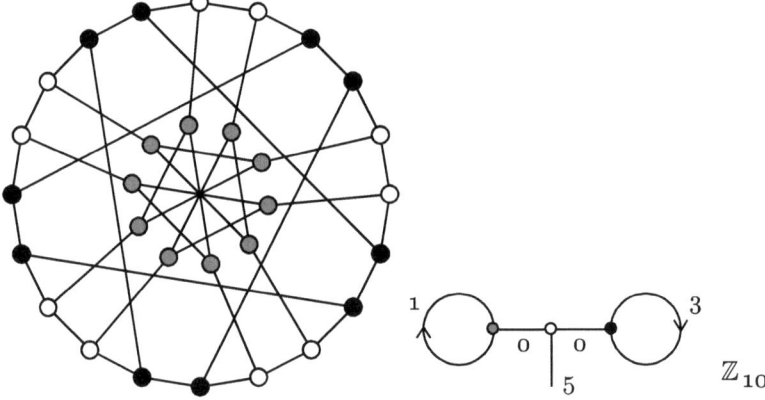

Fig. 3.36 The 8-cage with its voltage graph. This is also known as the Cremona–Richmond graph

3.6 Automorphisms of the Symmetric Group

Via the left regular action, every group **G** acts on itself *freely,* that is, with trivial stabilizers, and transitively and ditto with the right regular action. Both of these actions, while interesting and useful, do not completely respect the group structure of **G**; they do not act as *group automorphisms*. A group automorphism is a bijective group homomorphism so must preserve the identity, which the nontrivial permutations associated with the left and right regular actions do not.

There is, however, another action of **G** onto itself defined by conjugation:

$$\pi : \mathbf{G} \to \text{Aut}(\mathbf{G})$$
$$g \to [\pi_g(h) = g^{-1}hg]$$

3.6 Automorphisms of the Symmetric Group

The image of this homomorphism is called the *inner automorphism group* of **G**, denoted by Inn(**G**). Clearly an element is in the kernel of the homomorphism exactly when it commutes with every element of **G**.

So the inner automorphisms form a very nice class of automorphisms of **G**, since they borrow all their structure from **G** itself. By the same token, the noninner automorphisms must be the interesting ones.

Usually we do not expect the image of a group under a homomorphism to be normal, but in the case of Inn(**G**), it is, see Exercise 3.64. So we can define

$$\mathrm{Out}(\mathbf{G}) = \mathrm{Aut}\,(\mathbf{G})/\mathrm{Inn}(\mathbf{G}),$$

the *outer automorphism group* of **G**. There is a slight infelicity of language in that the elements of the outer automorphism group are not automorphisms but equivalence classes of automorphisms.

The question we want to examine in this section is whether or not Sym(n) has any interesting automorphisms, and for most values of n, the answer is "no."

Theorem 3.42. *If $n \neq 6$, Sym(n) has only inner automorphisms.*

Proof. We use the fact that conjugation does not alter the cycle structure of permutations: If τ is a permutation and

$$a_1 \xrightarrow{\sigma} a_2 \xrightarrow{\sigma} \cdots \xrightarrow{\sigma} a_n \xrightarrow{\sigma} a_1,$$

is one of the cycles of the permutation σ, then

$$\tau(a_1) \xrightarrow{\tau\sigma\tau^{-1}} \tau(a_2) \xrightarrow{\tau\sigma\tau^{-1}} \cdots \xrightarrow{\tau\sigma\tau^{-1}} \tau(a_n) \xrightarrow{\tau\sigma\tau^{-1}} \tau(a_1),$$

is a cycle of $\tau\sigma\tau^{-1}$.

Suppose now that Sym(n) has an automorphism, f. The automorphism f must map transpositions, which are of order two, to involutions. Moreover, since we have just seen that transpositions form a conjugacy class, f must map the transpositions to a conjugacy class of involutions, that is to say, a product of k disjoint transpositions for some fixed k. We know that the number of transpositions in Sym(n) is $\binom{n}{2}$. To count the number of k disjoint transpositions, we first choose the $2k$ elements, which can be done in $\binom{n}{2k}$ ways. Then we partition those $2k$ elements in pairs. To count this, the smallest of them must be paired with one of $2k - 1$ elements. Then the smallest of the remainder must be paired with one of the $2k - 3$ remaining, etc. So we get $\binom{n}{2k}(2k-1)(2k-3)\cdots 1$ products of k disjoint transpositions. So for f to be an automorphism, we must have that

$$\binom{n}{2} = \binom{n}{2k}(2k-1)(2k-3)\cdots 1$$

This is clearly impossible if $2 < 2k < n - 2$. We need only consider $2k = n - 2$, $2k = n - 1$, and $2k = n$.

If $2k = n - 2$, then $(2k - 1)(2k - 3) \cdots 1 = 1$, and so $k = 1$ and $n = 4$.

If $2k = n - 1$, then $n = 2k + 1$ and

$$\binom{2k+1}{2} = \binom{2k+1}{2k}(2k-1)(2k-3)\cdots 1$$

gives $k = (2k - 1) \cdots (1)$ and again $k = 1$, with $n = 3$.

Lastly, if $2k = n$, then

$$\binom{2k}{2} = \binom{2k}{2k}(2k-1)(2k-3)\cdots 1$$

gives $k(2k - 1) = (2k - 1)(2k - 3) \cdots 1$, or $k = (2k - 3) \cdots 1$, which is the desired exception, $k = 3$ and $n = 6$. So, if $n \neq 6$, then f must map transpositions to transpositions.

Now, if f maps transpositions to transpositions, we can apply an inner automorphism so that $h_\tau f : (0k) \to (0k)$ for all k. Since transpositions of the form $(0, k)$ generate $\text{Sym}(n)$ for all n, we conclude $h_\tau f$ is the identity; hence, f is inner. Thus, since only $\text{Sym}(6)$ can have automorphisms which map transpositions to nontranspositions, only $\text{Sym}(6)$ can have outer automorphisms. □

Antipodally Symmetric Perfect Matchings of the Icosahedron Graph

So, what about $\text{Sym}(6)$: Does it have any noninner automorphisms? We will construct all the inner automorphisms of $\text{Sym}(6)$ by considering perfect matchings of the edges of the icosahedral graph, that is, the 1-factors.

To each vertex in the icosahedral graph, there is exactly one "antipodal" vertex, the unique vertex at distance 3 away, so to each edge, there is exactly one "antipodal" edge. We want to consider those perfect matchings which contain three pairs of antipodal edges.

Let us examine how such perfect matchings are distributed in the graph of the icosahedron. The icosahedron graph is edge transitive, so we may start with any pair of antipodal edges and extend that pair to a perfect matching. Deleting all the edges incident with the endpoints of the chosen antipodal pair, we have a graph on the right of Fig. 3.37. There are three choices, indicated in Fig. 3.38.

Thus, there are three antipodally symmetric perfect matchings containing any given antipodal pair.

3.6 Automorphisms of the Symmetric Group

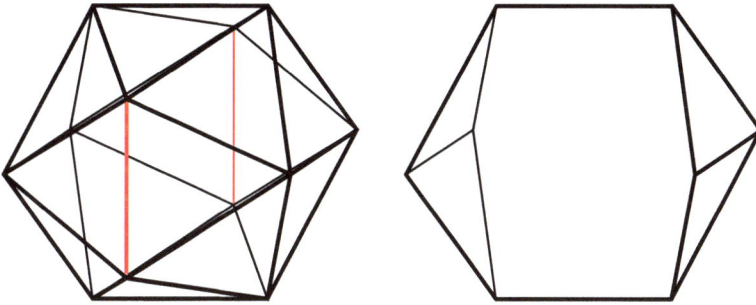

Fig. 3.37 Constructing a perfect matching on the graph of the icosahedron

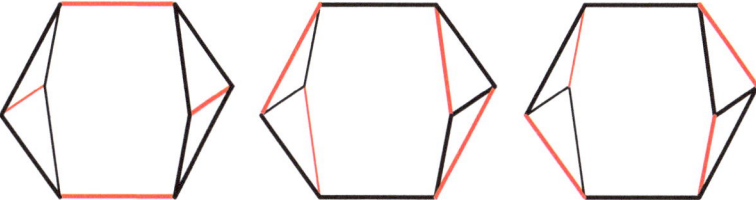

Fig. 3.38 Completing a perfect matching on the graph of the icosahedron

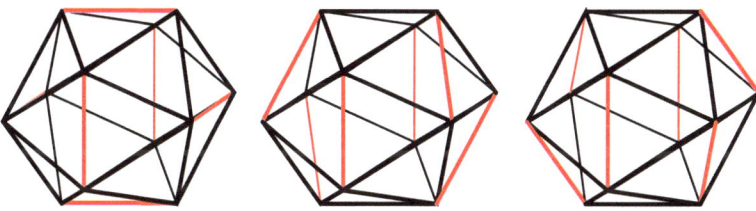

Fig. 3.39 The three antipodally symmetric perfect matchings containing the back and front edges

Cubic and Axial Matchings

The three perfect matchings belong to two combinatorially distinguishable types. In the perfect matching on the left of Fig. 3.39, which we call *cubic*, the 3-cycles not sharing an edge with the matching do not themselves share an edge. For the other two, in the geometrical icosahedron, the six lines of the edges of the perfect matching extend to two triples of coincident lines, see Fig. 3.40, such that the two points of coincidence define a line which passes through the center of two antipodal triangles. For this reason we call the second type of matching *axial*.

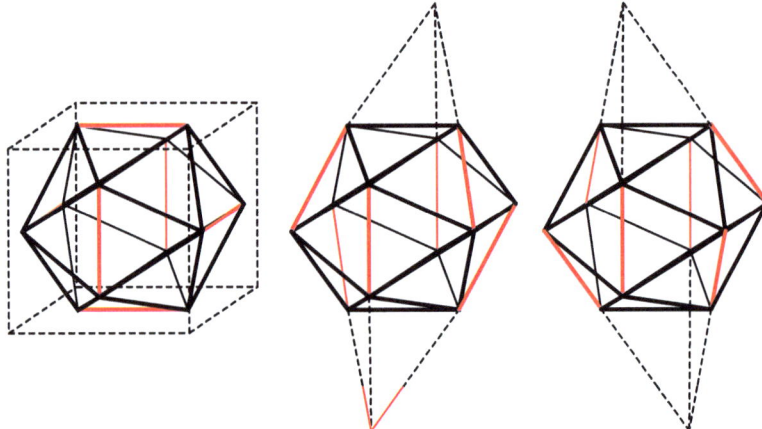

Fig. 3.40 A cubic and two axial matchings

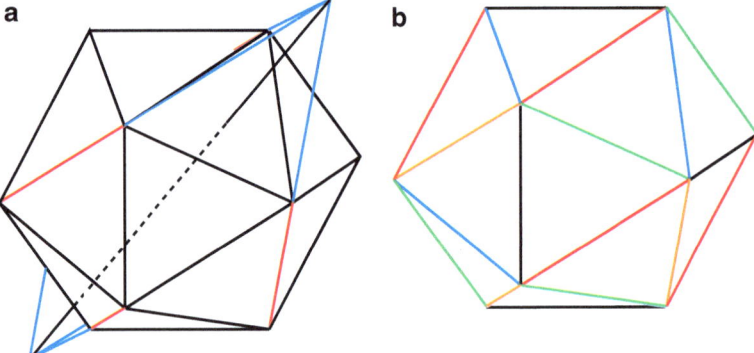

Fig. 3.41 Combining axial and cubic matchings

Perfect Matching Partitions

There are exactly five cubic perfect matchings contained in the edge set E of the icosahedron graph, and they partition E. The edges of any cubic matching M in E are disjoint from the edges from any axial matching in E whose axial 3-cycles are one of the four pairs of antipodal 3-cycles in $E - M$, see Fig. 3.41a. In fact, the edges of the four axial matchings corresponding to these 4 pairs of antipodal 3-cycles of $E - M$ are pairwise disjoint and, together with M, partition E.

Since there are five perfect matchings of cubic type, we have six partitions of the edges of the icosahedron into five antipodally symmetric perfect matchings, one partition with all matchings of cubic type and the other five partitions each having one matching of cubic type and the other four of axial type. There can be no other.

3.6 Automorphisms of the Symmetric Group

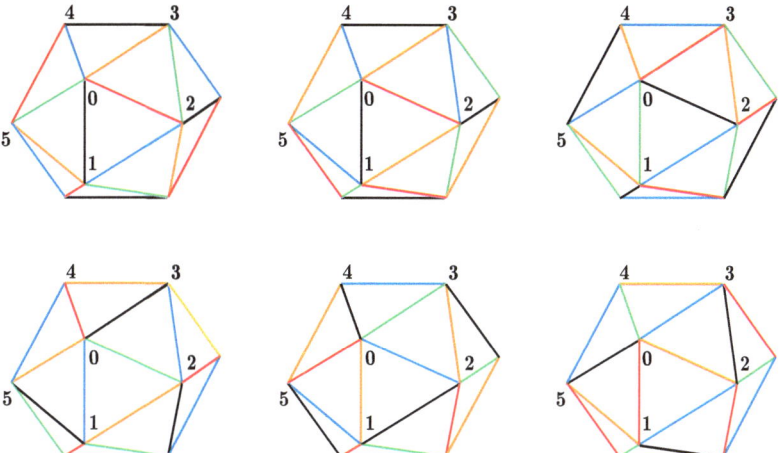

Fig. 3.42 Each color indicates a matching. Each graph indicates one of the six partitions of the edges of the icosahedron into 5 antipodally symmetric perfect matchings. The *top left* contains five matchings of *cubic* type; the others all have one matching of cubic type

Labeling the Matchings

Now let us label the antipodal pairs of vertices with $\{0, 1, 2, 3, 4, 5\}$. Each antipodally symmetric perfect matching, whether of cubic or axial type, will correspond to a partition of these labels into three two-subsets. So the black cubic matching in the top-left diagram of Fig. 3.42 is labeled by $\{\{0, 1\}, \{2, 5\}, \{3, 4\}\}$, which, to be compact, we will write as $01|23|45$. The 15 antipodally symmetric perfect matchings are in one-to-one correspondence with the 15 labels of the form $ij|lk|mn|$.

Now we can label each partition by a 5×3 array, five triples of pairs. The six partitions illustrated in Fig. 3.42 above are labeled

black	01|25|34	01|25|34	02|13|45
red	02|31|45	02|14|35	03|14|25
orange	03|42|51	03|12|45	04|15|23
blue	04|53|12	04|15|23	05|12|34
green	05|14|23	05|13|24	01|24|35
black	03|15|24	04|12|35	05|14|23
red	04|13|25	05|13|24	01|24|35
orange	05|12|34	01|23|45	02|15|34
blue	01|23|45	02|15|34	03|12|45
grccn	02|14|35	03|14|25	04|13|25

Let us now count the number of possible 5×3 arrays of pairs from $\{0, \ldots, 5\}$ with the properties that each digit $\{1, \ldots, 5\}$ occurs exactly once in each row and each pair ij occurs exactly once in the array. Each array has one triple with 01, and in that array, 2 is paired with $i \in \{3, 4, 5\}$. So we have three choices for that row $01|2i|jk$. Another row must have 02, and since j and k cannot be paired again, one of i and j is paired with 1, so there are six ways to choose the rows containing 01 and 02:

$$\begin{array}{ccc} 01|2i|jk & 01|2i|jk & 01|2i|jk \\ 02|1j|ik & 02|1j|ik & 02|1j|ik \\ 0i|??|?? & 0i|??|?? & 0i|1i|2k \\ 0j|??|?? & 0j|??|?? & 0j|1k|2j \\ 0k|??|?? & 0k|12|ij & 0k|12|ij \end{array}.$$

There is exactly one way to complete the array. Now, k will be paired with 1 and 2 in the third and fourth rows, so 1 and 2 must be paired with one another in the fifth. Then i and j must be paired with 1 and 2 in the fifth row, so the rest, realized above, is forced. This proves that the six partitions by antipodally symmetric perfect matchings given in Fig. 3.42 are the only ones possible. This also proves that any two partitions have exactly one matching in common.

Permuting the Partitions

If we permute the elements $\{0, 1, 2, 3, 4, 5\}$, then this induces a permutation of the labels of the partitions. Since the labels are in one-to-one correspondence with the partitions themselves, this gives us a homomorphism from the permutation group of $\{0, 1, 2, 3, 4, 5\}$ to the permutation group of the six partitions. We show that this homomorphism is an isomorphism. For this, we need only show that the homomorphism has trivial kernel. The kernel is a normal subgroup of Sym(6), that is, either Sym(6) itself; A_6, the alternating group of even permutations; or just the identity. If either Sym(6) or A_6 were the kernel, then every even permutation of $\{0, \ldots, 5\}$ would induce the identity permutation on the partitions. So let us check (123):

black	01|25|34	01|25|34		02|13|45
red	02|31|45	02|14|35		03|14|25
orange	03|42|51	$\to$ 03|12|45		04|15|23
blue	04|53|12	04|15|23		05|12|34
green	05|14|23	05|13|24		01|24|35
	$\uparrow$	$\swarrow$	$\nearrow$	$\downarrow$
black	03|15|24	04|12|35		05|14|23
red	04|13|25	05|13|24		01|24|35
orange	05|12|34	01|23|45	$\to$	02|15|34
blue	01|23|45	02|15|34		03|12|45
green	02|14|35	03|14|25		04|13|25

So (123) clearly does not induce the identity permutation on the partitions. Thus, the induced permutation on the partitions gives an isomorphism from the group of permutations of $\{0,\ldots,5\}$ to the group of permutations of the six partitions. Moreover, if we arbitrarily label the partitions with $\{0,\ldots,5\}$, we have an automorphism Sym(6) → Sym(6). Actually we have 6! automorphisms, one for each of the ways of labeling of the partitions.

Note that for these automorphisms, the three-cycle (123) has been mapped to a product of two disjoint three-cycles, that is, it is not an inner automorphism. In fact, none of these 6! automorphisms of Sym(6) are inner automorphisms; they are a coset of the inner automorphism group, an outer automorphism.

In the geometrical literature, the antipodally symmetric perfect edge matchings correspond to the synthemes of H. S. M. Coxeter, which he describes in the first of his *Twelve Geometric Essays,* [17], and the edge partitions are exactly his *pentads.* This section is adapted from that source.

Out(Sym(6)) ≅ $\mathbb{Z}_2$

Now that we have an outer automorphism, to complete the combination of Aut(S_6), we note that every automorphism must permute the conjugacy classes. By Exercise 3.65, every automorphism which permutes the transpositions (ij) must be inner, so the subgroup of inner automorphisms in Aut(S_6) must have index 2 and so the product of every two outer automorphisms is an inner automorphism. In particular, the square of any outer automorphism is an inner automorphism. We can express the automorphism group of Sym(6) as a semidirect product of the inner automorphism group, which is isomorphic with Sym(6) itself, with $\mathbb{Z}_2$:

$$\text{Aut}(S_6) = \text{Inn}(S_6) \rtimes \mathbb{Z}_2.$$

3.7 Exercises

Exercise 3.1. Show that the first group axiom implies that if $g_i \cdot g_j = g_i \cdot x$, then $x = g_j$ (left cancelation), as well as $g_i \cdot g_j = x \cdot g_j$ implies $x = g_i$ (right cancelation). Deduce the existence of a unique unit element e, satisfying $g \cdot e = e \cdot g = g$ for all $g \in \mathbf{G}$.

Exercise 3.2. Let g be an element of a group $\mathbf{G}$. Deduce from the group axioms the existence of a unique inverse g^{-1}, and prove that every group element commutes with its inverse.

Exercise 3.3. A more commonly used set of axioms to define a group is the following: A set $\mathbf{G}$ endowed with a binary operation $\cdot$ is called a group if the following axioms are satisfied: Closure axiom: $a, b \in \mathbf{G} \Rightarrow a \cdot b \in \mathbf{G}$. Associativity:

$(a \cdot b) \cdot c = a \cdot (b \cdot c)$ for all $a, b, c \in G$. Identity element: There exists an identity element, see Exercise 3.1. Inverse element: For each $a \in G$, there exists an inverse element, see Exercise 3.2.

Exercise 3.4. Give an example of an infinite group and an infinite subset which is closed under multiplication but does not form a subgroup.

Exercise 3.5. Let g be an element of order k in a group of order n. Show that k divides n.

Exercise 3.6. Determine whether or not three-dimensional nonzero vectors form a group under the cross product.

Exercise 3.7. Verify by the axioms the additive groups described in Example 3.2.
Show that the nonzero elements of $\mathbb{Q}[\sqrt{3}] = \{n + m\sqrt{3} \mid n, m \in \mathbb{Z}\}$ form a multiplicative group.

Is it true that the nonzero elements of any additive subgroup of the real numbers $\mathbb{R}$ form a multiplicative group?

Exercise 3.8. Identify which symmetries of the pyramid in Example 3.7 correspond to which complex transformations.

Exercise 3.9. Compute the Cayley graph of Sym(4) using the generators (1234), (234), and (14).
Are any of these generators redundant?
Draw the Cayley graph in the plane without edge crossings.

Exercise 3.10. Show that all elements of Alt(n) can be written as products of three-cycles, i.e. permutations that cyclically permute three elements and leave the other ones fixed.

Exercise 3.11. Show that a normal subgroup N of Alt(n) which contains one three-cycle contains all three-cycles.

Exercise 3.12. Prove that Alt(4) is the only alternating group which has a nontrivial normal subgroup.

Exercise 3.13. Show that the kernel of a group homomorphism $\phi : G \to H$ is a normal subgroup of H.

Exercise 3.14. Show that for each natural number n, there exists at least one group of order n.

Exercise 3.15. Let G be a finite group of order n that has only two subgroups. Prove that n is prime.

Exercise 3.16. Show that for a prime p, all groups of order p are isomorphic. Hence, up to isomorphism, there is only one group of order p.

Exercise 3.17. Show that the kernel of the homomorphism defined in Example 3.21 is generated by the words r^3, s^2, and $(rs)^3$ and their conjugates.

3.7 Exercises

Exercise 3.18. Draw the Cayley graph of the group acting on the 2×2 Escher stamps, see Example 3.37.

Exercise 3.19. Use the method of Example 3.36 to compute the number of three colorings of the faces of the cube up to reflective or rotational symmetry.

Exercise 3.20. Use the method of Example 3.36 to compute the number of three colorings of the edges of the cube up to reflective or rotational symmetry.

Exercise 3.21. Use the method of Example 3.36 to compute the number of three colorings of the vertices of the cube up to reflective or rotational symmetry.

Exercise 3.22. Use the method of Example 3.36 to compute the number of colorings of the cube with n colors.

How do you interpret the results as $n \to \infty$?

Exercise 3.23. Find representatives for each of the 23 types of 2×2 Escher stamp patterns, see Example 3.37.

Exercise 3.24. Determine the wallpaper group for each of the 23 stamp patterns.

Exercise 3.25. Show that the multiplicative group $\{1, -1, i, -i\} \subset C$ is isomorphic to $\mathbb{Z}_4$.

Exercise 3.26. Determine all subgroups of $\mathbb{Z}_n$.

Exercise 3.27. Let X be any of the three graphs: G_1, two triangles joined at a vertex; G_2, three triangles joined at a vertex; and G_3, two triangles joined by an edge.

1. Determine the (abstract) group of automorphisms $\text{Aut}(X)$.
2. Action of $\text{Aut}(X)$ on $V(X)$. What are the vertex orbits?
3. Action of $\text{Aut}(X)$ on $E(X)$. What are the edge orbits?

Exercise 3.28. Determine the group of symmetries of the prism Π_6.

Exercise 3.29. Determine the group of symmetries of the antiprism A_6.

Exercise 3.30. Determine the group of symmetries for the pyramid W_6.

Exercise 3.31. Determine the group of symmetries of the double pyramid B_6.

Exercise 3.32. Consider the transformations of the colored cube in Example 3.8. Compute $\mathbf{C}_{\text{rwb}}$, and show that it is normal in the group of symmetries of the cube.

Exercise 3.33. Consider the symmetries of the cube considered in Example 3.8. Can you find a subgroup of the group of symmetries of the cube which is not the group of color-preserving symmetries of some face coloring of the cube?

Is the same true of vertex colorings?

Exercise 3.34. In the truncated hexagonal prism of Example 3.9, color the top triangles red, blue, yellow, red, blue, and yellow and the bottom green, orange, purple, green, orange, and purple. Determine whether the red-preserving subgroup is normal.

Exercise 3.35. Consider the rotational symmetries of the ordinary hexagonal prism or, equivalently, its graph Π_6. There is a subgroup A consisting of all transformations which preserve the pair of antipodal edges $\{\{u_0, v_0\}, \{u_3, v_3\}\}$. There is also a subgroup B preserving the pair of antipodal 4-cycles $\{\{u_1, v_1, u_2, v_2\}, \{u_3, v_3, u_4, v_4\}\}$.

Show that A and B are isomorphic. Show that these subgroups are not conjugate. Is either of them normal?

Exercise 3.36. Show that the symmetric group Sym(n) can be generated by just two elements.

Exercise 3.37. Prove directly that each element of Sym(n) can either only be written as an even number of transpositions or only be written as an odd number of transpositions.

Exercise 3.38. Consider the Escher problem with the motif a square partitioned by a diagonal into a red and a green triangle.

1. Determine the abstract group of automorphisms and its Cayley graph.
2. What is the number of different patterns?
3. What is the number of different patterns if reflections are allowed?
4. What is the number of different patterns in the original Escher problem if reflections are allowed?

Exercise 3.39. Given a graph G, prove that any double cover over G is a regular cover.

Exercise 3.40. Find an example of a threefold cover that is not regular.

Exercise 3.41. Express the 3-prism graph as a sixfold cover over a pregraph on a single vertex.

Exercise 3.42. Let $\mathbb{Z}_n^k$ be an elementary abelian group. Let S be a set of generators with the following property. Each element is a 0-1 vector. They generate the whole group. S has the least number of elements with this property.

1. Show that $|S| = k$.
2. Show that there is an automorphism of the group mapping S to the standard generating set.

Exercise 3.43. Determine the Kronecker double covers of all generalized Petersen graphs GP(n, k).

Exercise 3.44. Show that GP(10, 2), the skeleton of the dodecahedron, arises as a double cover of the pentagonal embedding of the Petersen graph in the projective plane. Find the corresponding voltage assignment on GP(5, 2).

Exercise 3.45. Draw the Kronecker double cover of $G(10, 2)$.

3.7 Exercises

Exercise 3.46. Determine a $\mathbb{Z}_n$ covering over the handcuff graph $G(1, 1)$ that is not a generalized Petersen graph $G(n, r)$.

Exercise 3.47. Prove that the generalized Petersen graph $GP(n, k)$ is vertex transitive, if and only if either $k^2 = \pm 1 \mod n$ or else $n = 10$ and $k = 2$.

Exercise 3.48. Prove that C_n, K_n, and Q_n are all vertex transitive.

Exercise 3.49. Which complete multipartite graphs $K_{a,b}$, $K_{a,b,c}$, ... are vertex transitive?

Exercise 3.50. Prove that the Cartesian product of vertex-transitive graphs is vertex transitive.

Exercise 3.51. Prove that $\mathbb{Z}_3 \times \mathbb{Z}_3 \neq \mathbb{Z}_9$.

Exercise 3.52. Prove that $\mathbb{Z}_2 \times \mathbb{Z}_3 = \mathbb{Z}_6$.

Exercise 3.53. Determine all groups that have a cycle C_n as their Cayley graph.

Exercise 3.54. Show that PG $(8, 3)$ is a regular cover of the graph of the cube Q_3.

Exercise 3.55. Show that PG $(12, 5)$ is a regular cover of the graph of the cube Q_3.

Exercise 3.56. Show that PG $(24, 5)$ is a regular cover of the graph of the cube Q_3.

Exercise 3.57. Carry out a construction of the Cayley graph of the automorphism group of K_3 using the method of Sect. 3.4.6.

Exercise 3.58. Carry out a construction of the Cayley graph of the automorphism group of $K_{3,3}$ using the method of Sect. 3.4.6.

Exercise 3.59. Complete the construction of Cayley graph of the automorphism group of $GP(5, 3)$ started in Sect. 3.4.6.

Exercise 3.60. Show that the cube graph Q_4 is a cover of the cubic theta graph with voltages from $\mathbb{Z}_4$.

Exercise 3.61. Show that the prism graphs are rotagraphs.

Exercise 3.62. Show that the antiprism graphs are rotagraphs.

Exercise 3.63. Show that the Petersen graph cannot be written as a rotagraph.

Exercise 3.64. Show that for any group, the group of inner automorphisms is a normal subgroup of the group all automorphisms.

Exercise 3.65. Show that every automorphism of Sym(6) which permutes the transpositions (ij) must be inner.

Exercise 3.66. Determine the group of automorphisms of the Heawood graph. Hint: It is of cardinality 336.

Fig. 3.43 The line graph of the Gray graph

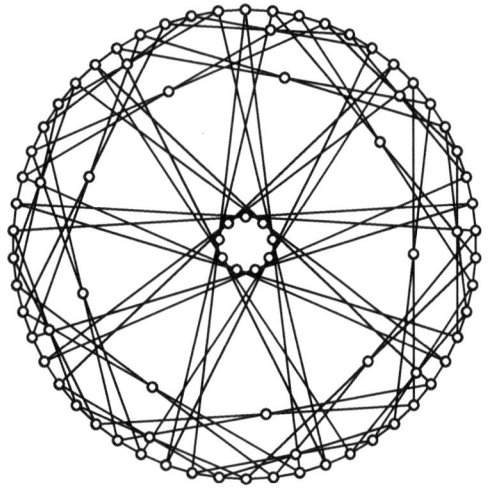

Fig. 3.44 A voltage assignment on $K_{3,3}$

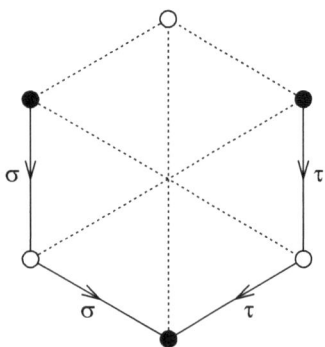

Exercise 3.67. Show that a vertex- and edge-transitive graph is 1/2-arc-transitive if and only if no automorphism of the graph interchanges the endvertices of some edge.

Exercise 3.68. Show that the girth of the Gray graph $\mathcal{G}$ is 8.

Exercise 3.69. Show that the two distance sequences of the Gray graph $\mathcal{G}$ are $(1, 3, 6, 12, 12, 12, 8)$ and $(1, 3, 6, 12, 16, 12, 4)$.

Exercise 3.70. Prove that the diameter of the Gray graph $\mathcal{G}$ is 6.

Exercise 3.71. Prove that there are a total of 81 octagons, that is, induced cycles of length 8, in the Gray graph and that each of its vertices is contained in 12 octagons and each of its edges is contained in 8 octagons.

Exercise 3.72. Prove that the order of the automorphism group of the Gray graph $\mathcal{G}$ is $1296 = 2^4 3^4$.

3.7 Exercises

Exercise 3.73. Show that the line graph
of the Gray graph, Fig. 3.43, is isomorphic to a a Cayley graph for the group $S(3) = \mathbb{Z}_3^3 \rtimes \mathbb{Z}_3$.

Exercise 3.74. Prove that the Gray graph $\mathcal{G}$ is a regular cover of $K_{3,3}$ with $\mathbb{Z}_3^2$ as the group of covering transformations by determining nonidentity voltages σ, τ from $\mathbb{Z}_3^2$ in Fig. 3.44.

Exercise 3.75. Verify that the incidence graph of the coset incidence structure of Example 5.15 is indeed the Desargues graph.

Chapter 4
Maps

4.1 Geometric Surfaces

4.1.1 Polyhedral Graphs

In Chap. 2, we saw several graphs which were obtained as the one-skeleta of polyhedra. Although such graphs lose some of the richness of the original object, e.g., the facial structure, the metric information, and the incidence angles, they do form an important class of graphs which we call *n-polyhedral graphs*. We say a graph is *polyhedral of dimension n* if it is a one-skeleton of a convex polyhedron in $\mathbb{R}^n$, with the dimension assumed to be three unless explicitly specified.

Using stereographic projection, we can associate to a convex polyhedron a specific drawing of its polyhedral graph in the plane; see Fig. 4.1. If the projection point is chosen to be a generic point outside the polyhedron but close to an interior point to some face, then the vertices will project to distinct points and the edges will project to noncrossing line segments. The resulting figure is called a Schlegel diagram. The faces of a Schlegel diagram are all convex polygons corresponding to the faces of the original polyhedron. The face of the polyhedron near the projection point corresponds to the cycle bounding the entire Schlegel diagram, the boundary of the *exterior face*. One may take the projection point to be on the surface of the polyhedron, which will result in that point projecting to infinity (see Fig. 4.2), and, again, all faces not incident to the projection point are projected to straight-sided convex polygons. For aesthetic reasons, the resulting drawing is often homeomorphically adjusted so that the angles are not too small and the vertices are not accumulated too much in the center of the figure, achieving a more Picasso-esque rendering which is also called a Schlegel diagram; see Fig. 4.3.

Stereographic projection implies that all polyhedral graphs are planar. Planarity is one of the essential qualities of polyhedral graphs. The following classical theorem of Steinitz states that the only other is 3-*connectivity*; see Sect. 2.4.4.

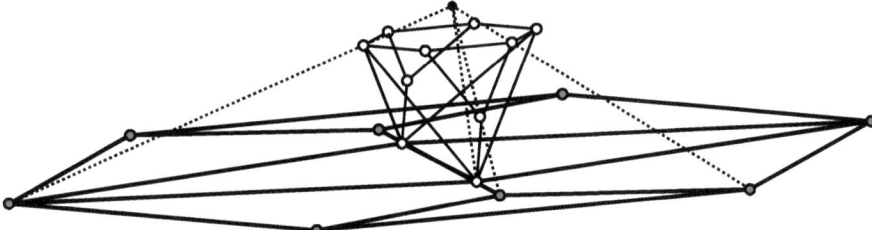

Fig. 4.1 Projecting a polyhedron onto a Schlegel diagram

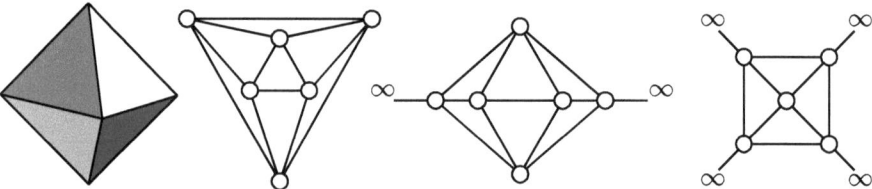

Fig. 4.2 Schlegel diagrams of an octahedron projected from outside a face center, from an edge center, and from a vertex

Fig. 4.3 Schlegel diagrams for the polyhedron of Fig. 4.1

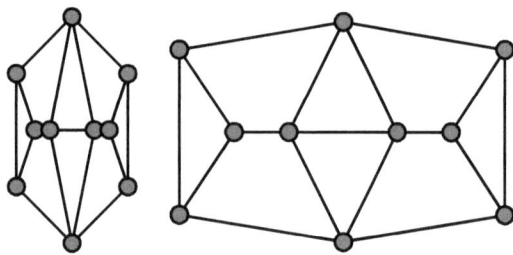

Theorem 4.1 (Steinitz [95]). *The one-skeleton of an arbitrary convex polyhedron is a planar 3-connected graph, and each planar 3-connected graph is polyhedral.*

Steinitz's theorem assures us that there is always a route back from a polyhedral graph to an actual polyhedron, but it will not be possible to recover the original polyhedron, even from a Schlegel projection. The combinatorial structure of the polyhedron, however, is recoverable just from the graph by Whitney's theorem [106] which states that the cyclic ordering of the edges and vertices around the faces of a drawing of a polyhedral graph *is* uniquely determined.

From Steinitz theorem and the projection argument, we also conclude that any 3-connected planar graph must have a planar drawing with convex faces and straight edges, a result known as Fáry's theorem, [100].

4.1 Geometric Surfaces

Fig. 4.4 A *stellated* cube and an infinite plane tiling

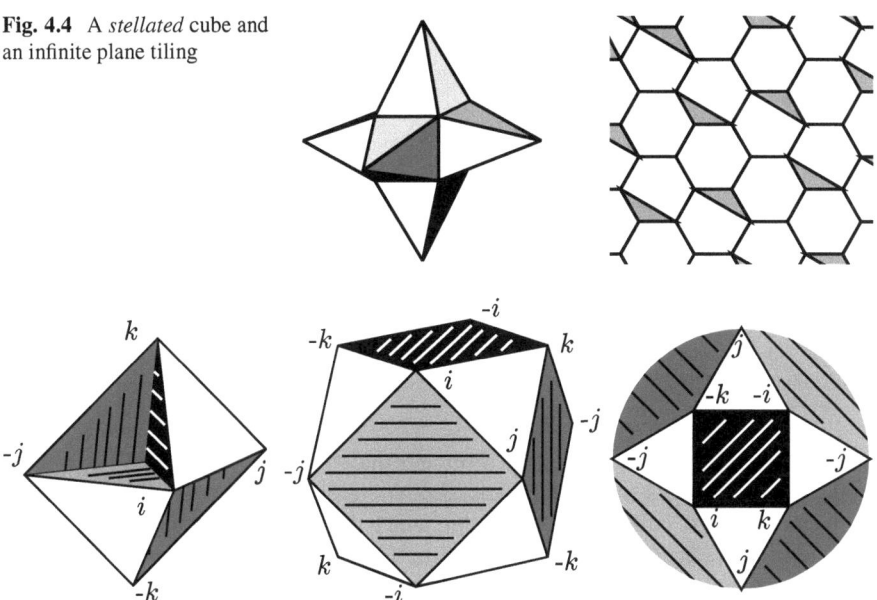

Fig. 4.5 The heptahedron, or tetrahemihexahedron, covered by the cuboctahedron

4.1.2 Polygonal Surfaces

The boundary of a convex polyhedron is a collection of convex polygons in space joined pairwise along their boundary edges. Such an object is called a *polygonal surface*. More generally, a polygonal surface is a collection $\{P_i\}$ of polygons in $\mathbb{R}^d$ whose boundaries are pairwise disjoint except that each edge of each polygon is the edge of precisely one other polygon and such that the resulting complex is connected.

Polygonal surfaces are general enough not only to represent nonconvex polyhedra and infinite plane tilings (see Fig. 4.4) but also abstract complexes, that is, complexes without a given embedding in Euclidean space.

Example 4.2. The tetrahemihexahedron. Start with $K_{2,2,2}$ embedded as the 1-skeleton of the regular octahedron (see Fig. 4.2), and consider the three planes which bisect the three main diagonals, that is, the segments joining the three pairs of nonadjacent edges. These three planes intersect the octahedron in three squares, which we consider as faces of a new complex. To these three squares, we add four of the eight triangular faces of the octahedron chosen such that no two of them meet along an edge. The resulting structure is called a *tetrahemihexahedron*, or a *heptahedron*, [103], and is pictured in Fig. 4.5. Each of the edges is shared by one triangle and one square, so the resulting complex is a polygonal surface with the same 1-skeleton as the regular octahedron.

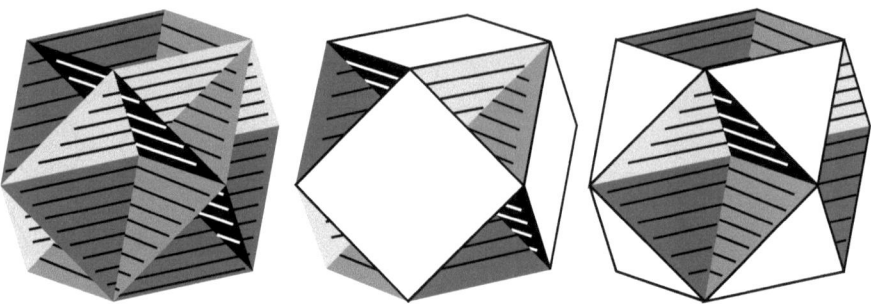

Fig. 4.6 The cubohemioctahedron, the octahemioctahedron, and the cuboctahedron share the same one-skeleton

Example 4.3. The cubohemioctahedron and the octahemioctahedron. The cuboctahedron also has a 1-skeleton which reoccurs as the skeleton of other polygonal surfaces. Given a cube, the four planes which bisect the diagonals intersect the interior of the cube in four regular hexagons, and the edges on the boundary of those hexagons form the graph of the cuboctahedron. Each edge of the cuboctahedron is incident to exactly one hexagon, so the four hexagons can be completed to a polygonal surface by adding the six squares of the cuboctahedron, forming the cubohemioctahedron, or the eight triangles, forming the octahemioctahedron; see Fig. 4.6.

The polygonal surfaces discussed so far are embedded in $\mathbb{R}^d$ and are properly regarded as geometric objects. More abstractly, we can consider a collection of polygons $\{P_i\}$ together with a perfect matching on the set of their oriented edges. If the 1-skeleton of the resulting complex is connected, the complex is again called a polygonal surface.

For example, in the degenerate position of the tetrahemihexahedron of Example 4.2, twisted and with lines of self-intersections, it is difficult to see the nature of the surface. Trying to reassemble the three squares and four triangles according to the edge matching in an untwisted position yields much frustration; however, if we take two indistinguishable copies of each polygon, the eight triangles and six squares may be readily reassembled in the form of a cuboctahedron with pairs of duplicated faces in antipodal position. Thus, the twisted surface represented by the tetrahemihexahedron is, in fact, the projective plane, which we may also see by slicing the labeled cuboctahedron in half and identifying opposite points on the boundary of one of the halves; see Fig. 4.5.

Two edges of a single convex polygon which is part of a polygonal complex in $\mathbb{R}^d$ may clearly not be matched to one another. For abstract polygonal surfaces, however, there is nothing to prevent this from happening. If we consider those polygonal surfaces with a single polygon, necessarily having an even number of sides, we obtain some of the classical descriptions of topological surfaces (see, for

4.1 Geometric Surfaces

Fig. 4.7 Some polygonal surfaces from a square

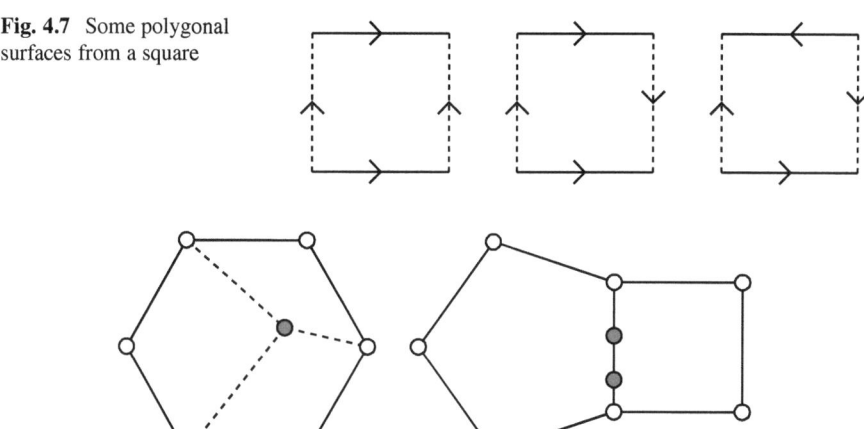

Fig. 4.8 A subdivision point in the interior of a face and two points of subdivision in the interior of an edge

Fig. 4.9 A barycentric subdivision

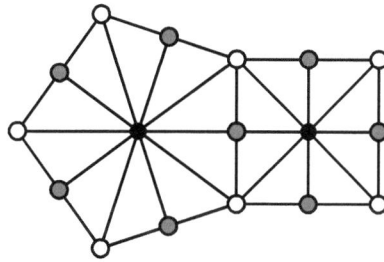

example, Fig. 4.7) in which the first square wraps up to form a torus, the second wraps to form the famous Klein bottle, and the third, noting that antipodal points on the square are brought together, yields the topological projective plane.

Surfaces in topology are distinguished up to *homeomorphism*, that is, a continuous transformation with a continuous inverse. For polygonal surfaces, a more combinatorial notion is a bijection between polygons preserving the edge matching, which is appropriate if we are studying the polygonal complexes themselves but not if we want to capture only the topology. For this, we need to subdivide the complexes.

A polygon may be subdivided by either splitting it in its interior or adding new vertices along its boundary compatible with its associated neighbor polygon; see Fig. 4.8. There is a piecewise linear bijection between a polygonal surface and its subdivision and hence between any two subdivisions of the same polygonal surfaces, and so the topological type is unchanged. A particular choice, called the *barycentric subdivision*, is to place one new vertex at the midpoint of every edge and one new vertex at the centroid of every polygon joined to every vertex along the polygon; see Fig. 4.9. Subdivision turns out to be a powerful tool to compare surfaces; see [84].

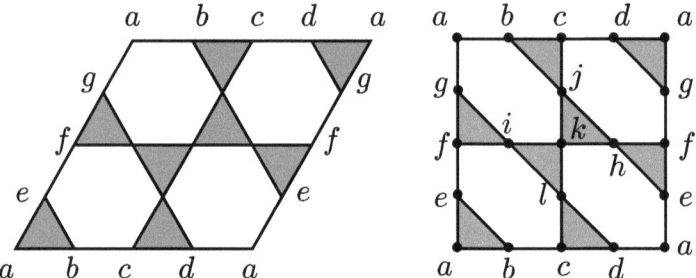

Fig. 4.10 The map of an octahemioctahedron in a torus with eight triangles and four hexagons

Fig. 4.11 The unfolded cubohemioctahedron

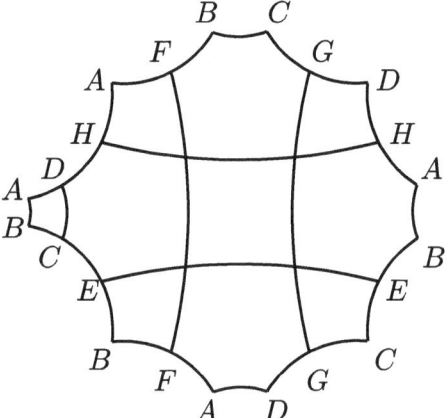

Theorem 4.4. *Two polygonal surfaces are homeomorphic if and only if they have polygonally equivalent subdivisions.*

To determine the character of the octahemioctahedron of Example 4.3, we can again try to take apart the polygons and reassemble them in an untwisted position. This time, there are two triangles and two hexagons at each vertex, so they will lie flat in the plane (see Fig. 4.10) where they are arranged in a parallelogram with the edges along the opposite sides matched, so the octahemioctahedron is polygonally equivalent to the subdivision of the torus made from one square, shown on the right.

For the cubohemioctahedron, unfolding is made more difficult because each vertex is incident to two hexagons and two squares, giving a total vertex angle of 390° if regular Euclidean polygons are used. One alternative is to alter the size and shape of the polygons, as in the rectangular torus in Fig. 4.10. Another, since this is such a symmetric example, is to use regular hyperbolic polygons of appropriate size; see Fig. 4.11. In either case, the overall shape of the surface when completely folded is not immediately apparent. See Exercise 4.4.

For more examples and details on complexes of polygons, see [82, 104].

4.1.3 The Topology of Polygonal Surfaces

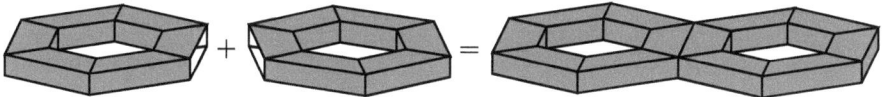

Fig. 4.12 The connected sum of two tori

Subdividing a polygonal surface does not alter its topological type, although it does introduce new vertices and edges and may introduce new faces. These new elements, however, are balanced, yielding a classical invariant, the *Euler characteristic*,

$$\chi(S) = |V| - |E| + |F|,$$

which was introduced for planar graphs in Sect. 2.2.11. From Theorem 4.4, we conclude that two polygonal surfaces which are homeomorphic must have the same Euler characteristic.

Suppose we are given two polygonal surfaces, S and S', containing two polygons P and P', respectively, such that P and P' have the same number of sides and such that no two edges of P or P' are matched. We can create a new polygonal surface, called the *connected sum*, $S \sharp S'$, of S and S' as follows. Let the edges of P be matched to edges $\{e_1, e_2, \ldots, e_n\}$ in S and those of P' matched to $\{e'_1, e'_2, \ldots, e'_n\}$ in S', in cyclic order around P and P'. The polygons of $S \sharp S'$ are those of S and S' excluding P and P', with the matching inherited from S and S', with e_i matched with e'_i. See Fig. 4.12 in which two polygonal tori are joined. Note that while $S \sharp S'$ is not necessarily defined, there exist subdivisions of S and S' for which the connected sum is defined.

The following is immediate.

$$\chi(S \sharp S') = \chi(S) + \chi(S') - 2.$$

Example 4.5. We can take two copies of the tetrahemihexahedra t and T (see Example 4.2) and join them along two quadrilaterals labeled $\{i, j, -i, -j\}$ and $\{I, J, -I, -J\}$. The result is a polygonal surface with eight triangles and four quadrilaterals with the topological type of a Klein bottle; see Fig. 4.13. If, on the other hand, we join the surfaces along the faces $\{i, j, k\}$ and $\{I, J, K\}$, then the connected sum is a surface with six triangles and six quadrilaterals, but it still has the topological type of the Klein bottle.

It is true, but not obvious (see Sect. 4.3), that the topological type of the connected sum only depends on the topological type of the summands and not on any of the choices made in its construction. In fact, all topological types of closed surfaces can be described as connected sums.

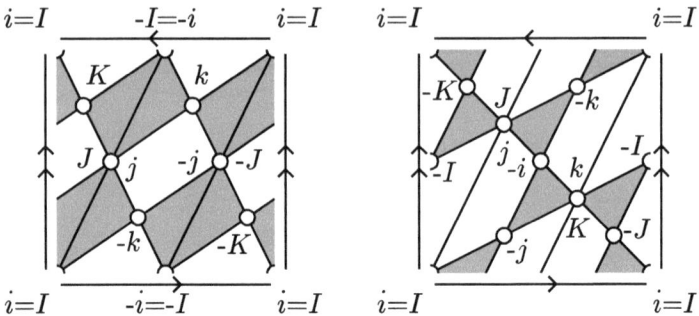

Fig. 4.13 The connected sum of two tetrahemihexahedra

Theorem 4.6. *Every closed surface has the topological type of either:*

1. *The sphere.* ($\chi(S) = 2$).
2. *A connected sum of n tori.* ($\chi(T^n) = 2(n-1)$).
3. *A connected sum of n projective planes.* ($\chi(P^n) = 2 - n$).

The surfaces S and T^n are two sided, or *orientable* surfaces, while P^n is one sided, or *nonorientable*. Therefore, two surfaces have the same topological type if and only if they have the same orientability and the same Euler characteristic.

4.2 Maps and Flags

4.2.1 A Graph Theoretical Approach to Surfaces

While polygonal surface is the most natural way to approach surfaces topologically, it is not the most transparent from the point of view of graph theory. For a polygonal surface, the faces and the edges are known a priori, while the vertices, the fundamental unit of graph theory, are an end product of the construction. In general, discrete and algebraic methods are central to the study of topology, and in particular, there are several methods to combinatorially describe surfaces. We follow the approach of Tutte, [99], which is equivalent to the one found, for instance, in [35].

The motivating example is the barycentric subdivision. If S is a polygonal surface with $|E|$ edges, then the barycentric subdivision $\mathbf{BS}(S)$ is a triangulated surface with $4|E|$ triangles. In the construction of $\mathbf{BS}(S)$ from the triangles, each triangle T is matched with three triangles which we denote by $\tau_0(T)$, $\tau_1(T)$, and $\tau_2(T)$. The notation is chosen so that T and $\tau_0(T)$ share an edge and face of S, T and $\tau_1(T)$ share a vertex and face of S, while T and $\tau_2(T)$ share a vertex and an edge of S; see Fig. 4.14. In other words, T and $\tau_i(T)$ share an object of dimension $i + 1$ and

4.2 Maps and Flags

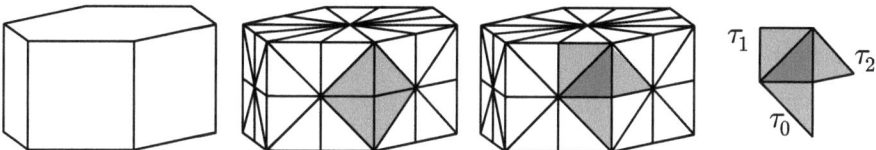

Fig. 4.14 Decomposition of the hexagonal prism into flags

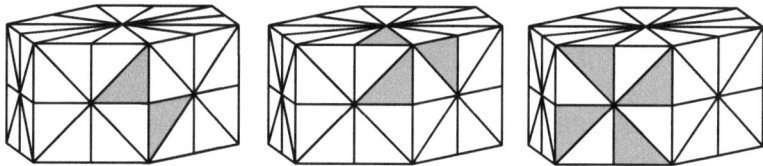

Fig. 4.15 Orbits under $\tau_2\tau_0$, $\tau_2\tau_1$, and $\tau_0\tau_1$

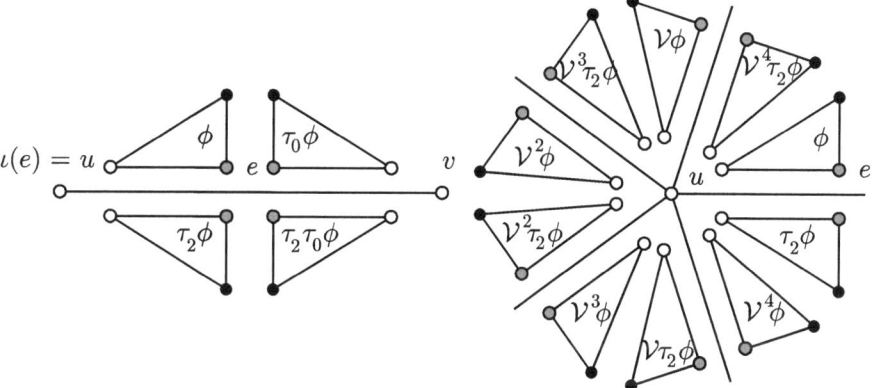

Fig. 4.16 Flags at an edge and at a vertex

object of dimension $i + 2$ of S, indices modulo 3. The perfect matchings τ_i can be regarded as permutations on the set of triangles in **BS**(S). The cycles in $\tau_2\tau_1$ give the cyclic orderings of the triangles around the faces of S, and the cycles of $\tau_0\tau_1$ give the cyclic orderings of the triangles about the vertices of S; see Fig. 4.15. From the point of view of the graph, the orbits of τ_0 and τ_2 are local to each edge, and the orbits of the product $\tau_2\tau_1$ are local to each vertex. This local information, τ_0, τ_2, and $\tau_2\tau_1$, carries the same information as in τ_0, τ_1, and τ_2, since $\tau_1 = \tau_2(\tau_2\tau_1)$.

Now, suppose we are given a connected graph G. We allow G to have both loops and multiple edges and, for notational convenience, assume each edge has an orientation, that is, a specified initial vertex $\iota(e)$. We define the set of *flags* to be $\Phi = E \times \mathbf{K}$, where **K** is the Klein four group, generated by involutions τ_0 and τ_2. There are four flags associated with each edge e, namely, $(e, 1)$ and (e, τ_2) are associated with the initial vertex of e, and (e, τ_0) and $(e, \tau_0\tau_2)$ are associated with the terminal vertex; see Fig. 4.16. The action $g(e, g') = (e, gg')$ defines an action of **K** on Φ, with τ_0 and τ_2 satisfying

1. $\tau_0^2 = \tau_2^2 = \text{Id}$.
2. $\tau_0 \tau_2 = \tau_2 \tau_0$.

To conclude the local picture for the drawing of G at a vertex v, consider the set Φ_v of all flags which are associated with a vertex v. There are two flags, ϕ and ϕ', for each half edge incident with v, and $\phi' = \tau_2(\phi)$. Define $\mathcal{V}_v$ to be a fixed point free permutation on Φ_v such that $\mathcal{V}_v$ is a product of two disjoint cycles $\mathcal{V}_v = L_v R_v$ with $\tau_2 L_v \tau_2 = R_v^{-1}$. Now define the *vertex permutation* $\mathcal{V}$ to be the product of all $\mathcal{V}_v$,

$$\mathcal{V} = L_{v_1} R_{v_1} L_{v_2} R_{v_2} L_{v_3} R_{v_3} \cdots L_{v_n} R_{v_n} = \mathcal{V}_{v_1} \mathcal{V}_{v_2} \mathcal{V}_{v_3} \cdots \mathcal{V}_{v_n},$$

where the order of multiplication is arbitrary since the cycles L_{v_i} and R_{v_j} act on disjoint sets of flags. The graph G together with its flag permutations is called a *map*, $\text{M} = \text{M}(G, \tau_0, \tau_2, \mathcal{V})$. We often write just $\text{M} = \text{M}(G, \mathcal{V})$.

Clearly we have

1. $\mathcal{V}\tau_2 = \tau_2 \mathcal{V}^{-1}$.
2. $\{\mathcal{V}^i \phi\} \cap \{\mathcal{V}^i \tau_2 \phi\} = \emptyset$.

Theorem 4.7. *Let* $\text{M}(G, \tau_0, \tau_2, \mathcal{V})$ *be a map. The permutation group generated by* τ_0, τ_2, *and* $\mathcal{V}$ *acts transitively on the flags.*

Proof. From the action of τ_0 and τ_2 alone, we know that the flags associated with any edge of G belong to one orbit. From the action of τ_2 and $\mathcal{V}$ alone, we conclude that the flags associated with any vertex of G belong to the same orbit, since on those flags, the action of $\mathcal{V}$ and $\mathcal{V}_v$ is the same. The flags of adjacent vertices in G belong to the same orbit because those of their connecting edge are in the same orbit. Therefore, the action is transitive by the connectivity of G. □

To every map $\text{M}(G, \tau_0, \tau_2, \mathcal{V})$, there is a naturally associated polygonal complex which consists of one triangle for each flag and matching rules defined by τ_0, τ_2, and $\tau_1 = \tau_2 \mathcal{V}$. Note that τ_1 is necessarily an involution since $\tau_2 \mathcal{V} \tau_2 \mathcal{V} = \mathcal{V}^{-1} \mathcal{V}$ by property (3) above. For convenience, we color the vertices of each triangle white, gray, and black and match the white–gray edges of ϕ and $\tau_2 \phi$ and the black–gray edges of ϕ and $\tau_0 \phi$ and match the white–black edges of ϕ and $\tau_1 \phi$. To establish that this does indeed define a polygonal complex, we need only check that the graph of the complex of triangles is connected, which follows immediately from Theorem 4.7. We have the following:

Theorem 4.8. *The flags of the map* $(G, \tau_0, \tau_2, \mathcal{V})$ *correspond to the triangles of a polygonal complex with matching rules* τ_2, τ_0, *and* $\tau_1 = \tau_2 \mathcal{V}$.

The flag complex defines a surface in which the graph G is naturally embedded, with the ordering of the flag triangles around each vertex corresponding to $\mathcal{V}$; the ordering of the flag triangles associated with each edge correspond to τ_0 and τ_2, and the faces of the map, making their late entrance, correspond to the disjoint cycles of $\mathcal{F} = \tau_0 \tau_1$, called the *face permutation*. Note that the embedding of a graph into a

4.2 Maps and Flags

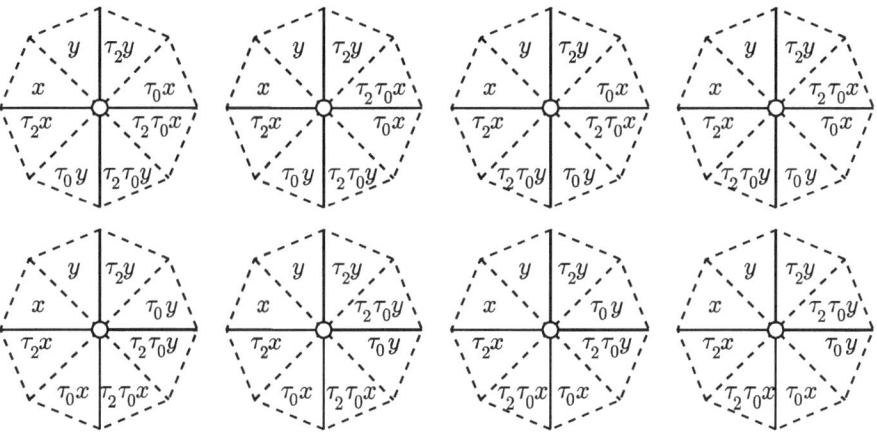

Fig. 4.17 Maps on the two-loop graph

surface such that its complement is homeomorphic to a disjoint union of open disks is considered a map from the topological, noncombinatorial point of view.

Example 4.9. The graph with a single vertex and two loops may be realized as a map with eight flags. How many ways may this be done? Somewhere in the ordering of the eight flags around the vertex, two flags belonging to different edges will be neighbors, call them x and y; see Fig. 4.17. So the eight flags are now labeled x, $\tau_0 x$, $\tau_2 x$, $\tau_2 \tau_0 x$ y, $\tau_0 y$, $\tau_2 y$, and $\tau_2 \tau_0 y$, with $y = \tau_1 x$. Once the flags x and y are placed, $\tau_2 x$ and $\tau_2 y$ are forced, so there are four possibilities if the edges alternate around the vertex and four possibilities if they do not.

We can pursue one further level of abstraction. If we have three fixed point free permutations τ_0, τ_2, and $\mathcal{V}$ on a set Φ such that

A1 $\tau_0^2 = \tau_2^2 = \mathrm{Id}$
A2 $\tau_0 \tau_2 = \tau_2 \tau_0$
A3 $\mathcal{V} \tau_2 = \tau_2 \mathcal{V}^{-1}$
A4 $\{\mathcal{V}^i \phi\} \cap \{\mathcal{V}^i \tau_2 \phi\} = \emptyset$
A5 τ_0, τ_2, and $\tau_0 \tau_2$ are fixed point free
A6 $\langle \tau_0, \tau_2, \mathcal{V} \rangle$ acts transitively on Φ,

then we can define a graph G whose vertices are the orbits of Φ under $\langle \tau_2, \mathcal{V} \rangle$ and whose edges are the orbits of Φ under $\langle \tau_0, \tau_2 \rangle$. The orbits of $\langle \tau_0, \tau_2 \rangle$ each have four elements and intersect either one or two orbits of $\langle \tau_2, \mathcal{V} \rangle$, defining the endpoints of the edge. It is not difficult to verify that this realizes the permutations of Φ as those of a map $\mathrm{M}(G, \tau_0, \tau_2, \mathcal{V})$ with 1-skeleton G.

Fig. 4.18 K_4 in the projective plane and its dual

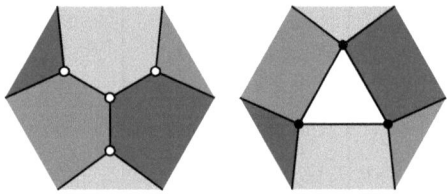

Fig. 4.19 K_4 and dual mapped onto the torus or Klein bottle

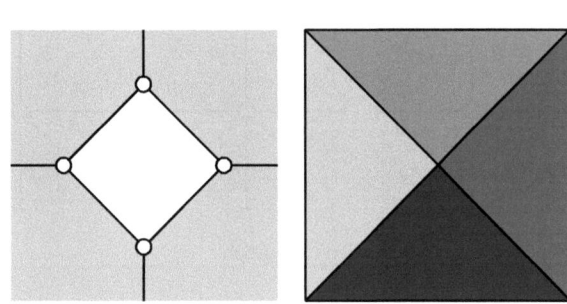

4.2.2 Dual Constructions

The face permutation $\mathcal{F} = \tau_0 \tau_1$ of a map $M = M(G, \tau_0, \tau_2, \mathcal{V})$ satisfies relations similar to that of $\mathcal{V}$ in the axiom system A1–A6.

Theorem 4.10. *If we have τ_0, τ_2, and $\mathcal{V}$ satisfying A1–A6 above, then $\mathcal{F}\tau_0 = \tau_0 \mathcal{F}^{-1}$ and $\{\mathcal{F}^i \phi\} \cap \{\mathcal{F}^i \tau_0 \phi\} = \emptyset$.*

Proof. For the first equation, $\mathcal{F}\tau_0 = \tau_1 = \tau_0(\tau_0 \tau_1) = \tau_0 \mathcal{F}^{-1}$.

For the second equation, assume to the contrary that $\mathcal{F}^m \phi = \tau_0 \phi$, and assume that ϕ is chosen so that m is minimal. If $m = 1$, we have $\mathcal{F}\phi = \mathcal{V}\tau_2(\tau_0 \phi) = \tau_0 \phi$, contradicting Axiom 4 for the element $\tau_0 \phi$. If $m \geq 2$, then $\mathcal{F}^m \phi = \tau_0 \phi$ implies $\mathcal{F}^{m-2}(\mathcal{F}\phi) = \mathcal{F}^{-1} \tau_0 \phi = \tau_0(\mathcal{F}\phi)$, contradicting the minimality of m. □

Using Theorem 4.10, given a map $M = M(\tau_0, \tau_2, \mathcal{V})$, we can interchange the roles of $\mathcal{V}$ and $\mathcal{F}$ to define a new map $\mathbf{Du}(M) = \mathbf{Du}(M)(\tau_2, \tau_0, \mathcal{F})$, called the *dual map*. Notice that τ_0 and τ_2 have also exchanged roles.

For a planar graph, which we regard as a map on the sphere, the 1-skeleton of $\mathbf{Du}(M)$ corresponds the usual dual graph, with one vertex in the center of each face and edges of M and $\mathbf{Du}(M)$ in one-to-one correspondence crossing transversely.

Example 4.11. The graph K_4 is the 1-skeleton of the tetrahedron map, whose dual is likewise the map of the tetrahedron. However, K_4 is the 1-skeleton of maps on other surfaces as well. If K_4 is embedded on the projective plane as in Fig. 4.18, then the dual map has three vertices with three pairs of doubled edges. The doubled edges do not correspond to faces with two edges; however, all faces are triangles. K_4 can also be embedded on the torus or Klein bottle; see Fig. 4.19 with one square

4.2 Maps and Flags

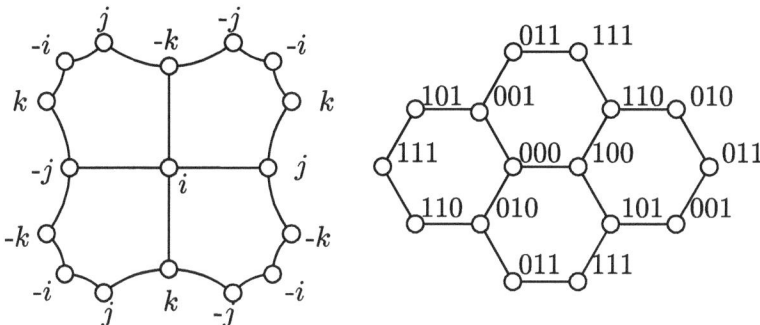

Fig. 4.20 The Petrie dual of a cube and octahedron

face and one octagonal face. In either case, the dual map has four triangles, and the dual graph has two vertices joined by four parallel edges, with two loops at one vertex.

Suppose we have a map $M = M(\tau_0, \tau_2, \mathcal{V})$. Let $\tau_0' = \tau_0 \tau_2$. Then the permutations $\tau_0', \tau_2, \mathcal{V}$ trivially satisfy the axioms for a map, **Pe**(M), called the *Petrie dual* of M. The correspondence is called a dual since, clearly, **Pe**(**Pe**(M)) = M. The face permutation $\mathcal{F}' = \mathcal{V}'\mathcal{E}' = \mathcal{V}\tau_0$ of **Pe**(M) gives a sequence of flags along a left–right alternating path in M, called a *Petrie walk*, hence the term Petrie dual. **Pe**(M) and M have the same 1-skeleton.

The Petrie dual of the cube has four hexagonal faces as does the Petrie dual of the Octahedron; see Fig. 4.20.

Alternating the dual and the Petrie dual eventually returns the original map,

$$\begin{array}{cccccc} \mathbf{Du} & \mathbf{Pe} & \mathbf{Du} & \mathbf{Pe} & \mathbf{Du} & \mathbf{Pe} \\ \tau_2 \longrightarrow \tau_0 & \longrightarrow \tau_0 & \longrightarrow \mathcal{E} & \longrightarrow \mathcal{E} & \longrightarrow \tau_2 & \longrightarrow \tau_2 \\ \tau_0 \longrightarrow \tau_2 & \longrightarrow \mathcal{E} & \longrightarrow \tau_0 & \longrightarrow \tau_2 & \longrightarrow \mathcal{E} & \longrightarrow \tau_0 \\ \mathcal{V} \longrightarrow \mathcal{V}\mathcal{E} & \longrightarrow \mathcal{V}\mathcal{E} & \longrightarrow \mathcal{V}\tau_0 & \longrightarrow \mathcal{V}\tau_0 & \longrightarrow \mathcal{V} & \longrightarrow \mathcal{V} \end{array}$$

so we have the equation

$$\mathbf{Du}(\mathbf{Pe}(\mathbf{Du}(\mathbf{Pe}(\mathbf{Du}(\mathbf{Pe}(M)))))) = M.$$

Rewriting the above relation $\mathbf{Du}(\mathbf{Pe}(\mathbf{Du}(M))) = \mathbf{Pe}(\mathbf{Du}(\mathbf{Pe}(M))) = \mathbf{An}(M)$ defines the *antipodal dual*. The vertices in the antipodal dual are the faces of **Pe**(**Du**(M)) and correspond to the Petrie walks in **Du**(M), that is, the set of Petrie walks in M itself. The faces of the **An**(M) are the vertices of **Pe**(**Du**(M)), which are the vertices of **Du**(M), i.e. the faces of M. The two maps in Fig. 4.20 are antipodal duals.

4.2.3 Flag Orbits and Orientability

There are $2|V|$ orbits of the flags Φ for map M under the vertex permutation $\mathcal{V}$ and, by Theorem 4.10, $2|F|$ orbits under the face permutation $\mathcal{F}$. We define the *edge permutation* $\mathcal{E}$ as $\mathcal{E} = \tau_0\tau_2$, which has $2|E|$ orbits. Locally, $\mathcal{V}, \mathcal{E}$, and $\mathcal{F}$ rotate the flags around the vertices, edges, and faces, respectively. Some relationships between $\mathcal{V}, \mathcal{E}, \mathcal{F}$, and τ_i are collected below:

$$\begin{aligned} \mathcal{V} &= \tau_1\tau_2, & \mathcal{E} &= \tau_0\tau_2, & \mathcal{F} &= \tau_1\tau_0, \\ \mathcal{V}\tau_2 &= \tau_2\mathcal{V}^{-1}, & \mathcal{E}\tau_2 &= \mathcal{E}\tau_2, & \mathcal{F}\tau_2 &= \tau_2\mathcal{E}\mathcal{F}^{-1}\mathcal{E}. \\ \mathcal{V}\tau_0 &= \tau_0\mathcal{E}\mathcal{V}^{-1}\mathcal{E}, & \mathcal{E}\tau_0 &= \mathcal{E}\tau_0, & \mathcal{F}\tau_0 &= \tau_0\mathcal{F}^{-1}. \\ & & \mathcal{F} &= \mathcal{V}\mathcal{E}. & & \end{aligned}$$

We see that each element in the flag permutation group $\langle \tau_0, \tau_2, \mathcal{V} \rangle = \langle \tau_2, \mathcal{V}, \mathcal{E}, \mathcal{F} \rangle$ can be written as $\tau_2 R$, where $R \in \langle \mathcal{V}, \mathcal{E}, \mathcal{F} \rangle$, so the subgroup $\langle \mathcal{V}, \mathcal{E}, \mathcal{F} \rangle$ has either one or two cosets, depending on whether $\tau_2 \in \langle \mathcal{V}, \mathcal{E}, \mathcal{F} \rangle$ or not.

If $\tau_2 \notin \langle \mathcal{V}, \mathcal{E}, \mathcal{F} \rangle$, then there are exactly two flag orbits under the action of $\langle \mathcal{V}, \mathcal{E}, \mathcal{F} \rangle$, called *orientation classes*, and we say the map is orientable. If, on the other hand, $\tau_2 \in \langle \mathcal{V}, \mathcal{E}, \mathcal{F} \rangle$, then $\langle \mathcal{V}, \mathcal{E}, \mathcal{F} \rangle$ acts transitively on Φ, and we say the map is nonorientable.

Theorem 4.12. *The map M is orientable if and only if there is a coloring of the flags with two colors so that ϕ and $\tau_i\phi$ receive different colors, for $i \in \{0, 1, 2\}$.*

Proof. If M is orientable, then color the two orbits of $\langle \mathcal{V}, \mathcal{E}, \mathcal{F} \rangle$ with different colors. If ϕ and $\tau_i\phi$ are colored the same, then, since $\tau_i\phi = \tau_2 R\phi$ for some $R \in \langle \mathcal{V}, \mathcal{E}, \mathcal{F} \rangle$, we would have $R\phi$ and $\tau_2 R\phi$ having the same color, a contradiction.

If, on the other hand, the flags can be bicolored so that ϕ and $\tau_i\phi$ receive different colors, then each of the elements $\mathcal{V}, \mathcal{E}$, and $\mathcal{F}$ preserves the colors of the flags, so $\langle \mathcal{V}, \mathcal{E}, \mathcal{F} \rangle$ must as well, and there must be two orbits. □

Another way to detect orientability is to consider the *flag graph*.

The flag graph has vertex set Φ with flags ϕ and ϕ' adjacent if $\phi' = \tau_i\phi$, for $i \in \{0, 1, 2\}$. An example of a flag graph can be seen in Fig. 4.38. The flag graph is trivalent and, in fact, is the union of the three perfect matchings induced by τ_0, τ_1, and τ_2. The following is immediate:

Theorem 4.13. *The map M is orientable if and only if the flag graph is bipartite.*

4.2.4 Map Projections

To depict a map obtained from a graph by choosing a particular clockwise order of semiedges around each vertex, we may start out by drawing each vertex, on a piece of paper, together with its half edges in the correct cyclic order with each side

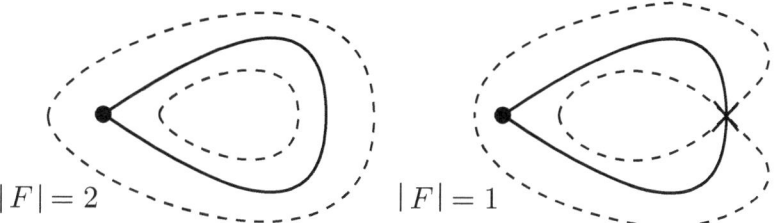

Fig. 4.21 A loop embedded on the sphere (*left*) and in the projective plane (*right*)

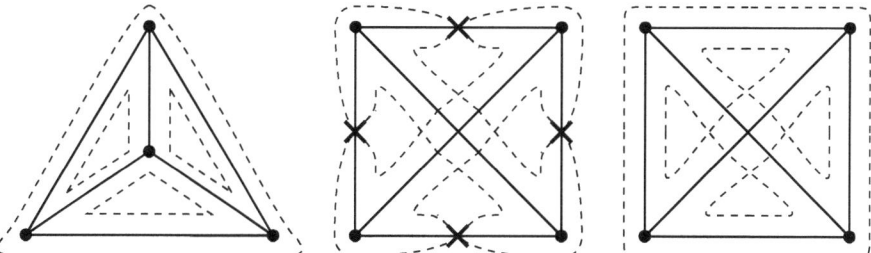

Fig. 4.22 Three map projections of the same graph K_4. The *left* one and the *central* one correspond to the embedding of K_4 in the sphere with four triangular faces; the *right* one corresponds to the embedding of K_4 in the torus

of a semiedge marked with a flag. Then connect corresponding semiedges by line segments, disregarding the perhaps unavoidable line crossings. If ϕ and $\tau_2\phi$ appear in clockwise order around v, but $\tau_2\tau_0\phi$ and $\tau_0\phi$ appear in counterclockwise order around w, mark the edge corresponding to $\{\phi, \tau_2\phi, \tau_0\phi, \tau_2\tau_0\phi\}$ with a little cross in the middle. The resulting drawing is called a *map projection*. It allows us in small cases to easily deduce the set of faces from the diagram: We start at one side of an edge, staying on the same side of the edge until we reach a vertex where we turn to follow the next edge as prescribed by the corner. However, if an edge is marked by a cross, we use the cross to switch to the other side and proceed as before. We keep walking on the same side (using all crosses as they come along) until we reach the starting point again. The edges we have traveled along are the boundary edges of the corresponding face (Figs. 4.21–4.23).

4.3 The Classification of Surfaces

In this section we will show how to incrementally alter a map M to yield a canonical normal form representing its topological type.

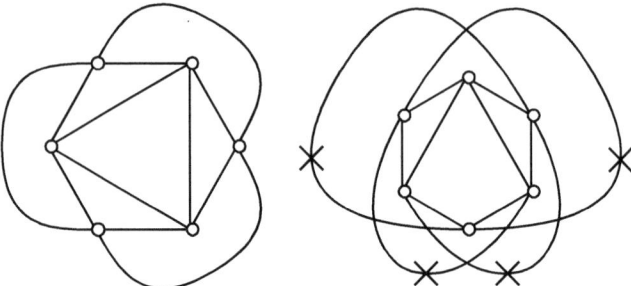

Fig. 4.23 An octahedron (8 faces) and a tetrasemihexahedron (four triangles and three squares)

4.3.1 Vertex Splitting and Edge Contraction

Given a map M on a graph G, the flags which are associated to a vertex v belong to two conjugate cycles in the factorization of $\mathcal{V}$, and since $\mathcal{V}$ satisfies (3) and (4) in Sect. 4.2.1, they must be in the form

$$(x, \mathcal{V}x, \ldots, \mathcal{V}^k x, \mathcal{V}^{k+1} x, \ldots, \mathcal{V}^n x)(\mathcal{V}^n \tau_2 x, \ldots, \mathcal{V}^{k+1} \tau_2 x, \mathcal{V}^k \tau_2 x, \ldots, \mathcal{V} \tau_2 x, \tau_2 x)$$

We can create a new map by introducing a new variable z, together with its attendant flags $\tau_0 z$, $\tau_2 z$, and $\mathcal{E} z$, and splitting the cycles compatibly by replacing the first cycle with the two cycles

$$(x, \mathcal{V}x, \ldots, \mathcal{V}^k x, y)(\mathcal{E} y, \mathcal{V}^{k+1} x, \ldots, \mathcal{V}^n x)$$

and its conjugate by

$$(\mathcal{V}^n \tau_2 x, \ldots, \mathcal{V}^{k+1} \tau_2 x, \tau_0 y)(\tau_2 y \mathcal{V}^k \tau_2 x, \ldots, \mathcal{V} \tau_2 x, \tau_2 x)$$

in the factorization of $\mathcal{V}$. The new permutations τ_0, τ_2, and $\mathcal{V}$ clearly define a map whose graph has exactly two vertices replacing the single vertex v and one new edge connecting the new vertices corresponding to z, and we say that the new map is the result of a *vertex split*. The inverse operation is called an *edge contraction*. Note that if an edge is a loop, then there are no distinct cycles to be joined, so contraction of a loop is undefined. Also, since a map must have at least one edge to have any flags, edge contraction is not defined for the map consisting of exactly one edge.

Geometrically, vertex splitting only alters the local picture of the map near v (see Fig. 4.24) and apparently does not change the number of faces of the map, which may be verified directly from the effect on $\mathcal{F}$ (see Exercise 4.14), so the Euler characteristic is unchanged. Intuitively it is clear that vertex splitting does

4.3 The Classification of Surfaces

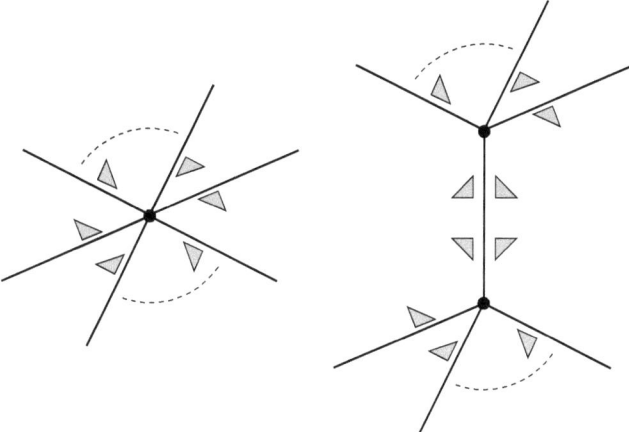

Fig. 4.24 Vertex split/edge contraction

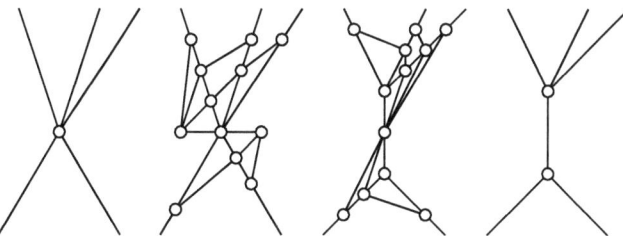

Fig. 4.25 Combinatorially equivalent common subdivision

not, in fact, alter the topological type of the map. For a proof, we note that the polygonal complexes of the two maps have combinatorially equivalent subdivisions; see Fig. 4.25.

4.3.2 Reduction to a Unitary Map

The first step toward the normal form is to reduce a given map M to one that has exactly one vertex and exactly one face, which may be done by a sequence of edge contractions on M, to reduce the number of vertices, or on M*, to reduce the number of faces. There are two base cases. If the reductions result in a map whose graph and dual graph consist only of loops, we call this base case a *unitary map*. For a unitary map, the local picture of half edges meeting at a vertex in an assigned order tells the whole story, and the polygonal complex associated with a unitary map can be regarded as starting out with all the triangles meeting a vertex and forming a polygon

Fig. 4.26 The base cases: A polygon with boundary edges identified in pairs and a one-edge sphere map

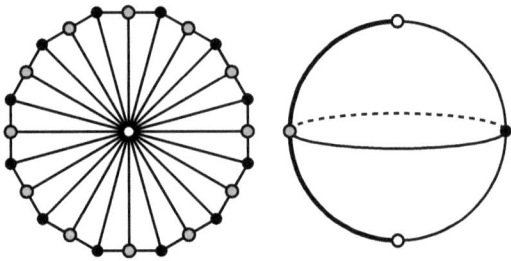

with $|4E|$ sides, some pairs of which will be identified, as with the typical picture of the torus; see Fig. 4.26. The permutation $\mathcal{V}$ on a unitary map consists of exactly two disjoint conjugate cycles.

The other base case is not a unitary map but contains exactly one edge with two distinct vertices and no other loops at all. Since the flags are associated with the edges, no edge contraction can be performed. In this case, the vertex permutation must be $(x, \mathcal{E}x)(\tau_0 x, \tau_2 x)$ with the topological type of the Sphere; again see Fig. 4.26.

4.3.3 Assembling Crosscaps

The flags of each edge of a unitary map $\{x, \tau_0 x, \tau_2 x, \mathcal{E}x\}$ are distributed among the two disjoint cycles of $\mathcal{V}$ either by

$$(x, A, \mathcal{E}x, B)(\tau_2 B, \tau_0 x, \tau_2 A, \tau_2 x)$$

or

$$(x, A, \tau_0 x, B)(\tau_2 B, \mathcal{E}x, \tau_2 A, \tau_2 x)$$

with A and B representing sequences of flags. In the latter case, we say that the flags $\{x, \tau_0 x, \tau_2 x, \mathcal{E}x\}$ form a *crosscap*. If the flags x and $\tau_0 x$ occupy successive positions in the cycle, i.e. A or B is empty, then we say that the crosscap is *assembled*.

We can inductively assemble all the crosscaps. If $\mathcal{V}$ has an unassembled crosscap, it is of the form $(x, A, \tau_0 x, B)(\tau_2 B, \mathcal{E}x, \tau_2 A, \tau_2 x)$ where A and B represent nonempty sequences of flags. A vertex split at z gives

$$(x, A, z)(\mathcal{E}z, \tau_0 x, B)(\tau_2 B, \tau_2 \tau_0 x, \tau_0 z)(\tau_2 z, \tau_2 A, \tau_2 x)$$

which is equal to $(A, z, x)(\mathcal{E}x, \tau_0 z, \tau_2 B)(B, \mathcal{E}z, \tau_0 x)(\tau_2 x, \tau_2 z, \tau_2 A)$, so contracting x gives $(A, z, \tau_0 z, \tau_2 B)(B, \mathcal{E}z, \tau_2 z, \tau_2 A)$, a unitary map where the crosscap z is assembled, while the ordering of the flags in A and B is undisturbed.

4.3.4 Assembling Handles

If the flags $\{x, \tau_0 x, \tau_2 x, \mathcal{E}x\}$ associated to an edge of a unitary map do not form a crosscap, so $\mathcal{V}$ is of the form

$$(x, P, \mathcal{E}x, Q)(\tau_2 P, \tau_0 x, \tau_2 Q, \tau_2 x)$$

then, if all the crosscaps are assembled, there must be a second flag y so that $\mathcal{V}$ is of the form

$$(x, A, y, B, \mathcal{E}x, C, \mathcal{E}y, D)(\tau_2 D, \tau_0 y, \tau_2 C \tau_0 x, \tau_2 y, \tau_2 B, \tau_2 A, \tau_2 x).$$

Suppose to the contrary that each flag z in P is either part of an assembled crosscap contained in P or both z and $\mathcal{E}z$ are contained in P. Let us consider the action of $\mathcal{F} = \mathcal{V}\mathcal{E}$ on a flag z of P. If z is part of an assembled crosscap, $\mathcal{V}z = \tau_0 z$ and $\mathcal{V}\mathcal{E}z = \tau_2 z$, then $\mathcal{F}z = \mathcal{V}\mathcal{E}z = \tau_2 z$, and $\mathcal{F}^2 z = \mathcal{V}\tau_0 z$, so either $\mathcal{F}^2 x$ is a flag of P or, if we are at the end of P, it is $\mathcal{E}x$. On the other hand, if z is not part of an assembled crosscap, then by assumption $\mathcal{E}z$ is also in P, and so $\mathcal{F}z = \mathcal{V}\mathcal{E}z$ is either a flag of P or $\mathcal{E}x$, but $\mathcal{F}\mathcal{E}x = \mathcal{V}x$, so the entire orbit of $\mathcal{F}$ does not contain x or $\tau_2 x$, contradicting the fact that M is a unitary map.

So, for unitary maps with crosscaps assembled, the flags which are not crosscaps come in interlocked pairs:

$$(x, A, y, B, \mathcal{E}x, C, \mathcal{E}y, D)(\tau_2 D, \tau_0 y, \tau_2 C \tau_0 x, \tau_2 y, \tau_2 B, \tau_2 A, \tau_2 x)$$

The set of flags $\{x, \tau_0 x, \tau_2 x, \mathcal{E}x, y, \tau_0 y, \tau_2 y, \mathcal{E}y\}$ is called *handle*, and we say the handle is *assembled* if all but one of A, B, C, and D are empty. It is possible for x to belong to more than one handle; however, an assembled handle can have no flags in common with any other handle or crosscap.

Once the crosscaps are assembled, if any, the handles can be assembled one by one as follows. If one of the cycles of $\mathcal{V}$ is given by $(x, A, y, B, \mathcal{E}x, C, \mathcal{E}y, D)$, then splitting after $\mathcal{E}x$ gives $(x, A, y, B, \mathcal{E}x, z)(\mathcal{E}z, C, \mathcal{E}y, D)$, which equals $(B, \mathcal{E}x, z, x, A, y)(\mathcal{E}y, D, \mathcal{E}z, C)$, so contracting gives $(\mathcal{E}z, C, B, \mathcal{E}x, z, x, A, D)$. Now splitting after w gives $(\mathcal{E}z, C, B, \mathcal{E}x, z, w)(\mathcal{E}w, x, A, D)$, which equals $(z, w, \mathcal{E}z, C, B, \mathcal{E}x)(x, A, D, \mathcal{E}w)$. Now, contracting at x yields the cycles $(z, w, \mathcal{E}z, C, B, A, D, \mathcal{E}w) = (\mathcal{E}w, z, w, \mathcal{E}z, C, B, A, D)$ in which the handle $\{z, \tau_0 z, \tau_2 z, \mathcal{E}z, w, \tau_0 w, \tau_2 w, \mathcal{E}w\}$ is assembled, and the ordering in A, B, C, and D, which contains any previously assembled handles or crosscaps, is undisturbed.

4.3.5 Crosscaps Canceling Handles

In the previous section we have seen how to use edge contraction and vertex splitting to transform any map M into a unitary map with assembled handles and crosscaps. If such a map has both handles and crosscaps, we now show that it is equivalent to one which has no handles at all.

Table 4.1 Normal forms of maps

S	cycle of $\mathcal{V}$	$\chi(S)$
Sphere	$(x, \mathcal{E}x)$	2
g tori	$(x_1, y_1, \mathcal{E}x_1, \mathcal{E}y_1, \ldots, x_g, y_g, \mathcal{E}x_g, \mathcal{E}y_g)$	$2-2g$
g crosscaps	$(x_1, \tau_0 x_1, x_2, \tau_0 x_2, \ldots x_g, \tau_0 x_g)$	$2-g$

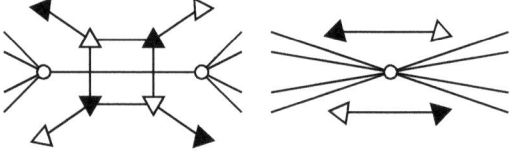

Fig. 4.27 Contracting an edge preserves a flag bipartition

If all handles and crosscaps of M are assembled and both occur, then there must be one assembled handle next to one assembled crosscap, and $\mathcal{V}$ is of the form

$$(x, \tau_0 x, y, z, \mathcal{E}y, \mathcal{E}z, A)(\tau_2 A, \tau_2 \mathcal{E}z, \tau_2 \mathcal{E}y, \tau_2 z, \tau_2 y, \tau_2 \tau_0 x, \tau_2 x)$$

in which case the sequence of moves

$$(\mathcal{E}y, \mathcal{E}z, A, x, \tau_0 x, y, z)(\tau_2 z, \tau_2 y, \tau_2 \tau_0 x, \tau_2 x, \tau_2 A, \tau_2 \mathcal{E}z, \tau_2 \mathcal{E}y)$$
$$(\mathcal{E}y, \mathcal{E}z, A, x, w)(\mathcal{E}w, \tau_0 x, y, z)(\tau_2 z, \tau_2 y, \tau_2 \tau_0 x, \tau_0 w)(\tau_2 w, \tau_2 x, \tau_2 A, \tau_2 \mathcal{E}z, \tau_2 \mathcal{E}y)$$
$$(w, \mathcal{E}y, \mathcal{E}z, A, x)(\mathcal{E}x, \tau_0 w, \tau_2 z, \tau_2 y)(y, z, \mathcal{E}w, \tau_0 x)(\tau_2 x, \tau_2 A, \tau_2 \mathcal{E}z, \tau_2 \mathcal{E}y, \tau_2 w)$$
$$(w, \mathcal{E}y, \mathcal{E}z, A, \tau_0 w, \tau_2 z, \tau_2 y)(y, z, \mathcal{E}w, \tau_2 A, \tau_2 \mathcal{E}z, \tau_2 \mathcal{E}y, \tau_2 w)$$

transforms the handle and crosscap into three unassembled crosscaps, which can subsequently be assembled, as we have seen. This can be done for each handle.

4.3.6 Normal Forms

We now have that each map is equivalent, by a sequence of vertex splits and edge contractions, either to a single edge considered as a map on a sphere or to a unitary map consisting either of g assembled handles or g assembled crosscaps; see Table 4.1.

The map of the sphere and the maps corresponding to g assembled handles are all orientable, and those of the assembled crosscaps are nonorientable; see Exercise 4.16. Moreover, orientation only depends on the topological type of a map, since vertex splitting and edge contraction preserve any bicoloring of the flags; see Fig. 4.27.

The complex of g assembled handles has Euler characteristic $2-2g$, and g is called the *orientable genus* of M, so if M and M′ have a different number of handles, they are of distinct topological types.

The complex of g assembled crosscaps has Euler characteristic $2-g$, and g is called the *nonorientable genus* of M, so if M and M′ have a different number of crosscaps, they are of distinct topological types.

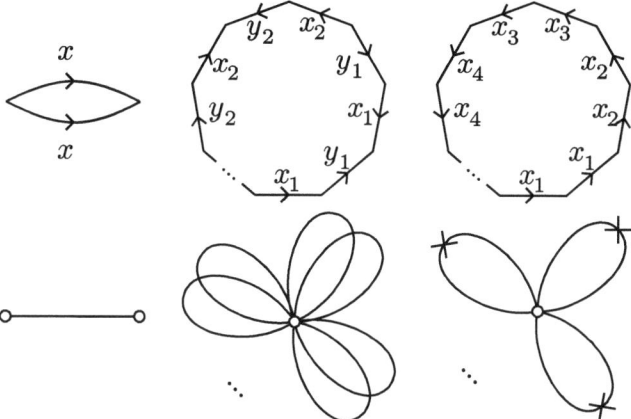

Fig. 4.28 Map projection and polygon models of canonical normal forms

Collectively, maps corresponding to the sphere, the assembled crosscaps, and the assembled handles all represent distinct topological types, which must be all the possible topological types of maps, and we say they are the *canonical normal forms* of maps; see Fig. 4.28.

The canonical normal forms actually represent the maps as a connected sum of g tori, if the orientable genus is g, or a connected sum of g projective planes, if the nonorientable genus is g. This establishes Theorem 4.6.

4.4 Operations on Maps

4.4.1 Uniform Flag Operations

The symmetry groups of the Platonic polyhedra act transitively on the vertices. Through each vertex, there is a rotation axis, and the symmetry group also acts on these axes and hence, on the collection of planes normal to these axes, a constant distance d from the center of the polyhedra. One can add these planes to the collection of planes defining the half-spaces whose intersection forms the polyhedron. If d is slightly less than the distance from the center to a vertex of the Platonic solid, then this operation cuts off a small pyramid at each vertex, and we say the polyhedron has been *truncated*. As d is decreased, the amount of volume removed increases, and at certain specified distances, the Archimedean solids make their appearance. See Fig. 4.29 in which the cube is transformed into the truncated cube, the cuboctahedron, the truncated octahedron, and, lastly, the octahedron, which is the dual of the cube. We can perform an analogous construction on the axes of the edge rotations (see Fig. 4.30) or a combination. The truncated solids

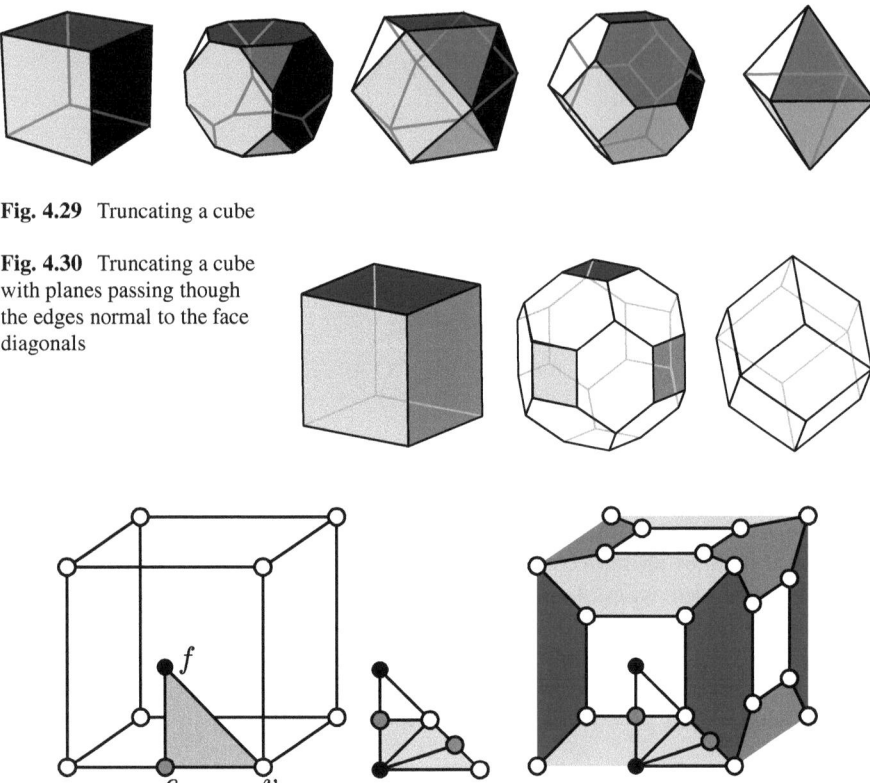

Fig. 4.29 Truncating a cube

Fig. 4.30 Truncating a cube with planes passing though the edges normal to the face diagonals

Fig. 4.31 Subdividing the flags of the cube map C to obtain **Cha**(C), the chamfered cube

must exhibit the at least the same symmetries as the original solid. In this section, we wish to examine analogous constructions on maps.

Given a map M, we have an associated polygonal complex in which the faces are triangles corresponding to M, whose vertices we will color, as before, white, gray, and black depending on whether they correspond to a vertex, edge, or face of M, with the same coloring scheme used on the flag graph. We have seen that the dual map corresponds to an operation on the flags of a map, interchanging the roles of the vertices and the faces, that is, a recoloring of the flags in the flag graph interchanging black and white. Since this operation is universal across the flags, any symmetries of the map are preserved. In the operation described in Fig. 4.31, each flag is to be divided into four subflags. This operation is called *chamfering* [26] and is one in the series of the so-called *Goldberg operations* [36]. For now we will use it to illustrate a convenient notation for flag subdivision operations. Each flag triangle of the polygonal complex of the map has three vertices, which we have colored white, gray, and black. The vertices of the subdivided flags are also colored white, gray, and back, with the only restriction being that the new gray vertices must have

4.4 Operations on Maps

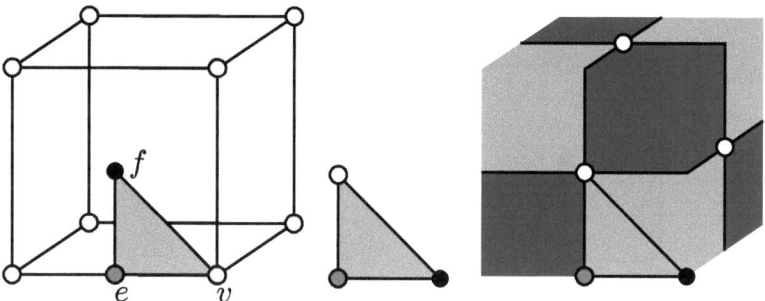

Fig. 4.32 A typical flag of the cube map C and its recoloring to produce **Du**(C)

degree four in the assembled subdivided complex and the original white and black vertices will not be recolored gray. If we regard the original triangle as a triangle in $\mathbb{R}^n$, with vertices v, e, and f, then the coordinates of the subdivision are an affine combination $\alpha v + \beta e + \gamma f$, with $0 \leq \alpha, \beta, \gamma \leq 1$, and $\alpha + \beta + \gamma = 1$. For each subdivision triangle we form a matrix whose rows are the affine coordinates of its vertices v, e, and f with respect to the triangle it subdivides. For the chamfering operation, there are four matrices:

$$M_1 = \begin{bmatrix} 1 & 0 & 0 \\ 3/4 & 0 & 1/4 \\ 0 & 1 & 0 \end{bmatrix}, \quad M_2 = \begin{bmatrix} 1/2 & 0 & 1/2 \\ 3/4 & 0 & 1/4 \\ 0 & 1 & 0 \end{bmatrix},$$

$$M_3 = \begin{bmatrix} 1/2 & 0 & 1/2 \\ 0 & 1/2 & 1/2 \\ 0 & 1 & 0 \end{bmatrix}, \quad M_4 = \begin{bmatrix} 1/2 & 0 & 1/2 \\ 0 & 1/2 & 1/2 \\ 0 & 0 & 1 \end{bmatrix}.$$

If the polygonal complex is actually embedded in $\mathbb{R}^n$, then we can use the matrices to determine the coordinates of the subdivided flags by multiplying M_i by the matrix whose rows are the coordinates of v, e, and f of the original flag.

$$\begin{bmatrix} v_1 \\ e_1 \\ f_1 \end{bmatrix} = \begin{bmatrix} 1 & 0 & 0 \\ 3/4 & 0 & 1/4 \\ 0 & 1 & 0 \end{bmatrix} \begin{bmatrix} v \\ e \\ f \end{bmatrix}$$

If the flag is not subdivided at all, then, since the white and black vertices may not be recolored as gray, there is only the identity matrix, which represents the identity operation, and the matrix $M = \begin{bmatrix} 0 & 0 & 1 \\ 0 & 1 & 0 \\ 1 & 0 & 0 \end{bmatrix}$ which represents the dual operation, **Du**(M), also denoted by M*, with which we are already familiar; see Fig. 4.32.

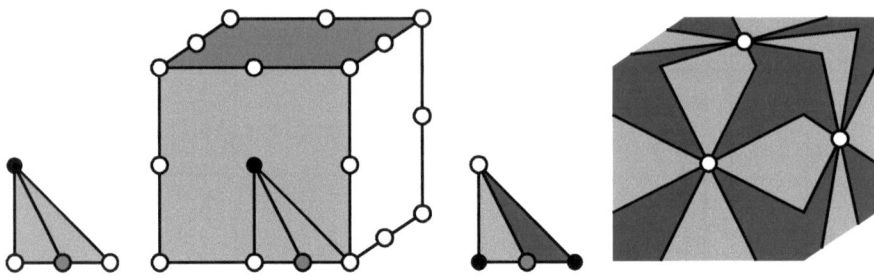

Fig. 4.33 **Su1**(C) and **Du**(**Su1**(C))

If we want to divide each triangle into two subtriangles, it must be by a new edge connecting one of the vertices of the triangle to a new vertex introduced along the opposite edge, and the new vertex must be colored gray in the result. There are six possibilities. One, in which the new gray vertex is introduced along the gray–black edge of each flag, is *one-dimensional subdivision*, which subdivides each edge in the map with a vertex of valence two. There are two matrices,

$$M' = \begin{bmatrix} 1 & 0 & 0 \\ 1/2 & 1/2 & 0 \\ 0 & 0 & 1 \end{bmatrix}, \quad M'' = \begin{bmatrix} 0 & 1 & 0 \\ 1/2 & 1/2 & 0 \\ 0 & 0 & 1 \end{bmatrix}$$

with the subdivided map having the same number of faces as the original; see Fig. 4.33. Dual to this is the subdivision scheme **Du**(**Su1**(M)), with matrices

$$M' = \begin{bmatrix} 0 & 0 & 1 \\ 1/2 & 1/2 & 0 \\ 1 & 0 & 0 \end{bmatrix}, \quad M'' = \begin{bmatrix} 0 & 0 & 1 \\ 1/2 & 1/2 & 0 \\ 0 & 1 & 0 \end{bmatrix}$$

obtained by interchanging the first and third rows of the matrices for **Su1**. The operation **Du**(**Su1**(M)) introduces *digons*, that is, faces bounded by only two edges.

If each flag triangle is cut in two with the new gray vertex introduced along the gray–white edge, the results are the same as applying the dual operation before the previous cases, giving **Su1**(**Du**(M)) and **Du**(**Su1**(**Du**(M))); see Fig. 4.34. The operation **Du**(**Su1**(**Du**(M))) is also known as *parallelization*, **Pa**(M), which replaces each edge by two parallel edges. The matrices for **Du**(**Su1**(M)) and **Du**(**Su1**(**Du**(M))) are obtained from those of **Su1**(M), **Su1**(**Du**(M)) by interchanging the first and third columns. In the next theorem we formalize the general method for iterating uniform flag subdivisions:

Theorem 4.14. *Let* $\{M_1, \ldots, M_m\}$ *be the matrices defining operation S, and let* $\{N_1, \ldots, N_n\}$ *define operation T. Then the composite operation ST is defined by the mn product matrices*

$$\{M_1 N_1, M_1 N_2, \ldots, M_1 N_n, M_2 N_1, \ldots, M_m N_n\}.$$

4.4 Operations on Maps

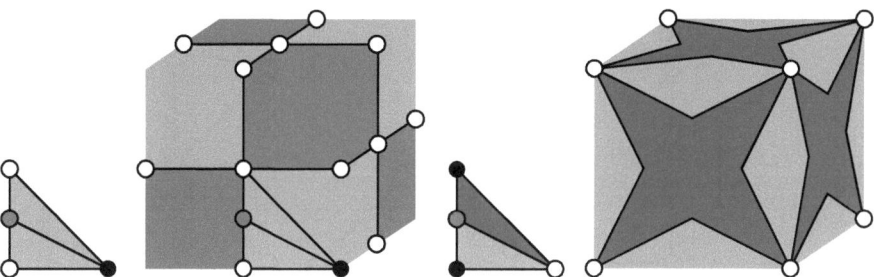

Fig. 4.34 Flag bisections along the vertex/edge incidence give either **Su1(Du(M))** or **Du(Su1 (Du(M)))** = **Pa(M)**

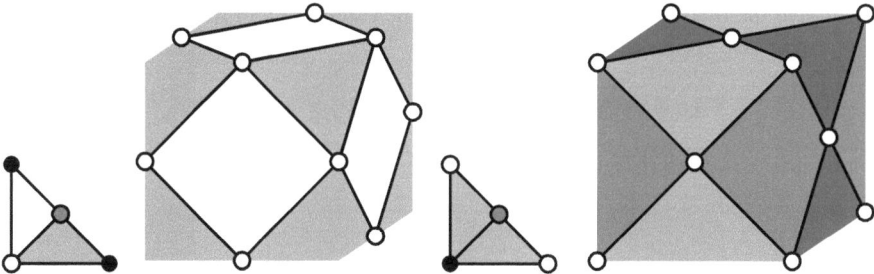

Fig. 4.35 **Me(M)** and **Du(Me(M))** = **An(M)**

4.4.2 The Medial

The previous two constructions correspond to choosing the new gray vertex to subdivide the edge of each flag having black and white endvertices. The *medial* subdivision also doubles the number of flags. It is illustrated by the relationship between the cube and the cuboctahedron in Fig. 4.29. The medial subdivision of a map M is denoted by **Me(M)**; see Fig. 4.35. Since the gray vertex is transformed to white, all vertices in **Me(M)** are 4-valent. Finally, **Du(Me(M))** is analogous to the operation of edge truncation on polyhedra illustrated in Fig. 4.30. It was studied in [69] under the name of the *angle transformation* and will therefore be denoted here by **An(M)**.

The matrices of the medial transformation are

$$M' = \begin{bmatrix} 0 & 1 & 0 \\ 1/2 & 0 & 1/2 \\ 1 & 0 & 0 \end{bmatrix}, \qquad M'' = \begin{bmatrix} 0 & 1 & 0 \\ 1/2 & 0 & 1/2 \\ 0 & 0 & 1 \end{bmatrix}.$$

If we interchange the first and third columns of M', we get M'', so we have **Me(Du(M))** = **Me(M)**. In terms of the flags of M, it is easiest to describe the

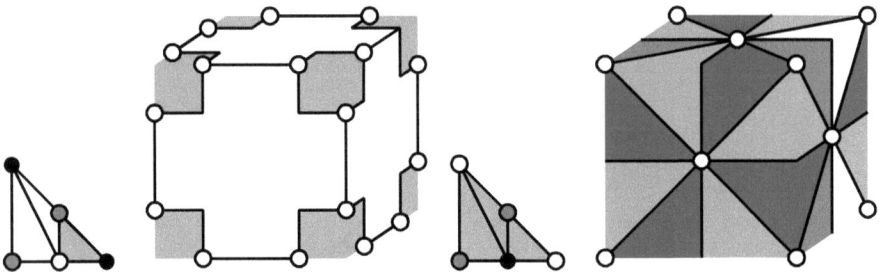

Fig. 4.36 The truncation, **Tr(M)**, and its dual **Du(Tr(M))**

change in the involutions τ_0, τ_1, and τ_2. Let the original set of flags be Φ, and let Φ' and Φ'' be two copies of this set.

$$\tau_0(x') = (\tau_1(x))'; \ \tau_1(x') = (\tau_2(x))'; \ \tau_2(x') = x''$$
$$\tau_0(x'') = (\tau_1(x))'; \ \tau_1(x'') = (\tau_0(x))''; \ \tau_2(x'') = x'$$

4.4.3 Truncation

The next group of four transformations is related to map truncation, **Tr(M)**, which for polyhedra corresponds to slicing off all the corners. The matrices for **Tr(M)** are given by

$$M' = \begin{bmatrix} 1/2 & 1/2 & 0 \\ 1/2 & 0 & 1/2 \\ 1 & 0 & 0 \end{bmatrix}, \ M'' = \begin{bmatrix} 1/2 & 1/2 & 0 \\ 1/2 & 0 & 1/2 \\ 0 & 0 & 1 \end{bmatrix}, \ M''' = \begin{bmatrix} 1/2 & 1/2 & 0 \\ 0 & 1 & 0 \\ 0 & 0 & 1 \end{bmatrix};$$

see Fig. 4.36.

The truncation operation on the dual map has been called by some chemists the *leapfrog transformation*, **Le(M)** = **Tr(Du(M))**, for instance, the leapfrog of the map Do of the dodecahedron, **Le(Do)**, the truncated icosahedron. Le(Do) has recently gone under the neologism "buckyball," since it is the structure of Buckminsterfullerene, a pure carbon molecule discovered in the 1980s. The dual of the leapfrog is the two-dimensional subdivision, **Su2(M)** = **Du(Tr(Du(M)))**, which places a new vertex in the interior of each face and joins that vertex to every vertex in the map to which that face is adjacent; see Fig. 4.37.

4.4 Operations on Maps

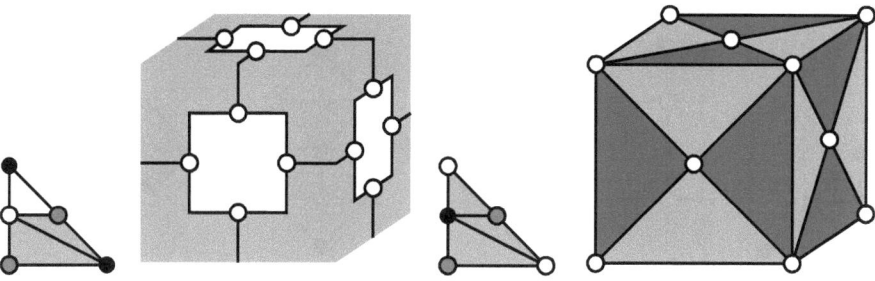

Fig. 4.37 Leapfrog, **Le**(M) = **Tr**(**Du**(M)), and its dual **Su2**(M) = **Du**(**Tr**(**Du**(M)))

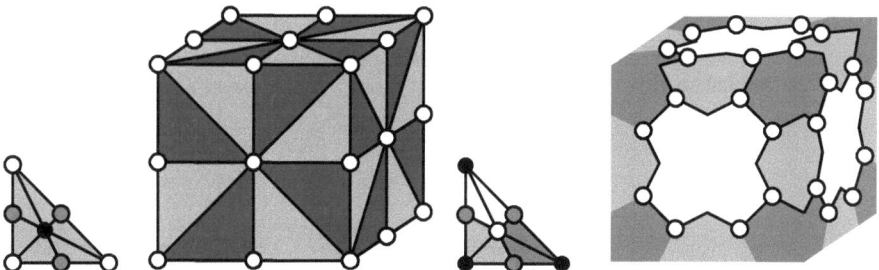

Fig. 4.38 BS(M) and Co(M)

4.4.4 Barycentric Subdivision and Combinatorial Map

The barycentric subdivision, BS(M), is defined by BS(M) = **Su2**(**Su1**(M)). It subdivides each flag into six pieces. Like the medial, its effect on the dual map is identical to its effect on the map, **BS**(**Du**(M)) = **BS**(M). The dual of BS(M) is called a combinatorial map, **Co**(M) = **Du**(**BS**(M)). **Co**(M) corresponds to the rhombitruncation of polyhedra, in which, once the vertices are truncated, the shortened original edges are then truncated into quadrilaterals; see Fig. 4.38. The vertices of **Co**(M) correspond to the flags of M, and two vertices in **Co**(M) are adjacent if an only if the corresponding flags in M share a face, so the graph of **Co**(M) is in fact the flag graph of M. **Co**(M) gives a natural embedding of the flag graph in the surface compatible with the map. This property will be useful later in Sect. 4.5.2.

4.4.5 Semiuniform Subdivisions

So far we have considered subdivisions which act uniformly on the flags. Since the flag graph consists of three perfect matchings, the flags naturally occur in pairs,

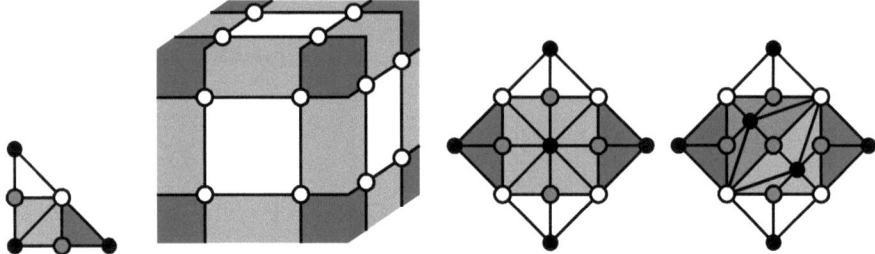

Fig. 4.39 $\mathbf{Me}^2(M)$, the first step in the snub operation, then a diagonal is introduced among the four subdivided flags of the original map

Fig. 4.40 The two snub cube maps, $\mathbf{Sn}_1(M)$ and $\mathbf{Sn}_2(M)$

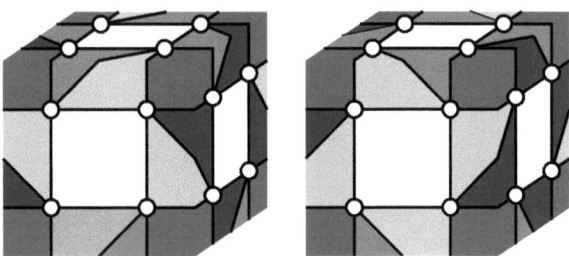

so it is reasonable to consider operations on pairs of flags, keeping in mind that any symmetry whose action on the flags is to interchange elements in that class may not be preserved.

If the map M is orientable, then there are exactly two orbits of flags under the action of $\langle \mathcal{V}, \mathcal{E}, \mathcal{F} \rangle$. It is natural to subdivide the flags in different orbits under different schemes. For each pair of schemes there will be two operations depending on which orbit is to be matched with which scheme. For the subdivided map, any orientation-preserving action will be preserved.

The most famous such subdivision is the snub operation, which starts with the square medial, $\mathbf{Me}^2(M)$ (see Fig. 4.39) and then, on the quadrilaterals introduced at the edges of M, one of the two diagonals is inserted, with the choice made consistently over M using the action of $\langle \mathcal{V}, \mathcal{E}, \mathcal{F} \rangle$; see Fig. 4.40. For the cube map, M, which has an orientation-reversing symmetry, $\mathbf{Sn}_1(M)$ is isomorphic to $\mathbf{Sn}_2(M)$, and we simply write $\mathbf{Sn}(M)$.

As with the other operations, it may happen that the symmetry group of $\mathbf{Sn}(M)$ is actually larger than that of M. A dramatic example is the tetrahedral map, T, for which snub $\mathbf{Sn}_1(T) \cong \mathbf{Sn}_2(T) = \mathbf{Sn}(T)$ is isomorphic to the map of the icosahedron.

We can also represent such operations as matrices, with two classes of matrices, one for each orientation class of flags.

4.4 Operations on Maps

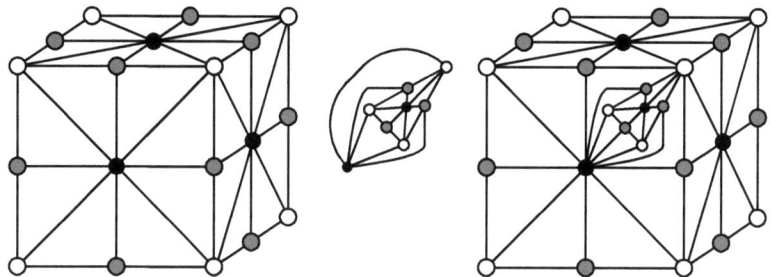

Fig. 4.41 The flag graphs of a vertex join of the cube map, with a triangle map

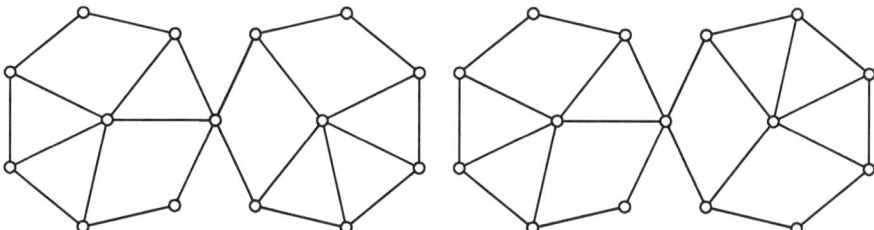

Fig. 4.42 Two ways to joining two maps at a vertex. Observe the sequence of vertex valences around the exterior face

4.4.6 Edge and Vertex Joins

We already saw how to join two maps along a face using the connected sum. We can also join maps using lower dimensional objects.

Suppose we have two maps M_i on disjoint sets of flags Φ_i, $i \in \{1, 2\}$, and let us choose specific cycles of the form $(x_1, \ldots, x_n)(\tau_2 x_n, \ldots, \tau_2 x_1)$ in the vertex permutation $\mathcal{V}_1$ for M_1 and $(y_1, \ldots, y_m)(\tau_2 y_m, \ldots, \tau_2 y_1)$ in the vertex permutation $\mathcal{V}_2$ for M_2. Then we can create a new map on $\Phi_1 \cup \Phi_2$ by taking for τ_0 and τ_2 the product of the corresponding permutations for M_1 and M_2. For $\mathcal{V}$, we again take the union of the flag cycles in the vertex permutations $\mathcal{V}_1$ and $\mathcal{V}_2$ except for the four specified above, which are amalgamated into the two cycles $(x_1, \ldots, x_n, y_1, \ldots, y_m)$ and $(\tau_2 y_m, \ldots, \tau_2 y_1, \tau_2 x_n, \ldots, \tau_2 x_1)$. It is easy to see that this defines a map which we call the *vertex join* of M_1 and M_2, denoted by $M_1 +_{v_1, v_2} M_2$, where v_1 and v_2 are the vertices corresponding to the amalgamated cycles. Topologically, the vertex join is the connected sum of the two surfaces. The effect of the vertex join on the flag graph is shown in Fig. 4.41. For maps, it is not enough to merely specify the vertices to be amalgamated or even to choose the face on which the connected sum is to be performed, since we can also combine the cycles via $(x_1, \ldots, x_n, \tau_2 y_m, \ldots, \tau_2 y_1)$ and $(y_1, \ldots, y_m, \tau_2 x_n, \ldots, \tau_2 x_1)$, which yields a map of the same topological type and with the same underlying graph; however, the maps will usually be distinct; see Fig. 4.42.

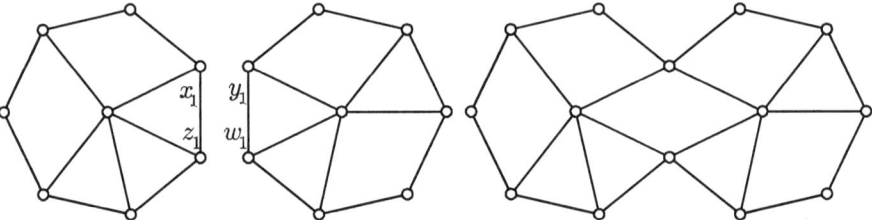

Fig. 4.43 $M_1 +_{x_1,y_1} M_2$, joining two maps along an edge

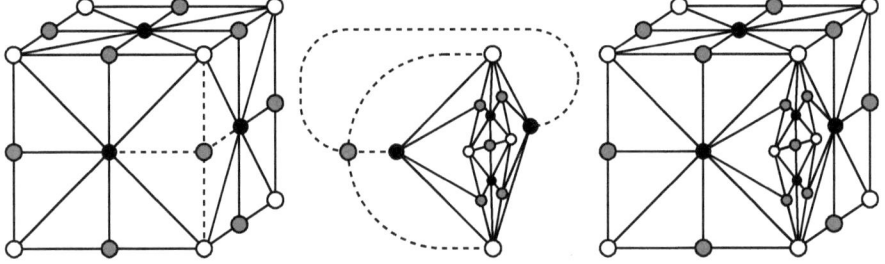

Fig. 4.44 The flag graphs of the cube map and the tetrahedron map under edge join

In a similar construction, if the vertex permutation $\mathcal{V}_1$ for M_1 contains the disjoint cycles

$$(x_1,\ldots,x_n)(\tau_2 x_n,\ldots,\tau_2 x_1)(z_1,\ldots,z_n)(\tau_2 z_n,\ldots,\tau_2 z_1)$$

with $\tau_0 x_1 = z_1$ and the vertex permutation $\mathcal{V}_2$ for M_2 contains the cycles

$$(y_1,\ldots,y_n)(\tau_2 y_n,\ldots,\tau_2 y_1)(w_1,\ldots,w_n)(\tau_2 w_n,\ldots,\tau_2 w_1)$$

with $\tau_0 y_1 = w_1$, then we can create a new $\mathcal{V}$ with the vertex cycles from $\mathcal{V}_1$ and $\mathcal{V}_2$ except for the eight specified, which are amalgamated into the four cycles

$$(x_2,\ldots,x_n,y_2,\ldots,y_m)(\tau_2 y_m,\ldots,\tau_2 y_2,\tau_2 x_n,\ldots,\tau_2 x_2)$$

and

$$(z_2,\ldots,z_n,w_2,\ldots,w_m)(\tau_2 w_m,\ldots,\tau_2 w_2,\tau_2 z_n,\ldots,\tau_2 z_2).$$

The flags $\{x_1, \tau_2 x_1, y_1, \tau_2 y_1, z_1, \tau_2 z_1, w_1, \tau_2 w_1\}$ no longer occur in $\mathcal{V}$, and we delete their cycles from τ_0 and τ_2 and eliminate $\{x_1, y_1, z_1, w_1, \tau_2 x_1, \tau_2 y_1, \tau_2 z_1, \tau_2 w_1\}$ from the list of flags, thereby erasing the amalgamated edge. The new map is called the *edge join* of M_1 and M_2, is denoted by $M_1 +_e M_2$, and corresponds topologically to the connected sum of M_1 and M_2 along two topological disks made up of the selected flags; see Fig. 4.43. The effect of the edge join on nearby elements can be seen in the effect on the flag graph; see Fig. 4.44 in which the four vertices incident to the amalgamated edge are also amalgamated.

4.4 Operations on Maps

Fig. 4.45 A Whitney twist, $M_1 +_{x_1,\tau_2 y_1} M_2$

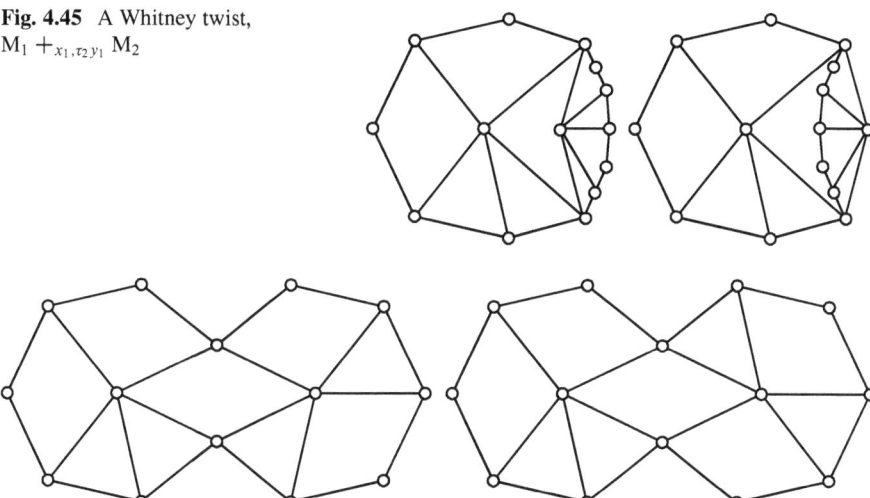

Fig. 4.46 A Whitney flip, $M_1 +_{x_1,\tau_0 y_1} M_2$

There are four possible edge joins of two maps along two specified edges, corresponding to the four ways of matching the flags representing the edges while respecting the action of τ_2 and τ_0. If we amalgamate the cycles as

$$(x_2, \ldots, x_n, \tau_2 y_m, \ldots, \tau_2 y_2)(y_2, \ldots, y_m, \tau_2 x_n, \ldots, \tau_2 x_2)$$

and

$$(z_2, \ldots, z_n, \tau_2 w_m, \ldots, \tau_2 w_2)(w_2, \ldots, w_m, \tau_2 z_n, \ldots, \tau_2 z_2)$$

the result differs from the previous by effectively twisting one of the summands over along the two vertices of attachment; see Fig. 4.45. This *Whitney twist*

$$M_1 +_{x_1, y_1} M_2 \longleftrightarrow M_1 +_{x_1, \tau_2 y_1} M_2$$

does not change the topological type of the map or the isomorphism class of the underlying graph; however, it does alter the dual graph; see [107].

Another amalgamation would be

$$(x_2, \ldots, x_n, \tau_2 w_m, \ldots, \tau_2 w_2)(w_2, \ldots, y_w, \tau_2 x_n, \ldots, \tau_2 x_2)$$

and

$$(z_2, \ldots, z_n, \tau_2 y_m, \ldots, \tau_2 y_2)(y_2, \ldots, y_m, \tau_2 z_n, \ldots, \tau_2 z_2).$$

which does alter the underlying graph of the result; see Fig. 4.46. The *Whitney flip*

$$M_1 +_{x_1, y_1} M_2 \longleftrightarrow M_1 +_{x_1, \tau_0 y_1} M_2$$

does not alter the isomorphism class of the dual graph, although it effects how the dual graph is mapped. Of course, the Whitney flip and twist are dual operations:

$$\mathbf{Du}(M_1 +_{x_1,y_1} M_2) = \mathbf{Du}(M_1) +_{x_1,\tau_0 y_1} \mathbf{Du}(M_2)$$

To complete the action of τ_0 and τ_2 on the possible edge joins, it is possible to flip and twist:

$$M_1 +_{x_1,y_1} M_2 \longleftrightarrow M_1 +_{x_1,\mathcal{E} y_1} M_2$$

We will use the edge and vertex joins, as well as the flips and twists, when we study the actions of maps.

4.5 Map Automorphisms

4.5.1 Regular Maps and Fundamental Regions

Given a map M on the set of flags Φ, we may specify any map automorphism of M either via the polygonal complex, as a permutation of the vertices, edges, and faces of M which preserve adjacency, or via the flag system, as a permutation of Φ which commutes with τ_2, τ_0, and $\mathcal{V}$.

A map automorphism that fixes any flag must fix its neighboring flags and, by connectivity, must be the identity. It follows that any two automorphisms f and g for which $f(x) = g(x)$ for some flag x must be equal. In other words, Aut (M) acts *freely* on the flags and $|\mathrm{Aut}\,(M)| \leq |\Phi|$. If, moreover, Aut (M) acts transitively on Φ, then Aut (M) acts *regularly*, so $|\mathrm{Aut}\,(M)| = |\Phi|$, and we say that M is a *regular map*.

It is a pitfall of this subject that, as the techniques of graph theory, group theory, and geometry have merged, two such wildly different meanings of "regular" have become firmly entrenched in the vocabulary to refer to important concepts. On the one hand, combinatorially, regularity refers to a collection of objects having equal numbers of incidences; on the other hand, algebraically, "regularity" refers to a collection of objects being equivalent under a group action. In this text we follow the convention that, in the combinatorial usage, the word regular never appears without a numerical qualifier, such as "4-regular," or "k-regular for some k." All undecorated uses of "regular" will be meant in the algebraic sense. Thus, the graph of a regular map will be $2|E|/|V|$-regular, and the dual graph will be $2|E|/|F|$-regular.

If M is a regular map on the sphere, so $\chi(M) = 2$, setting $2|E|/|V| = m$ and $2|E|/|F| = n$ gives

$$\frac{2|E|}{m} - |E| + \frac{2|E|}{n} = 2.$$

4.5 Map Automorphisms

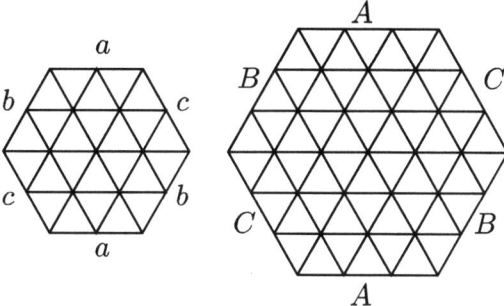

Fig. 4.47 Two regular torus maps

Since $|E|$ must be an integer, this equation severely restricts the possible values of the valences n and m;

$$1/n + 1/m > 1/2.$$

The possible solutions are $m = 2$ and $n \geq 2$ (n-cycles separating the sphere into two n-gonal faces), $m = 3$ and $n = 3$ (tetrahedron), $m = 3$ and $n = 4$ (octahedron), $m = 3$ and $n = 5$ (icosahedron), $m = 4$ and $n = 3$ (cube), $m = 5$ and $n = 3$ (dodecahedron), and $m \geq 6$ and $n = 2$ (two vertices connected by n edges forming n 2-gons).

Similarly, the regular maps on the plane with $1/n + 1/m = 1/2$ correspond to the tilings of the plane by squares, triangles, and hexagons. For $1/n + 1/m < 1/2$, it is no longer possible to represent the map by regular Euclidean n-gons. For example, for $n = 4$ and $m = 6$, we would have to fit six quadrilaterals around each vertex, which is not possible if the quadrilaterals are Euclidean squares but is possible if we take hyperbolic squares; see Fig. 4.20. In other words, for each $n, m > 1$, there is a regular map $[m, n]$ on the sphere, the Euclidean plane, or the hyperbolic plane with regular n-gons meeting m at a vertex.

For regular maps on other surfaces we have, for example, that four of the Platonic solids induce regular maps on the projective plane by identifying antipodal vertices in the flag graph. For regular maps on other surfaces, however, one cannot simply look at the local structure of the flag graph to determine the combinatorial class of the map. In Fig. 4.47, we have two combinatorially distinct regular maps on the torus. These both arise as orbit maps from the regular map $[6, 3]$ on the Euclidean plane under the automorphism group generated by three translations; see Fig. 4.48.

If M is a regular map, then Theorem 3.29 implies that the flag graph is a Cayley graph of Aut (M) where the generating set S has three elements, namely, the three group elements which map a given flag ϕ to $\tau_0\phi$, $\tau_1\phi$, and $\tau_2\phi$. More precisely, fix a particular flag ϕ, and associate to it the identity in Aut (M), and label each other flag ϕ' with the unique element of Aut (M) which maps ϕ to ϕ'. Then the action of $g \in$ Aut (M) on the flags associates with $f \in$ Aut (M), the flag associated with gf. Also, if $s_i\phi = \tau_i\phi$, so there is an edge in the flag graph from ϕ to $s_i\phi$, then, since g is an automorphism, there is an edge from $g\phi$ to $gs_i\phi$, as required.

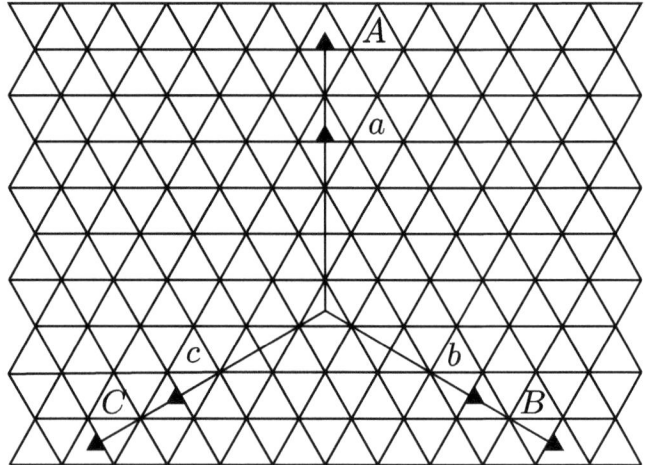

Fig. 4.48 Translations generating Aut ([6, 3])

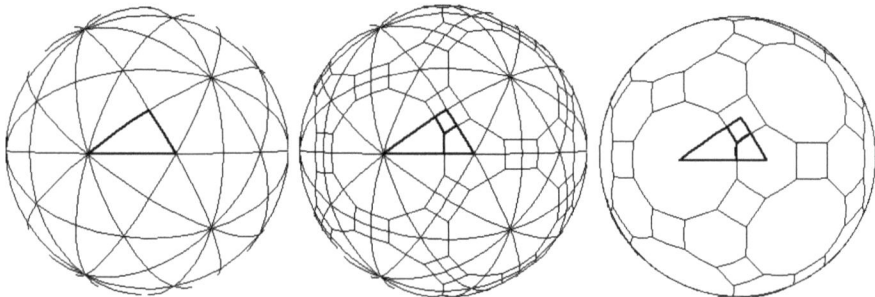

Fig. 4.49 Constructing a Cayley map for the icosahedron

Alternatively, the automorphism group of the map can be regarded as the subgroup of the permutation group on the set of flags Φ generated by τ_0, τ_1, and τ_2. In particular, every regular map has an automorphism group generated by three involutions.

We have already seen that the flag graph is the 1-skeleton of the map $\mathbf{Tr}(M) = \mathbf{Du}(\mathbf{BS}(M))$ and the action of Aut (M) on M corresponds to the action on the Cayley graph, with Aut (M) equal to the subgroup of Aut ($\mathbf{Tr}(M)$) which preserves the tricoloring of the edges by the generating set S. See Fig. 4.49 for the Cayley graph of the group of the icosahedron on the sphere. Since the one-skeleton of $\mathbf{Co}(M)$ is the Cayley graph of Aut (M) and the action on the Cayley graph extends to the faces, we say that $\mathbf{Co}(M)$ is the *Cayley map* of Aut (M).

The regular tetrahedron T considered as a self-dual regular map has Cayley map $\mathbf{Co}(T)$ such that Aut (T) is a proper subgroup of Aut ($\mathbf{Co}(T)$).

4.5 Map Automorphisms

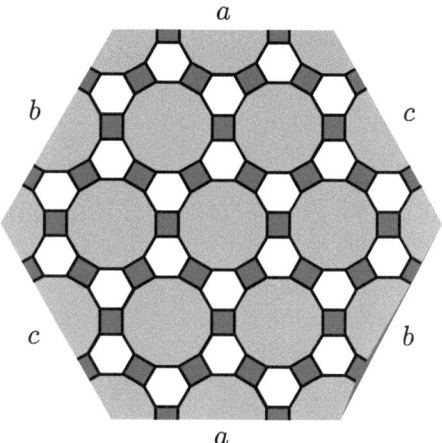

Fig. 4.50 A Cayley map on the torus

The Cayley map of the regular map M on the torus shown on the left of Fig. 4.47 decomposes the torus into hexagons, dodecagons, and squares; see Fig. 4.50. An important distinction must be made between Cayley maps on simply connected surfaces, that is, the sphere and the plane, and all other Cayley maps. The vertices in the Cayley map represent the group elements, and the paths in the Cayley map are words. Specifically, a path originating at the identity and terminating at g corresponds to an expression for g in the generating set $S = \{\tau_0, \tau_1, \tau_2\}$. Reading around any face F gives an expression w_F for the identity. For example, in the map in Fig. 4.50, there are three such expressions, $\{(\tau_0\tau_2)^2, (\tau_0\tau_1)^3, (\tau_1\tau_2)^6\}$. If there are two paths, w and w', from 1 to g, then $w'w^{-1}$ forms a closed path which splits into simple cycles. In the case of the plane or the sphere, the Jordan curve theorem implies that every simple closed curve bounds a topological disk, which for a map is a finite collection of faces in **Co(M)**. So, for planar Cayley maps, different expressions for g in the generating set can be reconciled using exclusively the identities $w_F = 1$ for representative faces and the trivial identity $aa^{-1} = 1$. We write

$$\text{Aut}(M) = \langle \tau_0, \tau_1, \tau_2 \mid \tau_0^2 = \tau_1^2 = \tau_2^2 = w_{F_1} = \cdots = w_{F_n} = 1 \rangle$$

For Cayley maps on other surfaces, however, there are cycles which do not bound disks and so are not expressible in terms of face words.

If the map M is not regular, then we can still construct a Cayley map; however, it will not be based on individual flags. The map **Sn**([4, 4]) obtained by taking the snub operation on the square lattice (see Fig. 4.51) is not regular, since there are five flag orbits. If we choose five contiguous representatives of these five orbits, then we can regard the five flags as subdivision of a pentagon, P. Moreover, every element of Aut (M) preserves the contiguity of these five, so the plane is tiled by the images of P forming a new map C and Aut (M) does act transitively on the faces of C.

Fig. 4.51 A fundamental region for **Sn**([4, 4])

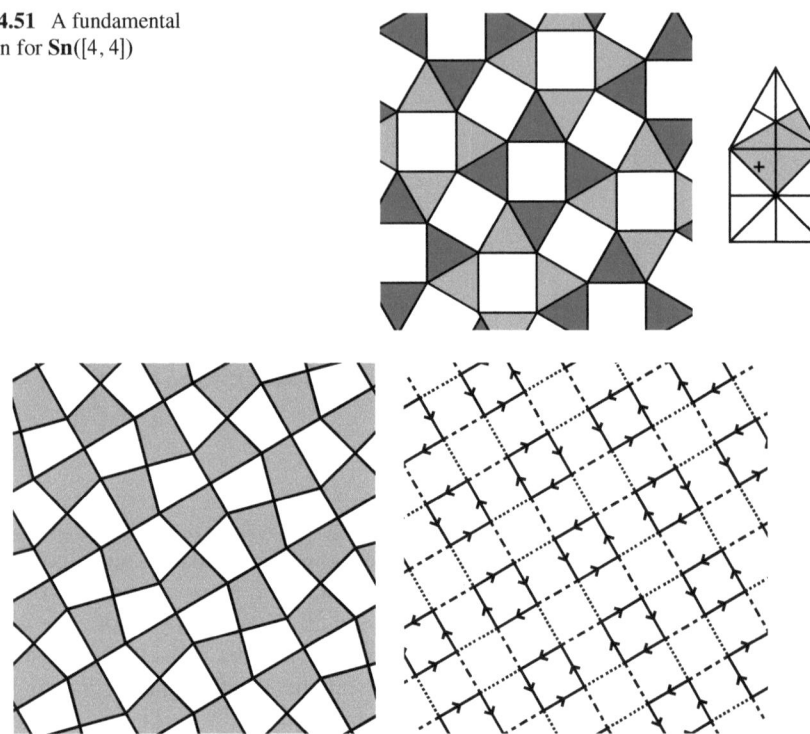

Fig. 4.52 The Cayley map of Aut(**Sn**([4, 4])

It is convenient to suppress the vertices of valence 2 in C and regard the faces as quadrilaterals (see Fig. 4.52), giving us a Cayley map via truncation, as before.

A *fundamental region* of a map M with respect to a subgroup of Aut (M) is a finite set of contiguous flags, that is, a set of flags which forms a connected subgraph of the flag graph **Co**(M), containing one flag from each orbit, and whose boundary is a simple cycle. As in the example, the orbits of the fundamental region give a new map having a common subdivision, namely, **Co**(M), with M on which Aut (M) acts regularly on the faces and whose dual is a Cayley map. It is possible to give a less strict formulation of the fundamental region. We are guaranteed the existence of a fundamental region if the map is on the sphere or the plane.

Theorem 4.15. *If M is a map whose topological type is that of the plane or the sphere and* **G** *is a subgroup of Aut* (M), *with finitely many face orbits, then there is a fundamental region F of M so that* **G** *acts regularly on the faces of the map derived from the orbits of F.*

Proof. **G** acts without fixed points on **BS**(M), whose triangles correspond to the flags of M. Choose a flag ϕ_1. If there is only one flag orbit, we are done. If not, there must be, by connectedness of the map, a flag ϕ_2 from a different orbit which is

4.5 Map Automorphisms

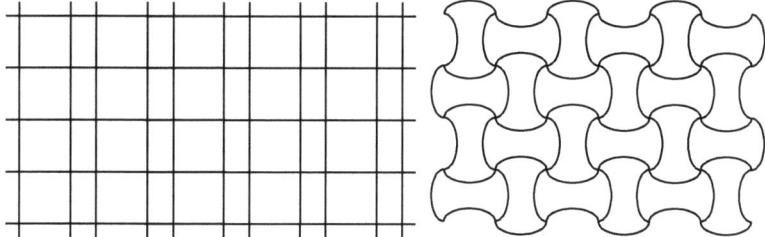

Fig. 4.53 Two combinatorially isomorphic but geometrically distinct tilings

contiguous to ϕ_1. So consider the orbits of the set $\{\phi_1, \phi_2\}$ in the same way. Either there is only one orbit, we select ϕ_3 contiguous to either ϕ_1 or ϕ_2. Continue to produce a fundamental region $F = \{\phi_1, \phi_2, \ldots\}$.

The set of cells contains one exterior boundary component, B. If there is another boundary component besides B, say B', then B includes B' as well as a set of cells C not in F. Therefore, there is a $g_1 \in \mathbf{G}$ such that $g_1(F) \subseteq C$. Continuing in this way produces $g_i F$, all of which are contained in the region enclosed by B, so the region enclosed by B contains infinitely many cells, a contradiction. □

Even if the map is not a regular map, once we have the fundamental region of M, we can express the action of Aut(M) in terms of the original flag permutations τ_i. Simply distinguish any flag in the fundamental region, and choose any path in Co(M) joining the distinguished flag to the corresponding flag in the faces of the Cayley map with which it shares an edge. For instance, in the example of Fig. 4.51, the distinguished flag is marked with a cross, and the permutations mapping it to the corresponding flag on the contiguous fundamental regions are $\{\tau_0\tau_1, \tau_1\tau_0, \tau_2\tau_1\tau_2, \tau_2\tau_1\tau_0\tau_2\tau_0\tau_1\tau_2\}$. So Aut(M) is the subgroup of Sym(Φ) generated by these four permutations.

4.5.2 Harmonious Maps

It surely has not escaped the reader's notice that, although a map is abstractly defined via a flag permutation system and realized only up to topological homeomorphism, our examples have been generally realized embedded on the sphere or Euclidean space with geodesic edges and regular polygonal faces and that the actions, which are only required to be combinatorial permutations, have been isometries.

If we have a map M realized as a polygonal complex $\mathcal{M}$ in a metric space X, then define Isom($\mathcal{M}$) to be the group of isometries of X which preserve the vertices, edges, and faces of M. In general Isom($\mathcal{M}$) $\subseteq$ Aut(M). For instance, the tilings indicated in Fig. 4.53 are both combinatorially isomorphic but have distinct metric automorphism groups. For a more dramatic example, since the hyperbolic plane $\mathbb{H}^2$ is homeomorphic to the Euclidean plane $\mathbb{E}^2$, any regular hyperbolic tiling gives rise

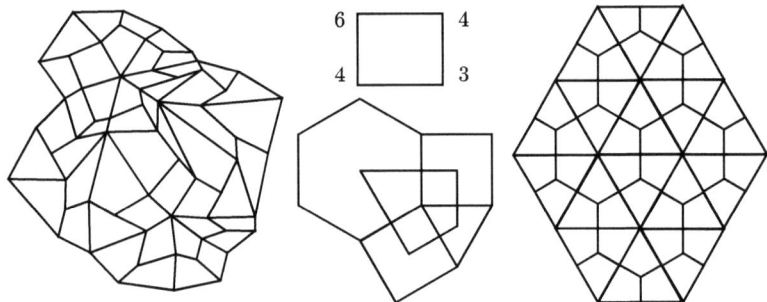

Fig. 4.54 Straightening part of an infinite tiling

to a rather unsymmetrical Euclidean tiling with large combinatorial automorphism group. In this case, Isom($\mathcal{M}$) is not even of finite index in Aut (M).

A complex $\mathcal{M}$ is *harmonious* if Aut M = Isom($\mathcal{M}$). Neither complex of Fig. 4.53 is harmonious; however, both are isomorphic to the square tiling [4, 4] which is harmonious. In [89], it was shown that every planar combinatorial map can be straightened to a harmonious map.

Theorem 4.16. *Let $\mathcal{M}$ be a complex of polygons piecewise linearly embedded on the sphere $\mathbb{S}^2$, the Euclidean plane $\mathbb{R}^2$, or the hyperbolic plane $\mathbb{H}^2$, and M is the associated map. If Aut M has finitely many face orbits, then there is a complex Ideal($\mathcal{M}$) of $\mathbb{S}^2$, $\mathbb{E}^2$, or $\mathbb{H}^2$ which is combinatorially isomorphic to $\mathcal{M}$ and such that every automorphism of M is expressed as an isometry of Ideal($\mathcal{M}$), so Isom(Ideal($\mathcal{M}$)) = Aut (M).*

Proof. By Theorem 4.15, the complex M has a fundamental region F. Every vertex of ∂F, the boundary of F, which belongs to more than two orbits of F we call a *corner point* of F, and we set the *multiplicity* of the corner point to be the number of orbits of F it intersects, and let the corner points of F be $x_1, \ldots, x_k$ reading counterclockwise around F, with multiplicities $m_1, \ldots, m_k$.

We want to construct Ideal(F), a polygon on $\mathbb{S}^2$, $\mathbb{R}^2$, or $\mathbb{H}^2$ such that the space is tiled by the shapes Ideal(F) via the action of Aut (M).

Let $\alpha = \sum (m_i - 2)$.

If $\alpha = 2$, then if we take, for each i, a regular Euclidean m_i-gon of unit side length. These regular polygons will just fit together cyclically about a common vertex in the Euclidean plane; see Fig. 4.54. Take the convex hull of the centroids of the k regular polygons as Ideal(F). If we use F for the realization of M as a polygonal complex, Ideal($\mathcal{M}$) is combinatorially isomorphic to $\mathcal{M}$ and is globally homeomorphic and locally isometric to $\mathbb{R}^2$, hence isometric to $\mathbb{R}^2$. Moreover, $f \in$ Aut (M) acts by permuting the elements in the orbit of Ideal(F), with each $f(gF) = fgF$ realized by a congruence, so f as a whole is realized by an isometry f on Ideal($\mathcal{M}$).

4.5 Map Automorphisms

If $\alpha > 2$, so that the Euclidean polygons fit around a vertex with overlap, choose instead regular hyperbolic polygons with equal side lengths. Their vertex angles will shrink as their common side length is expanded, so there is one choice of side length for which they will just fit and we proceed as above. In this case Aut $\mathcal{M}$ must be a hyperbolic group.

If $\alpha < 2$, then Euclidean polygons do not close, so choose regular spherical polygons whose vertex angles will expand as their common side length is expanded; again there will be a particular side length at which they just fit. □

We will say that the tiling $\mathcal{M}$ is *straightened* to Ideal($\mathcal{M}$). Note that each planar graph without loops and parallel edges admits a straight-line embedding in the plane, a result known as Fary theorem. In general it may not be possible to straighten the polygonal complex $\mathcal{M}$ to a straight-line tiling.

The straightening process is illustrated by Fig. 4.54, which is a tessellation of the Euclidean plane by irregular quadrilaterals where each quadrilateral may be taken to be the fundamental complex. They are individually straightened and then reassembled via the action of Aut (M). In general, different choices of fundamental region give nonisometric idealizations. For examples, see Figs. 4.61–4.63 which restraighten a harmonious (colored) tiling. See also [4, 91].

In the Euclidean and spherical cases, we have $\sum(m_i - 2) \leq 2$ which has only finitely many solutions with more than two vertex types, so it is possible to catalog all the possible discrete plane and spherical isometry groups. For the Euclidean plane, we obtain the so-called plane crystallographic groups, each of which has a subgroup of finite index generated by two noncollinear translations, from which it follows that each of these groups fixes a lattice in the plane. See Fig. 4.55 in which the lattice parallelograms, called *unit cells* by chemists, are shaded, lines of reflection are given by solid lines, and rotation centers are indicated by polygons. The groups are in two general classes, those that are symmetries of the square lattice [4, 4] and those which are symmetries of the hexagonal lattice [3, 6] and its dual [6, 3].

Since every Euclidean plane crystallographic group contains a translation subgroup which is free abelian of rank 2 and since this group does not act discretely on the hyperbolic plane, it follows from Theorem 4.16 that, for a plane map M with finitely many face orbits, we can say definitively whether it is intrinsically Euclidean, spherical, or hyperbolic.

For groups of the sphere, see Fig. 4.56, we get seven infinite families which are symmetries of prisms, as well as six groups which are the full symmetry groups, and the rotation symmetry groups of the platonic solids and the group $[3^+, 3]$, which is another subgroup of symmetries of the cube, having 24 elements and generated by the reflections in planes which are perpendicular bisectors of the edges, and rotations of 120° about the diagonals.

For more details on these groups, see [21] or [48].

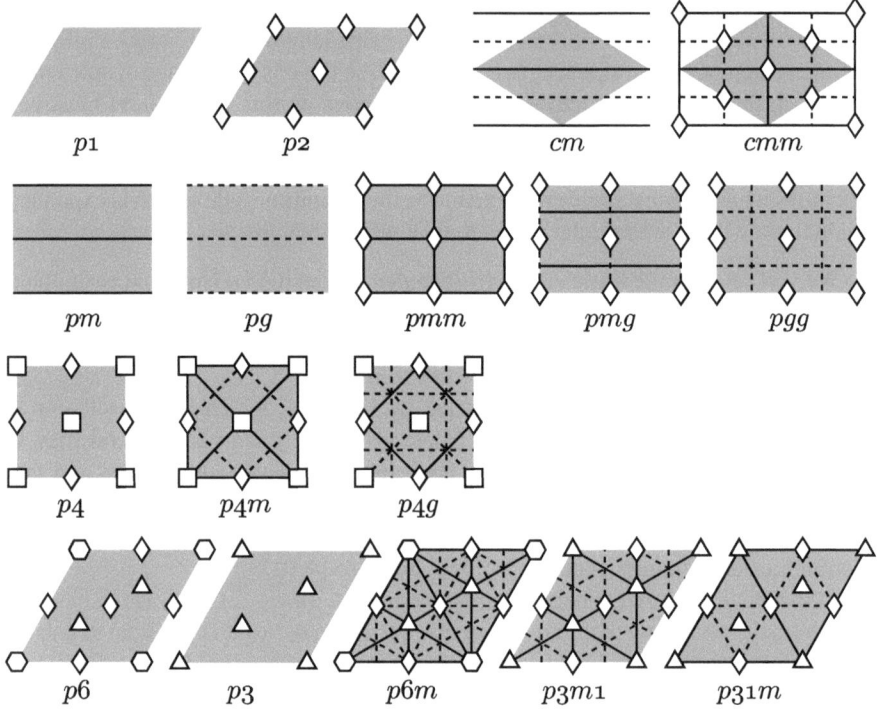

Fig. 4.55 The plane crystallographic groups

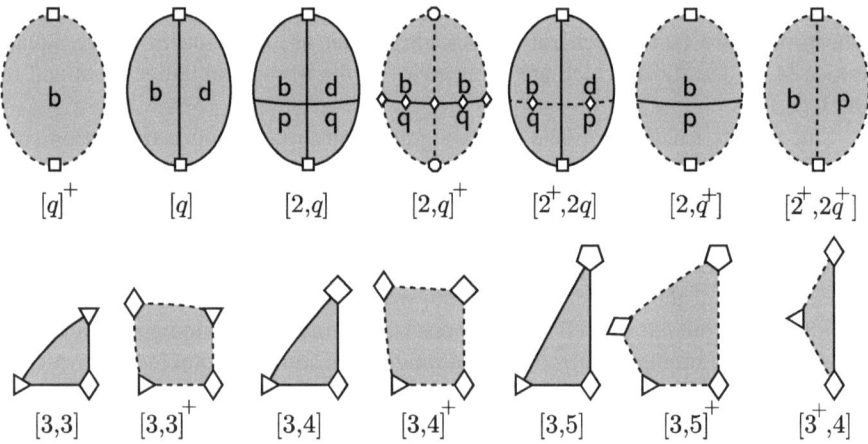

Fig. 4.56 The groups of the sphere

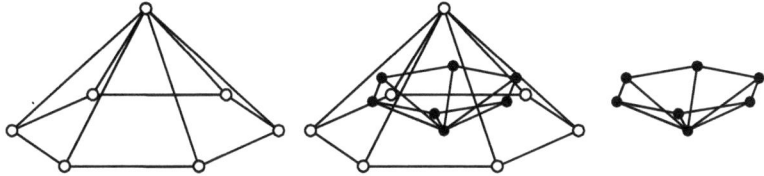

Fig. 4.57 A pyramid is self dual

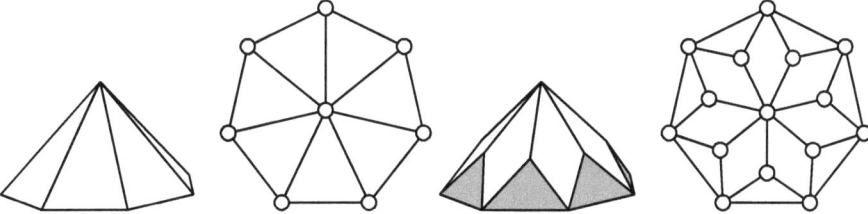

Fig. 4.58 A pyramid and hyperpyramid

Fig. 4.59 A self-dual graph with non-involutory self-duality

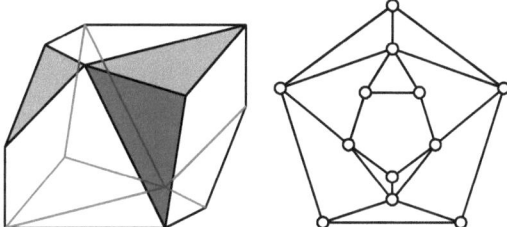

4.5.3 Self-dual Maps

We have defined dual objects for polyhedra, polygonal complexes, planar graphs, and maps. An object is said to be *self-dual* if it is isomorphic to its dual. The most famous self-dual object is the regular tetrahedron, with the vertices of two regular tetrahedra lying in dual position as the eight vertices of a cube divided according to the bipartition of the vertices of a cube. Any pyramid is self-dual; see Fig. 4.57.

The existence of several classes of self-dual polyhedral graphs, for instance, pyramids and hyperpyramids (see Fig. 4.58), was known to Kirkman [55], with their Schlegel diagrams known to graph theorists as wheels and hyperwheels. Other than these basic examples, self-duality was not understood very well until very recently, when it was realized by Grünbaum and Shephard [46] that the incidence-preserving correspondence between the vertices and faces of a self-dual polyhedron need not define an involution on the set $V \cup F$. The polyhedron of Fig. 4.59 is self-dual; however, the self-dual correspondences do not correspond to a mere interchange of vertices and faces.

With our study of maps, we have exactly the tools we need to study self-duality. We say a map M is *self-dual* if there is a map automorphism $f : \mathrm{M} \longrightarrow \mathbf{Du}(\mathrm{M})$.

Fig. 4.60 Superposition of a map and its dual

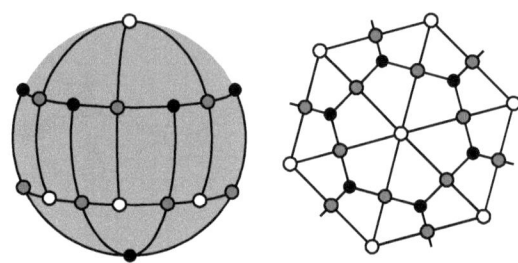

If such an f exists, then there must be a map automorphism $f_{\mathbf{Me}} : \mathbf{Me}(M) \longrightarrow \mathbf{Me}(\mathbf{Du}(M))$, and since $\mathbf{Me}(M) = \mathbf{Me}(\mathbf{Du}(M))$, $f_{\mathbf{Me}}$ is in fact a map automorphism, and so we have a map automorphism $f_{\mathbf{Me}^2} \mathbf{Me}(\mathbf{Me}(M)) \longrightarrow \mathbf{Me}(\mathbf{Me}(M))$. We can think of the map $\mathbf{Me}(\mathbf{Me}(M))$ as the superposition of M and $\mathbf{Du}(M)$, with a new vertex added wherever a dual pair of edges crosses. For ease of visualization, we will color the vertices of $\mathbf{Me}(\mathbf{Me}(M))$ which correspond to the original vertices of M with white, those corresponding to vertices of $\mathbf{Du}(M)$ with black, and those corresponding to crossed edge pairs with gray; see Fig. 4.60. Note that every face of $\mathbf{Me}(\mathbf{Me}(M))$ is a quadrilateral with two gray vertices separating a black and a white vertex and that every self-dual correspondence f induces an automorphism on $\mathbf{Me}(\mathbf{Me}(M))$ which interchanges the black and white vertices. These automorphisms will be called *self-dualities*. The set of self-dualities do not form a group, since the composition of two self-dualities preserves all the colors of the vertices of $\mathbf{Me}(\mathbf{Me}(M))$ and, hence, defines an automorphism of M; however, the union of white and black preserving and reversing automorphism does form a group, Dual(M), in which Aut (M) is a subgroup of index 2 and the self-dualities are the difference set Dual(M) − Aut (M) and comprise the other coset.

For the spherical complex of Fig. 4.60, which corresponds to the hexagonal pyramid, the group of automorphisms is the dihedral group of order 12 acting on the sphere as Aut (M) = [3]; see Fig. 4.56 with every transformation fixing both the north and south pole. For the black and white reversing automorphisms, there are six 180° rotations about axes through the equator, as well as six *rotary reflections*, rotations by 30°, 90°, 150°, 210°, 270°, and 330° followed by an equatorial reflection. The overall group is Dual(M) = $[2^+, 6]$. The ordered pair of groups (Dual(M), Aut (M)) = $([2^+, 6], [3])$ is called the *self-dual pairing* of M. We can use the pairings to classify the self-dual maps.

Theorem 4.17. *Every self-dual map M with finitely many face orbits is the map of a harmoniously self-dual complex.*

Our straightening process depends on the choice of fundamental region, so a self-dual tiling may have many different harmonious incarnations. This is illustrated in Figs. 4.61–4.63 where even the original tiling is harmonious. In Fig. 4.61, the first choice for the fundamental region has five corner points of multiplicities $\{4, 3, 4, 3, 3\}$, so in Fig. 4.62, we map it into a pentagon with angles $\{\pi/2, \pi/3, \pi/2, \pi/3, \pi/3\}$. The second choice in Fig. 4.61 specifies a fundamental

4.5 Map Automorphisms 147

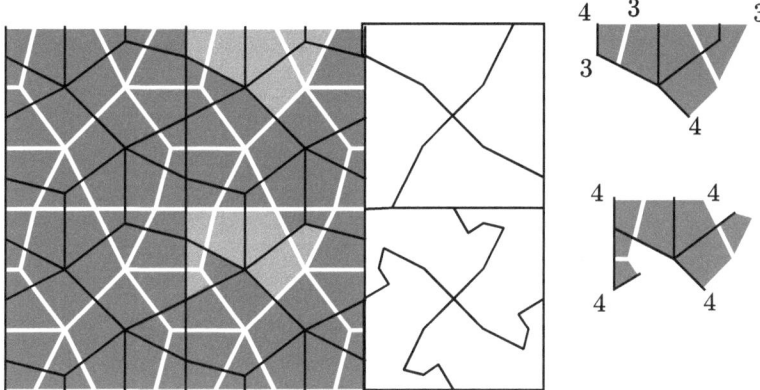

Fig. 4.61 Choosing a fundamental complex

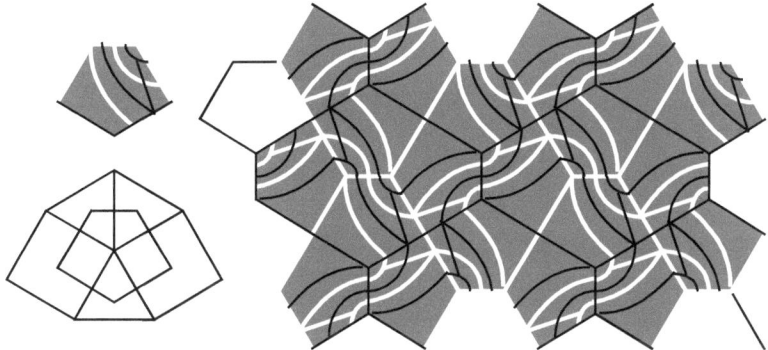

Fig. 4.62 Straightening a self-dual tiling

complex with 4 corner points each of multiplicity 4, so we map the fundamental complex into a square. We see in Fig. 4.63 that with some choices of fundamental complex, it is impossible to arrange the mapping into Ideal(X) so that the resulting tiling has straight edges or convex cells.

A doubly periodic self-dual tiling is necessarily Euclidean and has finitely many face orbits. Since every self-dual periodic tiling is isomorphic to a harmonious one, Dual($\mathcal{T}$) and Aut $\mathcal{T}$ are both one of the 17 plane crystallographic groups, and the pair (Dual($\mathcal{T}$), Aut $\mathcal{T}$) is one of the 46 2-*color groups* enumerated by Coxeter [19], with the color-preserving symmetries corresponding to Aut $\mathcal{T}$.

Once M is harmoniously embedded, we see that only rotations of order 2 and 4 are possible for self-dualities: The rotation center is necessarily at a vertex, or the center of an edge, or the center of a face of **Me**(**Me**(M)). Since the colors must be reversed, it must be at the center of a face, which is a quadrilateral with opposite corners gray; hence, the rotation is of order 2, or it must be at a gray vertex, which is incident to black and white vertices alternatingly, so the rotation must be of order 4.

Fig. 4.63 An alternative straightening

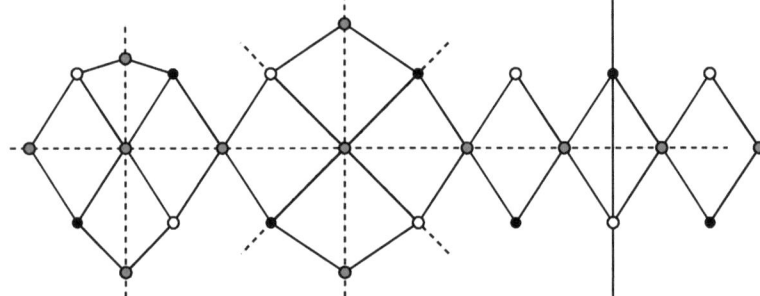

Fig. 4.64 Reflections on a straightened self-dual tiling

Similarly, the angles between reflection lines are restricted for self-dualities: The reflecting line of a self-duality must pass through the 1-skeleton only at gray vertices, bisecting the quadrilateral faces. All the reflecting lines passing through a point generate a dihedral group, and the point of intersection must be at the center of a quadrilateral, or at a gray vertex. If one reflection through a vertex is a self-duality, then all or half the reflections through that point are self-dualities. A gray vertex can have only two self-duality reflections, and a quadrilateral center can have only one. The possibilities are shown in Fig. 4.64. See Fig. 4.66 for an example.

Thus, by examining the generators of the plane crystallographic groups, we see that the groups $p6$, $p3$, $p6m$, $p3m1$, and $p31m$ do not occur as $\text{Dual}(\mathcal{T})$ in the self-dual pairing of a self-dual Euclidean tiling if Aut(M) is a proper subgroup. The groups $p3$ and $p6$ are generated by rotations of orders 3 and 6, none of which can be self-dualities. The group $p6m$ is generated by reflections in a right triangle with angle 60°, and $p3m1$ is generated by reflections in the sides of an equilateral triangle, and none of these generating transformations can be self-dualities. The group $p31m$ is generated by the reflections in the sides of an equilateral triangle and an order 3 rotation about its center, none of which can be self-dualities.

4.5 Map Automorphisms

Table 4.2 Self-dual pairings for doubly periodic maps

(p1,p1)				
(p2,p1)	(p2,p2)			
(cm,p1)	(cm,pg)	(cm,pm)		
(cmm,cm)	(cmm,p2)	(cmm,pgg)	(cmm,pmg)	(cmm,pmm)
(p4,p2)	(p4,p4)			
(p4g,cmm)	(p4g,p4)	(p4g,pgg)		
(p4m,cmm)	(p4m,p4m)	(p4m,pmm)		
(pg,p1)	(pg,pg)			
(pgg,p2)	(pgg,pg)			
(pm,cm)	(pm,p1)	(pm,pg)	(pm,pm)a	(pm,pm)b
(pmg,p2)	(pmg,pg)	(pmg,pgg)	(pmg,pm)	(pmg,pmg)
(pmm,cmm)	(pmm,p2)	(pmm,pm)	(pmm,pmg)	(pmm,pmm)

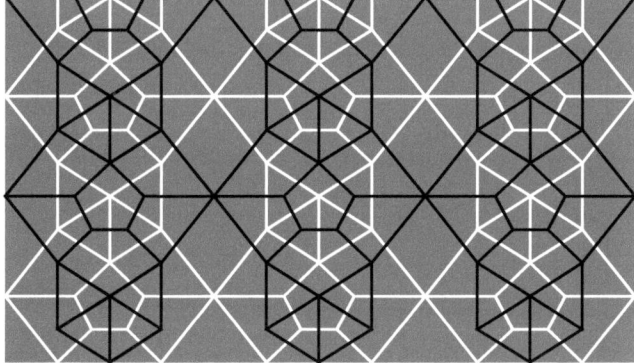

Fig. 4.65 $(pm, pm)a$

The 2-color groups which are pairings for self-dual tilings are listed in Table 4.2. Notice that the group p4g, which is generated by a 90° rotation and a reflection in a line not through the pole of the rotation, does not occur as Aut M for any self-dual pairing, even though it does occur on the list in [4] of possible symmetry groups of self-dual tilings. This is because any self-dual Euclidean tiling with Isom($\mathcal{T}$) = p4g has additional combinatorial automorphisms and can be further straightened to have symmetry group $p4m$, for instance, the tiling on the right in Fig. 4.53.

All pairings have harmonious incarnations with convex faces. As with Coxeter's color groups, the pairing (pm, pm) is not sufficient to determine the symmetry class, since the group pm has itself as a subgroup in two geometrically distinct ways, which are labeled $(pm, pm)a$ and $(pm, pm)b$. They can be distinguished by the fact that $(pm, pm)a$ has no self-duality reflection (see Fig. 4.65), while $(pm, pm)b$ does have a self-duality reflection (see Fig. 4.66).

Using similar methods one can show that self-dual maps on the sphere can either exhibit the symmetry of the dual tetrahedra embedded in the cube, with pairings $([3,4], [3,3])$, $([3,4]^+, [3,3]^+)$, or $([3^+,4], [3,3]^+)$, or that exhibited by the wheel, if there is no equatorial self-duality reflection, or the hyperwheel, if there is.

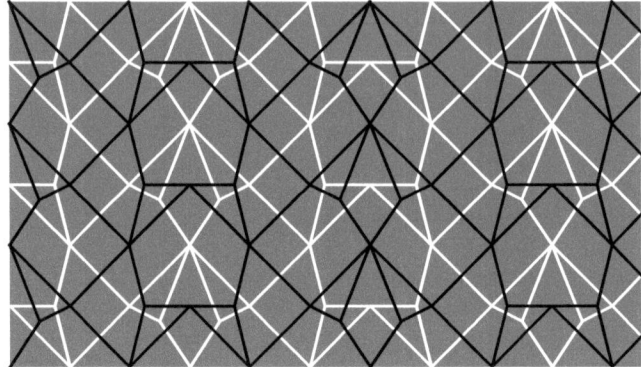

Fig. 4.66 $(pm, pm)b$

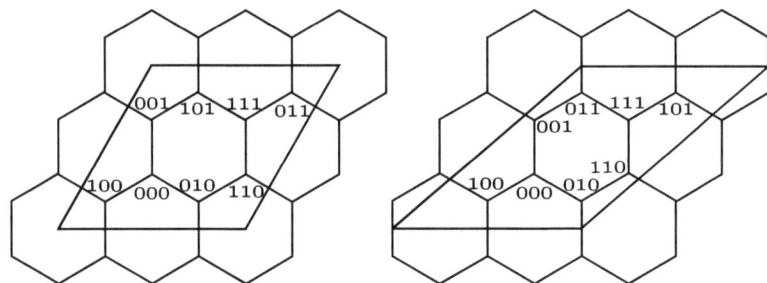

Fig. 4.67 Two inequivalent embeddings of the cube Q_3 on the torus ($H(2, 2, 1)$ and $H(2, 2, 0)$)

For more information of self-duality, see [4] for Euclidean plane tilings, [4,6,47, 54] for polyhedra, [3,89] for spherical maps, and [91] for planar maps.

A *metric self-duality* is a rigid motion of the plane which carries $\mathcal{T}$ into dual position, and examples are known [4] of combinatorially self-dual tilings which are not metrically self-dual.

4.5.4 Automorphisms of Planar Graphs of Low Connectivity

For graphs which are 3-connected, Whitney's theorem [106] says all its planar or spherical maps are combinatorially isomorphic. For such graphs, the automorphism group of the graph is the same as its map automorphism group, which, given the straightening theorem, can be studied with geometric methods alone. This is not the case for 3-connected graphs and an associated map on the torus; see Fig. 4.67.

If the graph is planar and has connectivity less than three, the faces of its planar realization are not uniquely defined, and there may be graph automorphisms which

4.5 Map Automorphisms

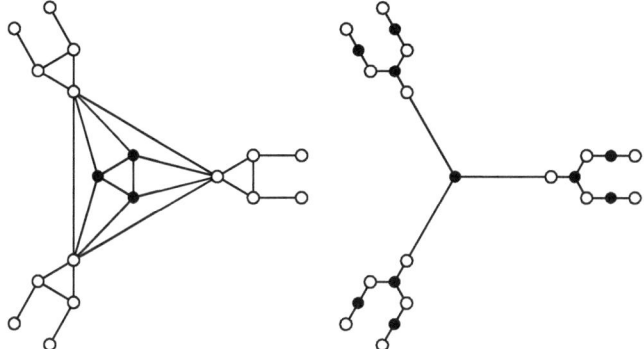

Fig. 4.68 The block-cutpoint tree

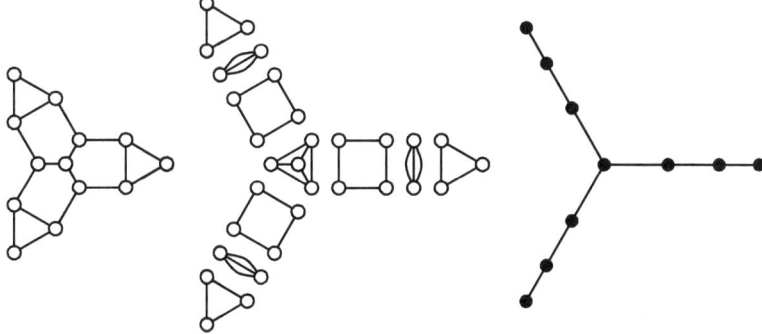

Fig. 4.69 The block-cutpoint tree

do not correspond to map automorphisms. Fortunately, in this case we have a combinatorial tool in the vertex and edge join of Sect. 4.4.6. Given a connected graph G, we can define a new graph $\mathcal{T}(G)$ whose vertices correspond to the cutpoints of G and the maximal 2-connected graphs of G, called *blocks*, with adjacency in $\mathcal{T}(G)$ corresponding to incidence in G. The graph $\mathcal{T}(G)$ is a tree, called the *block-cutpoint* tree (see Fig. 4.68), and any automorphism of G defines an automorphism of the block-cutpoint tree which permutes isomorphic blocks. So the automorphisms of a connected graph or map can be studied via the automorphisms of a tree, which are well understood, and automorphisms of 2-connected graphs, the blocks.

The block-cutpoint tree is a description of how to construct a connected graph G from 2-connected graphs via vertex joins. If G is two-connected, then it has a unique description via the edge join and a tree $\mathcal{T}_3(G)$, called the *3-block tree*. Each vertex of $\mathcal{T}_3(G)$ corresponds to a 3-*block*, which is either a 3-connected graph, a cycle, or a multilink, and each edge in $\mathcal{T}_3(G)$ corresponds to an edge join between two 3-blocks; see Fig. 4.69. Note that the 3-blocks are not subgraphs of G, since

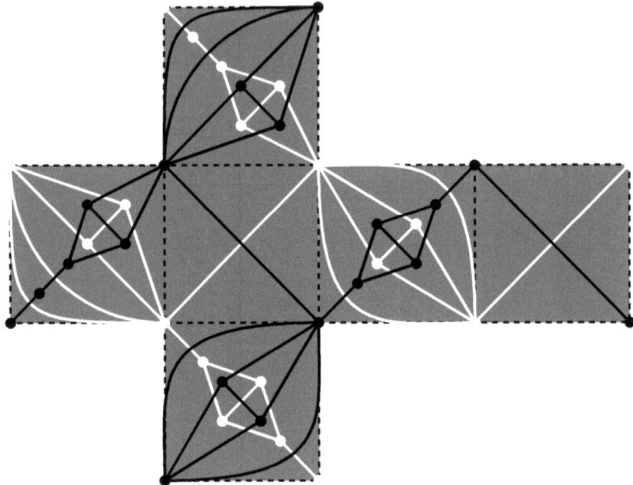

Fig. 4.70 A self-dual graph drawn on an unfolded cube with no corresponding self-dual map

in the edge join, the joined edge is erased; however, each 3-block has a subdivision which is isomorphic to a subgraph of G, so if G is planar, then all its 3-blocks will be planar. The 3-block tree is uniquely defined, and any group action of G induces an action on $\mathcal{T}_3(G)$, decomposing the action into an action on a tree, together with an action on 3-connected graphs. For more details see [23, 27, 28, 53, 101].

As an example of this method, consider a connected finite vertex-transitive graph. Its block-cutpoint tree is finite and so has a well-defined central element, which is either a vertex or an edge. The central element cannot be an edge, so it must be a block, which, since it is distinguished, must contain all the vertices of G, so G is actually 2-connected.

Theorem 4.18. *Let G be a simple 1-connected vertex-transitive graph. Then either G is 2-connected or its block-cutpoint tree is infinite.*

A similar argument shows

Theorem 4.19. *Let G be a simple 2-connected vertex-transitive graph. Then either G is a cycle, 3-connected, or its 3-block tree is infinite.*

For a planar graph G, we may now analyze the automorphisms of its 3-blocks via geometric methods and then piece these symmetries together combinatorially via its structure trees. This program has been followed for planar Cayley graphs [28] and self-dual graphs and matroids (see [89, 90]), where the construction of all self-dual graphs is described, even those such as that in Fig. 4.70 which have no associated self-dual map to be straightened.

4.6 Exercises

Exercise 4.1. Show that for any map M, the one-dimensional subdivision map **Su1**(M) is bipartite.

Exercise 4.2. A *fullerene* is a cubic planar graph whose faces are either hexagons or pentagons. Using the Euler formula for the sphere, prove that any fullerene has exactly 12 pentagons.

Exercise 4.3. Find an embedding of the octahedron graph on the torus. Is the embedding unique?

Exercise 4.4. Determine the topological type of the surface in Fig. 4.11 from the Euler characteristic.

Exercise 4.5. Show that Whitney's theorem does not hold for the torus. Hint: Check the embeddings in Fig. 4.67.

Exercise 4.6. Describe the operations of vertex splitting and edge contraction in M as operations on the dual map M*.

Exercise 4.7. Write the flag permutations of $M = M(G, \tau_0, \tau_2, \mathcal{V})$ in cycle notation for the tetrahexahedron map. Do the same for the octahedron, and find the permutations of the dual map in each case.

Exercise 4.8. Find an embedding of the octahedron on the torus. Is the embedding unique?

Exercise 4.9. Using the fact the Petrie walks are the orbits of $\tau_0 \tau_1 \tau_2$, list and trace all Petrie walks for the maps in Fig. 4.67.

Exercise 4.10. List and trace all Petrie walks for the maps in Figs. 4.5 and 4.6.

Exercise 4.11. For each of the maps in Example 4.9, determine the face permutation and the topological type of the resulting map.

Exercise 4.12. Show directly from the axioms on page 115 that $|\Phi| \equiv 0 \mod 4$.

Exercise 4.13. Decide whether or not the maps defined in Example 4.11 are the only maps with K_4 as a 1-skeleton.

Exercise 4.14. Show that the vertex split defined in Sect. 4.3 does not alter the number of faces of a map.

Exercise 4.15. Construct the flag graph of a unitary map consisting of k handles, and show that it is bipartite.

Exercise 4.16. So that a unitary map with g assembled handles is orientable, show that a unitary map with an assembled crosscap is nonorientable.

Exercise 4.17. Explain how a crosscap is related to the attachment of a Möbius band.

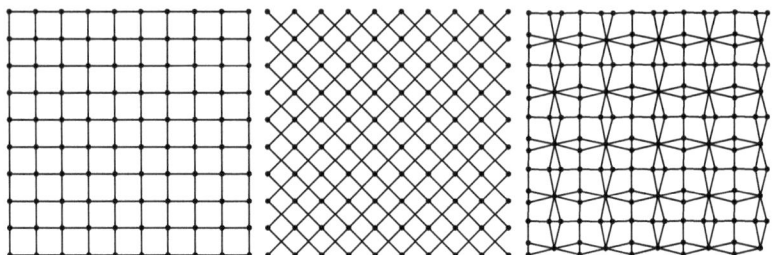

Fig. 4.71 The grid and its mysterious transformation

Exercise 4.18. Generalize the approach we took for Archimedean polyhedra in order to construct all uniform polyhedra from the tetrahedron by applying a series of operations on maps.

Exercise 4.19. Describe the operation on maps that transforms the squares of the grid on the left of Fig. 4.71 in order to obtain the map that looks locally like the one on the right of the same figure. How many matrices are needed for the description of this mysterious operation?

Exercise 4.20. Repeat the previous exercise by replacing the right map of Fig. 4.71 with the middle one.

Exercise 4.21. Show that the 1-skeleton of **Me**(M) is the line graph of the 1-skeleton of M.

Exercise 4.22. Show that the 1-skeleton of **Me**(M) is bipartite.

Exercise 4.23. Is it possible to derive octahemioctahedron from the tetrahedron by a series of map operations described in this chapter?

Exercise 4.24. Find the duals of the cube embedded on the torus as in Fig. 4.67. Are these duals isomorphic as maps? As graphs?

Exercise 4.25. Let M be a map. Define $\mathbf{Go}(M) := \mathbf{Du}(\mathbf{Me}(\mathbf{Tr}(M)))$. Draw $\mathbf{Go}(M)$ if M is locally a square grid.

Exercise 4.26. Calculate the topological types of the maps in Fig. 4.20.

Exercise 4.27. Compute all maps that can be obtained from the map of the cube using the dual and the Petrie dual.

Exercise 4.28. Generalize the matrix formulation to account for the snub operation.

Exercise 4.29. Show that $\mathbf{Co}(M)$ has a quadrilateral 2-factor.

Exercise 4.30. Draw all nine finite 3-connected planar edge-transitive maps.

Exercise 4.31. Determine all vertex-transitive fullerenes; see Exercise 4.2. Which of them are edge transitive?

4.6 Exercises

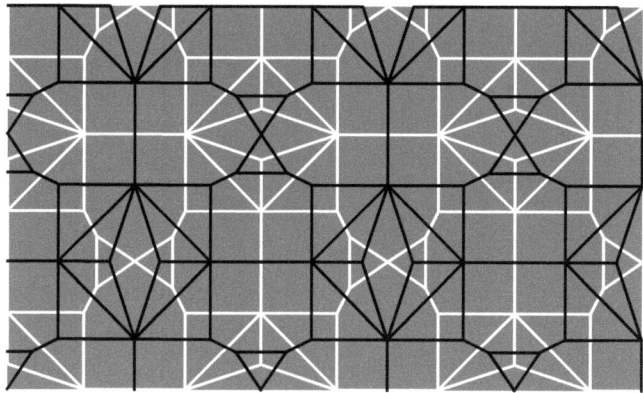

Fig. 4.72 What is the pairing?

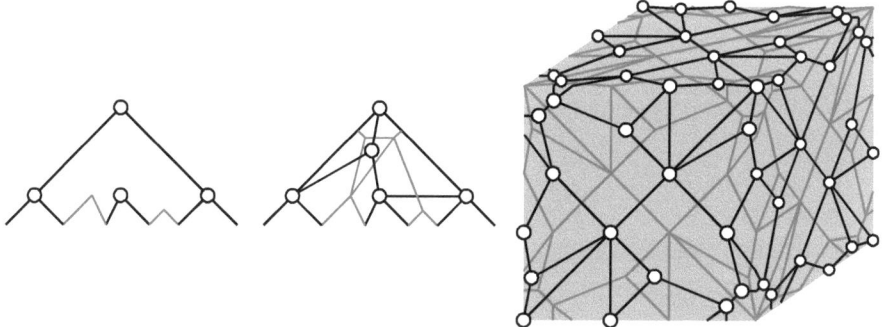

Fig. 4.73 A fundamental region for which groups?

Exercise 4.32. Explain how the seven frieze groups can be described by the four letters (aspects of the pattern): b, p, q, and d.

Exercise 4.33. Determine a Cayley graph of each of the seven frieze groups.

Exercise 4.34. Determine the crystallographic groups that may arise from the classical Escher problem.

Exercise 4.35. Which pairing is represented by the self-dual tiling of Fig. 4.72?

Exercise 4.36. To create a self-dual map with given pairing, one can start with a fundamental region of the groups; say using the Cayley graphs, draw an asymmetric dual pair in that region, and then, using the group action, patch the entire map together as illustrated in Fig. 4.73.

Using this method, find a self-dual map with pairings ($[3^+,4]$, $[3,3]^+$).

Exercise 4.37. Using the method of Exercise 4.36, find a self-dual tiling with pairing (p4g, pgg).

Chapter 5
Combinatorial Configurations

5.1 A Combinatorial Approach to Configurations

5.1.1 Incidence Structures

Let us formally recall the concepts introduced in Chap. 1. The most general framework with which we can describe the combinatorics of all the examples there is an incidence structure. An *incidence structure* S is a triple $S = (P, B, I)$ where P is the set of points, B is the set of blocks, $P \cap B = \emptyset$, and $I \subseteq P \times B$ is an incidence relation. In general there is no restriction on the number and types of incidences.

For example, the game of chess is played on an 8×8 grid of squares, which can be regarded as the points. At a given point in play, there is some subset of the 32 pieces, the blocks, scattered, presumably not at random, on the board, and we may say that piece p is incident to square s if p could move legally and occupy s.

Similarly, the nine squares of a tic-tac-toe grid may be regarded as points of an incidence structure, with the blocks being the various winning triples of three in a row. There are nine points labeled (i, j) for the square in the ith row and jth column, and eight blocks labeled r_i for row i, c_j for column j, and d_k for the two diagonals:

r_1	r_2	r_3	c_1	c_2	c_3	d_1	d_2
(0,0)	(0,1)	(0,2)	(0,0)	(1,0)	(2,0)	(0,0)	(2,0)
(1,0)	(1,1)	(1,2)	(0,1)	(1,1)	(2,1)	(1,1)	(1,1)
(2,0)	(2,1)	(2,2)	(0,2)	(1,2)	(2,2)	(2,2)	(0,2)

(5.1)

In the tic-tac-toe structure, while there are pairs of blocks which are incident to the same point, no two blocks are incident to exactly the same set of points. We say the tic-tac-toe structure is *simple*. It is not so easy to see if the chess structure is simple. A structure is *cosimple* if no two points are incident to the same set of lines. Of course, for nontrivial configurations of points and lines in the plane, simplicity and cosimplicity are obvious requirements for their incidence structures.

Many of the concepts we have considered for configurations are applicable to general incidence structures. An *incidence table* is a list, indexed on the number of blocks, such that the item on the list corresponding to block b is a list of those points p which are incident to b. The array in (5.1) is an incidence table for the tic-tac-toe incidence structure. The incidence table naturally highlights an ordering which has been imposed on the elements of P and B.

It is also valuable to encode an incidence structure as an *incidence graph*, $G(\mathcal{S})$, a bipartite graph with vertex set $P \cup B$, and the stipulation that point $p \in P$ is adjacent to block $b \in B$ if and only if p is incident to b in $\mathcal{S} = (P, B, I)$. The bipartition may be emphasized by coloring the vertices in P black and those in B white. The incidence graph generalizes directly the Levi graph introduced by Coxeter in [16] for configurations. For incidence structures, the incidence graph carries the complete information about the structure, so nothing whatever is lost by adopting an exclusively graph theoretical approach, in particular the language and definitions of graph theory.

The many definitions from graph theory carry over directly via the incidence graph to incidence structures. An incidence structure is *connected* if its Levi graph is connected. Otherwise, it is called disconnected or decomposable. The incidence structure corresponding to a subgraph of the Levi graph is a *substructure*. If necessary, we distinguish between those subgraphs which are induced and those which are not by denoting substructures as *strong* or *weak* respectively. Homomorphisms, isomorphisms, and automorphisms of the incidence graph become respectively homomorphisms, isomorphisms, and automorphisms of the corresponding incidence structures provided that they preserve the colors of the vertices, that is, the distinction between points and blocks. Morphisms which interchange the points and blocks are said to be *color reversing*, and a color-reversing automorphism is said to be a *self-duality;* see Sect. 4.5.3.

If necessary, to model the combinatorics of higher dimensional geometric structures, we consider an *incidence structure of rank k*, defined to be a $k+1$ tuple $\mathcal{S} = (P_1, \ldots, P_k, I)$ with $P_i \cap P_j = \emptyset$, and

$$I \subseteq \bigcup_{i<j} P_i \times P_j.$$

An incidence structure of rank k has been called a *rank k geometry*. The Levi graph of an incidence structure of rank k is defined analogously to be a multipartite graph, with vertices of each type colored separately.

Besides Coxeter's Levi graph, there have been several other graphs introduced in the study of configurations. The vertex set of the *Menger graph*, $M(\mathcal{S})$ is $\cup P_i$, and two vertices of $M(\mathcal{S})$ are adjacent if they are at distance 2 in the Levi graph. The restriction $M(\mathcal{S}, i)$ of $M(\mathcal{S})$ to the vertices of a given type i is called the *Menger graph of type i*.

5.1 A Combinatorial Approach to Configurations

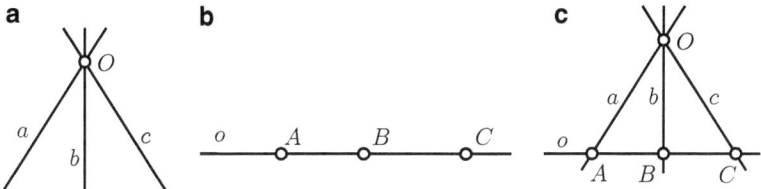

Fig. 5.1 (a) The pencil P_3, (b) its dual, (c) a near pencil

Related to the Menger graph is the *Grünbaum graph*, which has the same set of vertices as the Levi graph and in which two vertices are incident if either they are adjacent in the Levi graph or adjacent in the Menger graph. A set of points and blocks *splits* the incidence structure if they are independent in the Grünbaum graph, so there is no edge between them, and their deletion disconnects the Levi graph. Incidence structures for which no such set exists are called *unsplittable*.

An incidence structure is very general and requires none of the regularity of geometric configurations.

Example 5.1. Any graph gives rise to several incidence structures in a natural way. Clearly, one may regard the vertices as points, the edges as blocks, and incidence as graph incidence. Or one could take the p-cliques as points and q-cliques as blocks, with incidence defined via the subgraph relation.

Example 5.2. Any family of sets $\mathcal{F} \in \mathcal{P}(\mathcal{X})$ is an incidence structure with blocks $\mathcal{X}$, lines $\mathcal{F}$, and incidence relation $\in$.

Example 5.3. A set $L = \{l_1, l_2, \ldots, l_n\}$ consisting of a n distinct lines in the Euclidean plane $\mathbf{E}^2$ defines an incidence structure $\mathcal{L} = (P, L, I)$ called a *line arrangement* by letting $\mathcal{P}$ denote the set of points in $\mathbf{E}^2$ that are incident with at least two lines from L, and I is the point-line incidence in $\mathbf{E}^2$.

Example 5.4. A *pencil* $P(k)$ is an incidence structure with 1 point and k lines such that the point is incident with each line; see Fig. 5.1a. A *near pencil* $N(k)$ is an incidence structure with $1 + k$ points and blocks with a block L incident with k points and k blocks of size 2 connecting the k points incident with L to the remaining point. A pencil is unsplittable, while a near pencil is splittable. The dual of the pencil $P(k)$ is a single line incident to k points (see Fig. 5.1b) while the near pencil is self-dual. In fact, the near pencil has an involutory self-duality, that is, the self-duality is an automorphism of order 2 on the Levi graph. We say the near pencil is *autopolar*.

As the previous example suggests, incidence structures will be valuable not only to describe objects more general than configurations but to describe their substructures, which may lack regularity and so not be regarded as configurations in their own right.

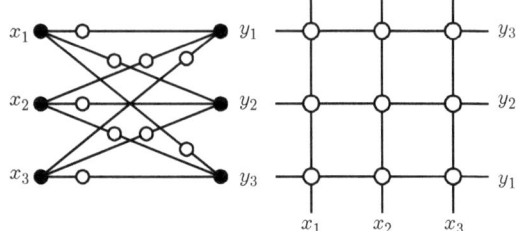

Fig. 5.2 The subdivision graph of $K_{3,3}$ gives a configuration which is resolvable into parallel classes

If the incidence table is modified so that instead of simply listing for each block b the vertices to which it is incident, we record a column vector of length $|V|$ placing a 1 or 0 in the row corresponding to v depending on whether or not v is incident to b. These column vectors form the *incidence matrix* of the incidence structure S. If we order the vertices of the Levi graph with all the black (point) vertices first and the white (block) vertices thereafter, then, setting to be N the incidence matrix with respect to this ordering, the adjacency matrix of the incidence graph (see Sect. 2.1) is given by

$$\begin{pmatrix} \mathbf{0} & \mathbf{N} \\ \mathbf{N}^T & \mathbf{0} \end{pmatrix}.$$

Note that entry (i, j) in the matrix $N \cdot N^T$ gives the number of blocks common to points i and j, and the (i, i) entry gives the number of blocks incident with i. The matrix $N^T \cdot N$ has the dual conclusion.

Associated with each incidence structure $S = (P, B, I)$ is a *dual structure* $S^* = (B, P, I')$ where $(b, p) \in I'$ if and only if $(p, b) \in I'$. As we have seen in the case of polyhedra and maps, the dual structure contains the same combinatorial information with the roles of the points and the blocks interchanged. The incidence matrix of S^* is N^T, and

$$\begin{pmatrix} \mathbf{0} & \mathbf{N}^T \\ \mathbf{N} & \mathbf{0} \end{pmatrix}$$

gives the adjacency matrix of its incidence graph.

In analogy with the Euclidean case, two blocks in an incidence structure are said to be *parallel* if they have no points in common, and a *parallel class* of blocks is a set of parallel blocks that partition the point set. An incidence structure is said to be *resolvable* if its blocks can be partitioned into parallel classes. The Pappus configuration (see Sect. 1.1.3) is a simple example of a resolvable incidence structure.

Example 5.5. Consider the subdivision graph of $K_{3,3} = (V, E)$; see Sect. 2.4. $S(K_{3,3})$, like any subdivision graph, is bipartite with bipartition $V \cup E$. The graph $S(K_{3,3})$ is the subdivision graph of a graph which is itself bipartite, $V = \{x_1, x_2, x_3, y_1, y_2, y_3\}$, with x_i adjacent to y_j for all $i, j \in \{1, 2, 3\}$. If we regard V as the blocks of an incidence structure and E as the points, then the bipartition of V is a resolution of the graph into two parallel classes; see Fig. 5.2.

5.1 A Combinatorial Approach to Configurations

Fig. 5.3 Generalized hexagon on 63 points and lines as drawn by Andreas Schroth. It is the dual of the one from the next figure

Example 5.6. A *generalized n-gon* is an incidence structure N (or rank 2 geometry) whose incidence graph satisfies the following axioms:

- N1 The distance between any two vertices of N is at most n.
- N2 For any two vertices at distance $k < n$, there is a unique path of length k.
- N3 Each vertex is incident with at least three lines.

In Fig. 5.3, we have a generalized hexagon on 63 points, which is the fewest possible. Since n-gons are self-dual, one might expect self-duality for generalized n-gons, but this example is not self-dual. However, its dual is also a generalized hexagon; see Fig. 5.4.

5.1.2 Coset Incidence Structures

Let G be a group and let $\{G_1, G_2, \ldots, G_k\}$ be a family of subgroups of G with cosets xG_t, $t \in \{1, 2, \ldots, k\}$. An incidence structure of rank k is obtained by taking the elements of type t, $t \in \{1, 2, \ldots, k\}$ to be the cosets xG_t. Two cosets

Fig. 5.4 Generalized hexagon on 63 points and lines as drawn by Andreas Schroth, [88]. It is the dual of the one from the previous figure

are mutually incident, $xG_t \sim yG_s$, if and only if $xG_t \cap yG_s \neq \emptyset$. This structure is called *rank k coset incidence structure* for the group G with respect to the subgroups $\{G_1, G_2, \ldots, G_k\}$. The subgroups are called *parabolic subgroups*, and their intersection is known as the *Borel subgroup*. If parabolic subgroups do not generate the group, then the incidence structure is not connected.

Example 5.7. Consider the three subgroups of the eight element quaternion group **H** generated by i, j, and k. The subgroups $\langle i \rangle$, $\langle j \rangle$, and $\langle k \rangle$ are rank 4, index 2, cyclic subgroups containing three types of elements, so the coset incidence structure has six points, with two elements of each type: $\langle i \rangle$ and $\mathbf{H} - \langle i \rangle = \langle i \rangle'$ of type i, $\langle j \rangle$ and $\langle j \rangle'$ of type j, and $\langle k \rangle$ and $\langle k \rangle'$ of type k. Of these six cosets, only the three pairs of cosets of the same type are disjoint and hence nonincident in the incident structure or nonadjacent in the incidence graph. So the incidence graph is the graph of the octahedron, with antipodal points having the same type. See Fig. 5.5a. If we add as a fourth subgroup the subgroup $\langle -1 \rangle$, with cosets $\{-i, i\}$, $\{-j, j\}$, and $\{-k, k\}$, then we add four new vertices of a new type. Since $\{-1, 1\}$ is contained in all of the subgroups $\langle i \rangle$, $\langle j \rangle$, and $\langle k \rangle$ and hence is disjoint from all the other cosets, the corresponding new elements will be each of valence three in the incidence graph,

5.1 A Combinatorial Approach to Configurations

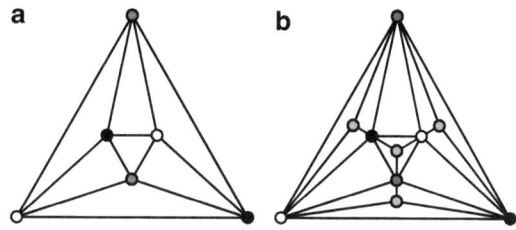

Fig. 5.5 Quaternions and some of their subgroups form a rank 3 incidence structure and rank 4 incidence structure

attached to the endpoints of four faces of the octahedron which do not share an edge, one of which is $\{\langle i \rangle, \langle j \rangle, \langle k \rangle\}$, which is incident to the new point $\langle -1 \rangle$. See Fig. 5.5b. This new incidence structure is of rank 4.

5.1.3 Lineal Incidence Structures

As we have seen, an incidence structure is no more or less than a bipartite graph with a prescribed bipartition. The types of incidence structures that actually arise as collections of points and lines in a geometric plane are a much more restricted class. There is no known combinatorial characterization of such incidences structures. Even for the Euclidean plane, there is no set of axioms to characterize linear arrangements which were introduced in Example 5.3. One obvious necessary condition we can specify is that there is in the Euclidean plane a unique line through any two points. We shall call an incidence structure *lineal* if any two points belong to at most one common block. (Lineal incidence structures have sometimes been called *semilinear* or *planar* incidence structures.)

Of course, all the incidence structures of all linear arrangements are lineal. On the other hand, it is not hard to find geometrically based nonlinear incidence structures. The points and planes of the octahedron are nonlineal, and it is not hard to show that it requires the deletion of at least four blocks to become lineal, in which case the structure is not only lineal but isomorphic to the linear arrangement corresponding to the complete quadrangle; see Sect. 1.1.5. The Miquel configuration (see Sect. 1.1.2) is not lineal. The Fano configuration (see Sect. 1.1.1) is lineal, but we have seen that it does not correspond to a Euclidean linear arrangement.

The incidence graph is bipartite, so every cycle has an even number of edges; moreover, the incidence graph is simple, so the incidence graph either has a four-cycle or has girth at least six. Two points being incident to two blocks correspond to a 4-cycle in the incidence graph, so we have the following:

Proposition 5.8. *An incidence structure is lineal if and only if its Levi graph has girth at least 6.*

This immediately implies the following:

Corollary 5.9. *The dual of a lineal incidence structure is lineal.*

5.1.4 Regularity of Incidence Structures

Let $\mathcal{S} = (P, B, I)$ be an incidence structure. The *degree of a point* $p \in P$ is the number of blocks to which it is incident. The *degree of a block* $b \in B$ is the number of points to which it is incident. The degrees of the points and the blocks are synonymous with their valences in the incidence graph. A *point regular incidence structure* is a structure in which all points have the same degree, r. A *block regular incidence structure* is a structure in which all blocks have the same degree, k. An incidence structure which is point and block regular is called *regular of type* $(|P|_r, |B|_k)$. For a regular incidence structure, we have

$$|I| = |P|r = |B|k$$

Note that the tic-tac-toe structure is block regular but not point regular. The points and planes of an icosahedron form a regular incidence structure of type $(12_5, 20_3)$ with 60 incidences. The vertices and edges of regular maps also form a regular incidence structure, although note the difference in the usage of the word regular.

The incidence structure is regular if and only if its incidence graph is bipartite and semiregular. A regular incidence structure is said to be *symmetric* or *square* if $|P| = |B|$, in which case necessarily $r = k$, and its type will merely be noted $|P|_r$.

The generalized hexagons of Figs. 5.3 and 5.4 are both regular of type (63_3).

5.1.5 Definition of Combinatorial Configurations

A lineal incidence structure which is point and block regular is called a *combinatorial configuration*. The blocks of a combinatorial configuration will generally be referred to as "lines", and the incidence graph will be called the *Levi graph*. It is common to denote by v the number of points in the configuration, b the number of lines, r the vertex degree, and k the block degree. As before, we will denote its type by (v_r, b_k), or sometimes $[r, k]$, and for regular configurations (v_k, v_k) merely by (v_k). Unless otherwise stated, all our configurations are connected.

Example 5.10. The only connected (v_1) combinatorial configuration is (1_1), and the only connected (v_2) combinatorial configurations are the incidence structures of polygons with at least three points since the incidence structure of the digon is not lineal.

Proposition 5.11. *An incidence structure is a (v_r, b_k) configuration if and only if its Levi graph is semiregular with valences r and k and has girth at least 6.*

An incidence structure is a (v_r) configuration if and only if its Levi graph is regular of valence r and has girth at least 6.

Combinatorial configurations are still a very general construct.

5.2 Combinatorial (v_3) Configurations of Small Order

Table 5.1 Properties of configurations and their Levi graphs

Configuration $(P, \mathcal{B})$	Levi graph $L(P, \mathcal{B})$
Incidence structure	Bipartite graph
n_r, b_k	Semiregular, girth ≥ 6
n_3	Bipartite, cubic, girth ≥ 6
Points	Black vertices
Lines (blocks)	White vertices
No triangles	Girth ≥ 8
Self-dual	$\gamma \in \text{Aut}(L)$ swaps black and white
Indecomposable	Connected
Incidence matrix	Biadjacency matrix
Point transitive	Black vertices in the same orbit under $\text{Aut}(L)$
Line transitive	White vertices in the same orbit under $\text{Aut}(L)$
Flag transitive	Edge transitive
Flag-transitive, self-dual	Arc transitive
Point transitive, self-dual	Vertex transitive
Cyclic	Haar graph of girth ≥ 6

Example 5.12. Each k-regular graph without loops and multiple edges can be viewed as an (n_k, b_2) configuration of points and lines. Vertices correspond to points and edges correspond to lines. The corresponding Levi graph is the subdivision graph of the original graph.

As with incidence structures, nothing is lost by adopting a completely graph theoretical point of view. Table 5.1 shows the correspondence between properties of configurations and their Levi graphs.

5.2 Combinatorial (v_3) Configurations of Small Order

While generically chosen lines in the plane will meet at one point with probability 1, three arbitrarily chosen lines will be concurrent with probability 0. Likewise, two points are always collinear, but three generic points are not. So, it is very natural to ask for a configuration of points and lines with the property that on every line, there are exactly 3 points and through every point, there are exactly three lines, that is, to study (v_3) configurations. As we have seen, combinatorially (v_3) configurations are characterized by the fact that their Levi graphs are connected, bipartite, 3-regular graphs of girth at least 6. For a small number of vertices, it is tractable to examine how many such nonisomorphic graphs exist.

Theorem 5.13. *There is, up to isomorphism, exactly one bipartite 3-regular graph of girth at least 6 on 14 and on 16 vertices.*

There is no such graph on fewer than 14 vertices.

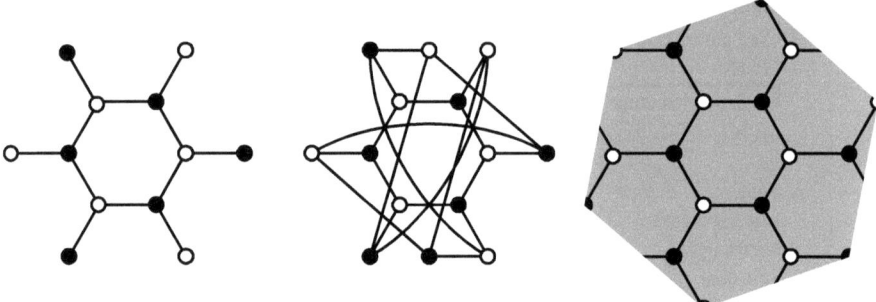

Fig. 5.6 The only Levi graph of a (7_3) configuration, embedded on a torus represented as a regular hexagon with opposite sides identified

Proof. Let G be a 3-regular bipartite graph on n vertices. A black vertex must be adjacent to three distinct white vertices since we assume G is simple. These three white vertices in turn must be adjacent to six distinct black vertices, all different from the initial black vertex, if G is required to have girth at least 6. Therefore, G must have at least seven black vertices and, by symmetry, seven white vertices, so $n \geq 14$.

If $n = 14$, G must have girth equal to 6, otherwise, by the above argument, we get $n > 14$. Consider a 6-cycle in G. The vertices of this hexagon must be adjacent to six distinct other vertices of G. These new vertices come in pairs of distance 5. The induced subgraph of G on the 12 vertices considered so far must contain the 3 edges whose endpoints have distance 5 in the hairy hexagon (leftmost graph in Fig. 5.6) since the two vertices of G not yet considered must both be adjacent to all three hexagon hairs of the correct color, so there is only one way to complete the graph; see Fig. 5.6.

We pause to remark at this point that the two vertices attached to a plane regular hairy hexagon will result in many edge crossings but not if they are placed on the vertices of an enclosing regular hexagon in which opposite sides have been identified. Moreover, the three edges joining the hairs may now be rerouted across the interiors of the identified edges, giving an embedding of the Levi graph in the torus, which has a combinatorial automorphism of order 6 rotating the hairy hexagon. Straightening the map using this automorphism and Theorem 4.16, we obtain a regular hexagon embedding in the hexagonal torus pictured on the right of Fig. 5.6.

If $n = 16$, G must also be of girth 6 and contain a hairy hexagon, but now there are 4 vertices outside the hairy hexagon. Let us consider the induced subgraph of G on these extra vertices. The requirement of 3-regularity and girth 6 implies that the extra vertices induce at most 3 edges, and this induced subgraph is either empty, a single edge, a path of length 2 plus a single vertex, a pair of independent edges, or a path of length 3. The induced subgraph cannot be empty or a single edge since then one black vertex would be adjacent to three white vertices of the hairy hexagon, and

5.2 Combinatorial (v_3) Configurations of Small Order

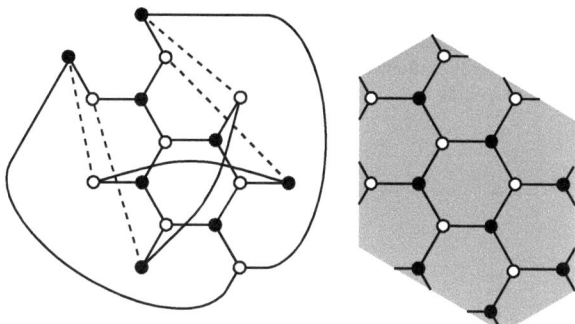

Fig. 5.7 The only Levi graph of an (8_3) configuration, embedded on a torus represented as an irregular hexagon with opposite sides identified

the other would be adjacent to at least two of them forming a 4-cycle. The induced subgraph cannot be a path of length two since both of its endpoints would have to be adjacent to at least one common vertex in the hairy hexagon, again forming a 4-cycle. It can also not be a path of length 3 since then no two vertices of the hairy hexagon could have two neighbors in the induced subgraph without forming a 4-cycle, so the hairy hexagon must have all three hairy diagonals. The endpoints of each hairy diagonal must now be adjacent to two nonadjacent vertices of the different colors in the induced subgraph, and it only has one such pair.

This leaves the pair of independent edges as the only possibility. The subgraph induced by the hairy hexagon contains two hairy diagonals. The two vertices of degree 1 of this induced subgraph must be adjacent in G to both of the extra vertices of the correct color. Now, the girth requirement forces the endpoints of the hairy diagonal to be adjacent to different components of the independent edge set; see Fig. 5.7.

Again, we can take a planar hairy hexagon, and, except for the last four edges, we can extend to a planar embedding and then remove the crossings by embedding in a hexagon with opposite sides identified. After straightening, the fundamental hexagon can no longer be regular since a rotation of the planar hairy hexagon does not extend to an automorphism of the Levi graph; see Fig. 5.7. □

Theorem 5.14. *There are, up to isomorphism, exactly three bipartite 3-regular graphs of girth at least 6 on 18 vertices.*

Proof. In the case of a bipartite graph with 18 vertices, we have, as before, a hairy hexagon which is an induced subgraph except for possibly three hairy diagonals.

If there are three hairy diagonals, then their endpoints are connected with the remaining six vertices by three edges, so they in turn are joined to one another by six edges, which must contain a cycle and thus, to avoid quadrilaterals, must be arranged in a hexagon. There is, up to isomorphism, one graph of this type as pictured in Fig. 5.8.

If there are two hairy diagonals, then the remaining six vertices have five edges among them, and since these five edges have no cycles, they must comprise a tree, and the two vertices of the hairy hexagon which are not joined by a hairy diagonal

Fig. 5.8 A combinatorial (9_3) configuration with a hairy hexagon with three hairy diagonals

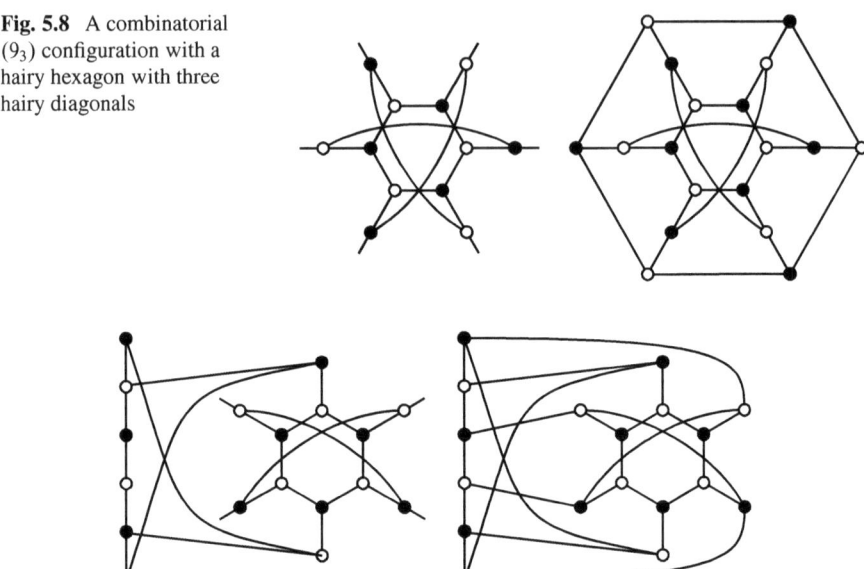

Fig. 5.9 A combinatorial (9_3) configuration with a hairy hexagon with two hairy diagonals

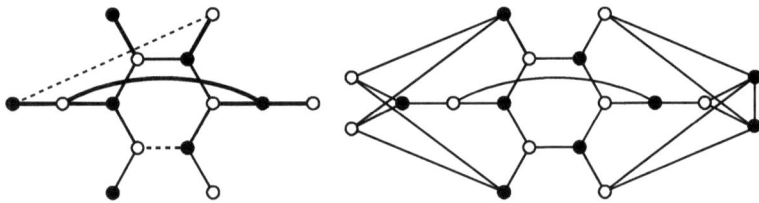

Fig. 5.10 There is no combinatorial (9_3) configuration having hairy diagonals in which every hairy hexagon has at most one hairy diagonal

must make a cycle with this tree and so must join the tree at two vertices of the same color at a distance of four or more apart. Thus, the tree can only be a path of length five (see Fig. 5.9) and the remaining edges, to avoid quadrilaterals with the hairy diagonals, must be arranged up to symmetry as on the left of Fig. 5.9.

The next case to consider is that of the hairy hexagon having one hairy diagonal, and to avoid duplication, we may assume that no hairy hexagon in the Levi graph has two or more hairy diagonals. In this case, the subgraph induced by the hairy hexagon and the third neighbors of the endpoints of the hairy diagonal is, in fact, an induced subgraph since any added edge would create a hairy hexagon with two hairy diagonals; see Fig. 5.10. Since this subgraph is induced, the 12 required incidences are among the incidences of the remaining four vertices, so 3-regularity forces the graph to be completed as in the right of Fig. 5.10, introducing several quadrilaterals. So this case gives no new examples.

5.2 Combinatorial (v_3) Configurations of Small Order

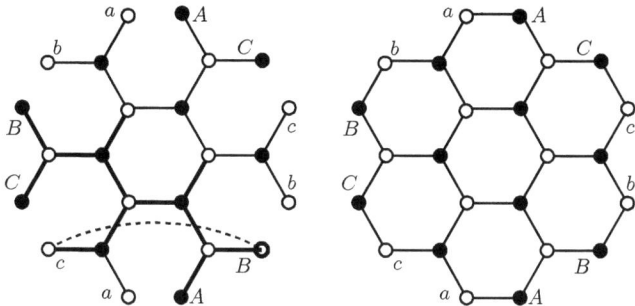

Fig. 5.11 Constructing the Levi graph of a combinatorial 9_3 configuration in which every hairy hexagon is an induced subgraph

Lastly, it may happen that all hairy hexagons are induced subgraphs. If any vertex x not in the hairy hexagon is adjacent to three free vertices, then there are three incidences yet to be accounted for these hairs, so one of the two other vertices of the same color as x must be adjacent to two, forming a quadrilateral. So each of the vertices not in the hairy hexagon is adjacent to at most two, and since there are only six of them, it must be exactly two, and the graph must appear as in Fig. 5.11, where the pairs of labeled vertices are to be identified and three edges remaining unaccounted for. Now, there cannot be an edge between c and B since that would induce a hairy diagonal to a hairy hexagon. Similarly, c and A cannot be adjacent, so the edge at c must be to C, and similarly, a and A must be adjacent and b and B, so if there are no hairy diagonals, the graph must be presented as on the right in Fig. 5.11. □

As is suggested by the proof, the Levi graph of the (9_3) configuration depicted in Fig. 5.11 has a natural embedding on the torus represented as a hexagon with opposite sides identified; see Fig. 5.12a. This is the Levi graph of the Pappus configuration, symmetrically drawn in Fig. 1.10.

The other combinatorial (9_3) configurations whose Levi graphs were constructed above correspond to the other two point-line configurations indicated in Fig. 1.10 and realized in $\mathbb{R}^2$. Its construction via the hairy hexagon also suggests a representation as a regular hexagonal tiling on a surface.

In Fig. 5.12, we have the Levi graph of the configuration $(9_3)_1$, that is, the Pappus configuration, embedded on a torus, with a fundamental region consisting of a hexagon whose opposite sides are identified. The Levi graphs of the other two combinatorial (9_3) configurations constructed above are shown embedded on the torus in Fig. 5.12b, and on the Klein bottle; see Fig. 5.13. It is left as an exercise, which of these has a hairy hexagon with two hairy diagonals, as well as which of these Levi graphs corresponds to the geometric configurations in Fig. 1.10.

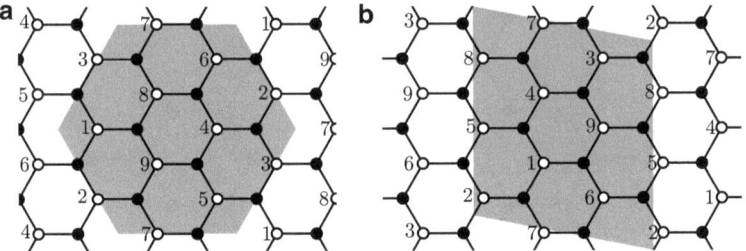

Fig. 5.12 Torus embeddings of (9_3) configurations

Fig. 5.13 A (9_3) configuration embedded in the Klein bottle

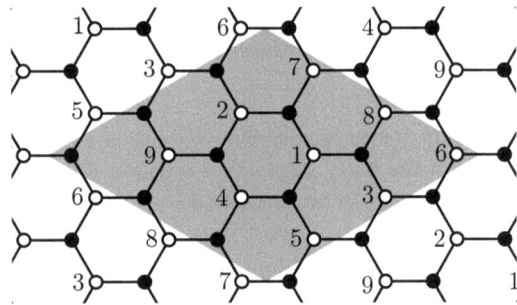

5.3 Classical Configurations

We have seen that there are unique combinatorial configurations of type (7_3) and (8_3) and 3 configurations of type (9_3), one of which is the classical Pappus configuration. There are ten configurations of type (10_3). One example is the Desargues configuration. Another example is pictured below; see Fig. 5.14.

As n increases, it becomes easier and easier to construct configurations of type (n_3) and less common for them to display large automorphism groups. There are 31 configurations of type (11_3) and 229 of type (12_3). These facts were established more than a 100 years ago by von Sterneck, [24, 25], although he missed one (12_3) configuration. Only recently, it has been shown [97] that all these configurations admit realizations in 3-D using integer coordinates. For more information, see, for instance, [37, 39–41]. Algorithmic aspects are covered in [10] and [97].

With the hundreds of relatively small interesting configurations from which to choose, we will consider a not at all random selection of classic configurations.

5.3.1 The Fano Plane

The Fano plane, named for the Italian mathematician Gino Fano, was already encountered in Chap. 1 as a configuration of 7 points and 7 lines. It was drawn

5.3 Classical Configurations 171

Fig. 5.14 A (10_3) not isomorphic to the Desargues configuration

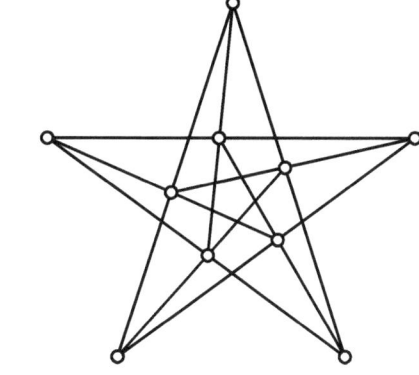

Fig. 5.15 The Fano configuration with 7 points and 7 lines. The seventh line is drawn as a circle

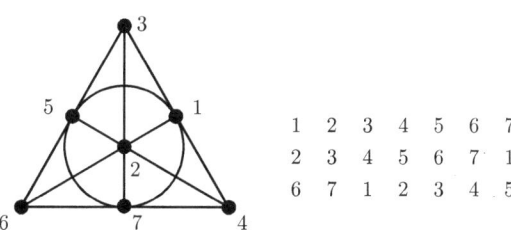

in the plane with a defective Euclidean representation via an equilateral triangle with all three altitudes and the inscribed circle and the incenter; see Fig. 5.15. While having no representation with straight Euclidean lines, it is nevertheless a lineal configuration, so its cubic Levi graph is of girth 6. In fact, with a mere 14 vertices, the Levi graph is the 6-cage described in Sect. 3.5, the Heawood graph.

The Fano plane has $\binom{7}{3} = 35$ triples of points of which 7 are collinear, so it has 28 nondegenerate triangles which correspond to the 28 different 6-cycles in the Levi graph. Any color-preserving automorphism of the Levi graph which fixes the points of any 6-cycle must fix as well all their six distinct neighbors, and the remaining two points, one of each color, must also be fixed. So we may count the automorphisms of the Levi graph by computing the orbit of the labeled 6-cycles. First, we simply count the labeled 6-cycles. There are three black vertices to choose, 7 ways to choose the first, 6 ways to choose the second, and, discounting the one point collinear with the first two choices, 4 ways to choose the third, giving $7 \cdot 6 \cdot 4 = 168$ labeled 6-cycles.

Now, we show that all these cycles are in the same orbit. Start with any color-preserving isomorphism between two 6-cycles. The vertices of the six-cycle and their six neighbors, which are all distinct since the graph is 3-regular, leave only two vertices, each of different color, and each must by 3-regularity be adjacent to all the three neighbors of the 6-cycle of the relevant color. The 6-cycle neighbors must each have one more edge, and since the Levi graph has no 4-cycles, the neighbor of each vertex of the 6-cycle must be adjacent to the neighbor of the vertex directly across on the 6-cycle. So our isomorphism between the labeled 6-cycles extends to an automorphism of the graph, and the two labeled 6-cycles are in the same orbit.

Fig. 5.16 The Heawood graph: given the outer 6-cycle, the points distance 1 from the outer 6-cycle, and the two vertices of distance 2 from the outer 6-cycle

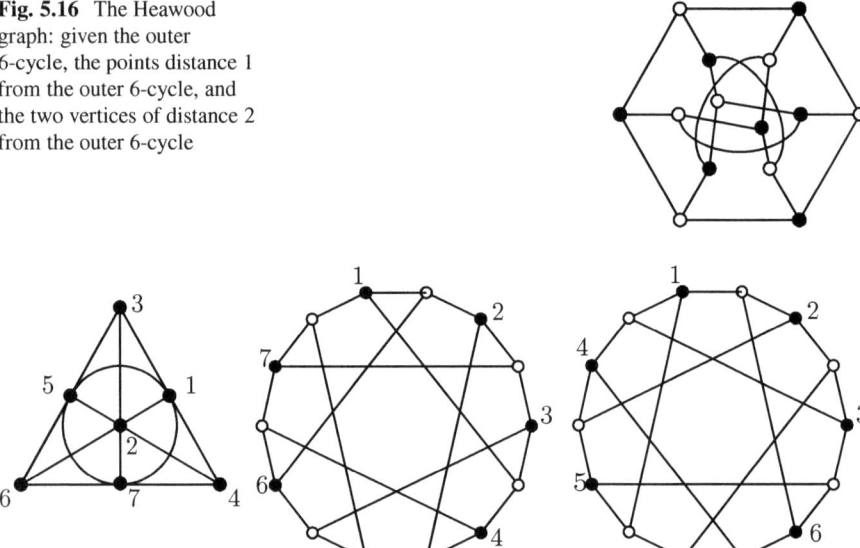

Fig. 5.17 The Hamilton cycles 1234567 and 1236754 represent the two orbits of 7-cycles in the Levi graph of the Fano configuration

This argument gives us another view of the Levi graph (see Fig. 5.16) and, in fact, essentially proves that the Levi graph of the Fano plane is the unique 6-cage.

We have established that the automorphism group of the Fano configuration has 168 elements.

Note that the Menger graph of the Fano plane is K_7, with automorphism group Sym(7), indicating a loss of combinatorial information as compared with the Levi graph.

Let us construct a Hamilton cycle of the Levi graph of the Fano plane. A Hamilton cycle cannot contain consecutively three black vertices corresponding to collinear points since the white vertex corresponding to their connecting line would occur twice in the cycle. So, since the automorphism group acts transitively on the labeled 6-cycles, we can assume our Hamilton cycle starts with black vertices 1, 2, and 3 as in Fig. 5.17 and conclude that the next black vertex cannot be 7 since it is collinear with 2 and 3 and cannot be 5 since the next three vertices would then have to be collinear. The choices are 4 and 6, both of which continue uniquely to give us two Hamiltonian cycles which are in different orbits under the automorphism group of the Levi graph. Each of the Hamilton cycles is stabilized by $\mathbb{Z}_7$ generated by the permutation (1234567) and (1236754) respectively of the black vertices, so the orbit of each has $168/7 = 24$ elements. Also, therefore, the automorphism group has 48 different 7-cycles in two conjugacy classes, one for (1234567) and one for (1236754).

5.3 Classical Configurations

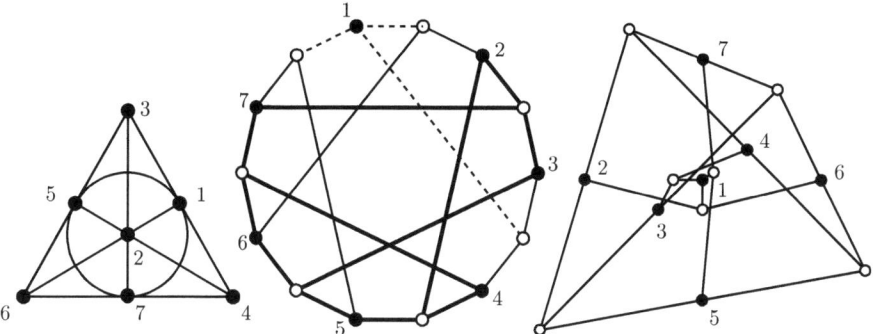

Fig. 5.18 Another view of the Heawood graph valuable in analyzing the vertex stabilizers

The action of $\mathbb{Z}_7$ via (1234567) on the Heawood graph is the action that gives the symmetry of the incidence table in Fig. 5.15 which identifies the Fano configuration as a 7-gon which inscribes and circumscribes itself. It is also the action which leads to the representation of the Heawood graph as the covering graph of the theta graph-assigned voltages in $\mathbb{Z}_7$ in Fig. 3.34 of Chap. 3.

We can use this action to refocus our examination of the automorphism group. As with any Levi graph, since there are no 4-cycles, we can specify a color-preserving automorphism by giving its action on the black vertices alone. Since $\mathbb{Z}_7$ acts transitively via (1234567) on the black vertices, we can find all the other color-preserving automorphisms of the Heawood graph by considering the stabilizer of a black vertex. The stabilizer of 1 acts as a group of automorphisms on the subgraph of the Levi graph of all vertices of distance 2 or more from 1. This graph is isomorphic to $S(K_4)$, the subdivision graph of K_4; see Fig. 5.18. This copy of $S(K_4)$ is connected to each of the three neighbors of 1 at the subdivision vertices of a pair of antipodal edges of K_4. Since every automorphism of K_4 preserves these pairs, the stabilizer of the vertex 1 may be identified with the group of symmetries of a regular tetrahedron. So we have by conjugacy class in the stabilizer of 1: the identity; 8 vertex rotations with each conjugate to (376)(245); 3 antipodal edge rotations with each conjugate to (26)(34); 6 reflections with each conjugate to (23)(46); and 6 90° rotary reflections on antipodal edge axes with each conjugate to (2364)(57).

So the color-preserving automorphism group of the Heawood graph is the subgroup of Sym(n) generated by

$$\{(1234567), (376)(245), (26)(34), (23)(46), (23)(46)(57)\}.$$

Note that some generators are redundant.

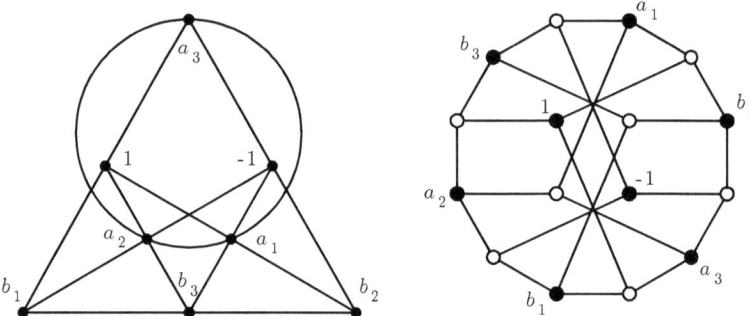

Fig. 5.19 The Möbius–Kantor configuration and its Levi graph

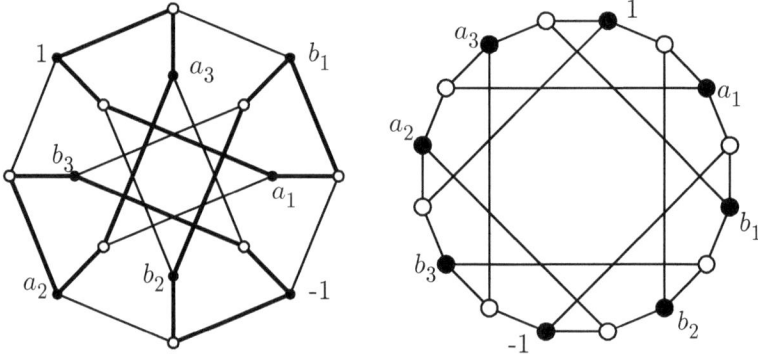

Fig. 5.20 Two drawings of the Levi graph of the Möbius–Kantor configuration

5.3.2 The Möbius–Kantor Configuration

We have seen that there is exactly one combinatorial configuration of type (8_3). It is called the Möbius–Kantor configuration. Like the Fano configuration, it has no representation in the Euclidean plane. It can be considered abstractly as having eight points with labels $\{1, a_1, a_2, a_3, -1, b_1, b_2, b_3\}$ and organized with two lines $\{a_1, a_2, a_3\}$ and $\{b_1, b_2, b_3\}$ as well as the six lines $\{1, a_i, b_{i+1}\}$ and $\{-1, a_i, b_{i-1}\}$, indices modulo 3; see Fig. 5.19. It is clear from both the incidence structure and the figure that the stabilizer of 1 under the automorphism group, which must perforce also to stabilize -1 since that is the unique point not sharing a line with 1, is the dihedral group of order 6, generated by $(a_1 b_1)(a_2 b_3)(a_3 b_2)$ and $(a_1 b_2)(a_2 b_1)(a_3 b_3)$.

To complete the computation of the automorphism group of the Levi graph, we must determine the orbit of 1. We already know that the generalized Petersen graph GP(8, 3) is cubic, bipartite, and of girth at least 6, so since there is exactly one (8_3) configuration, it must be the Levi graph of the Möbius–Kantor configuration; see Fig. 5.20. Just as for the Heawood graph, the Levi graph of the Fano configuration

5.3 Classical Configurations

Fig. 5.21 The Levi graph of the Möbius–Kantor configuration as a subgraph of the hypercube

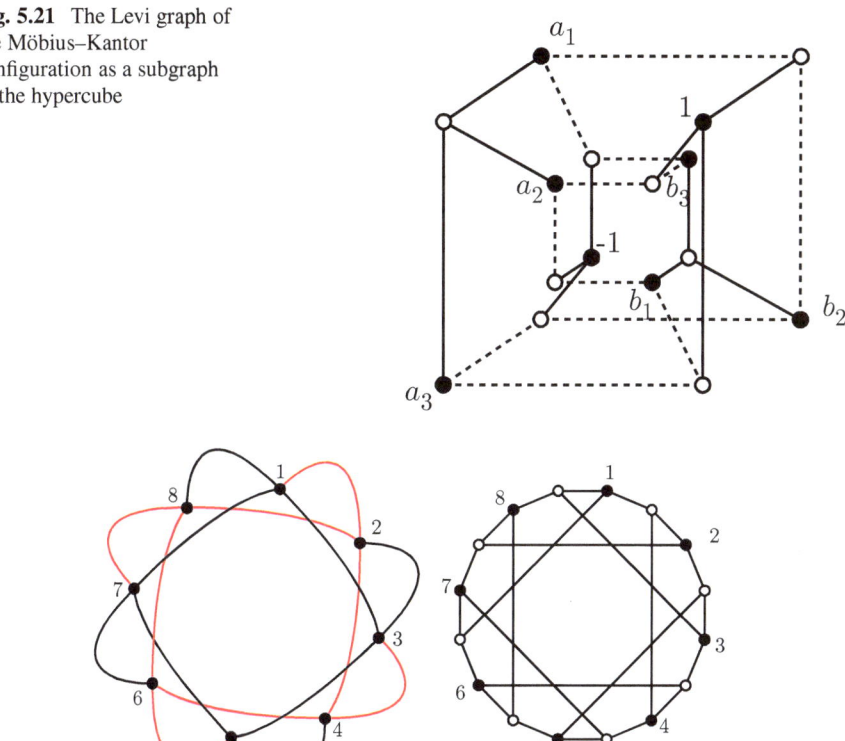

Fig. 5.22 The Möbius–Kantor configuration and its Levi graph

GP(8, 3) has a Hamiltonian cycle, and this cycle defines an action of $\mathbb{Z}_8$ on the Levi graph, generated by $(1a_1b_3b_2-1b_1a_3a_2)$.

So the automorphism of the configuration is generated by $(1a_1b_3b_2-1b_1a_3a_2)$, $(a_1b_1)(a_2b_3)(a_3b_2)$ and $(a_2a_3)(b_2b_3)$ and has order $8 \cdot 12 = 48$.

The number 48 is suggestive of the automorphism group of the cube, which is Sym(4), but Sym(4) has no eight-cycle. There is connection to the cube, [16]. The sixteen-vertex Levi graph is 3-regular. The graph of the hypercube Q_4 is 4-regular with sixteen vertices, so it is not surprising that the Levi graph of the Möbius–Kantor configuration can be viewed as Q_4 minus a 1-factor; see Fig. 5.21.

This generalized Petersen graph GP(8, 3) is one of the exceptional examples which is arc transitive (see Sect. 3.4) so the Möbius–Kantor configuration is also self-dual, as well as autopolar; see Sect. 5.4.

Considering the representation of the Levi graph as a generalized Petersen graph, the inner and outer 8-cycles may be regarded as the Levi graphs of two quadrilaterals, with the spoke vertices implying that the quadrilaterals inscribe one another; see Fig. 5.22.

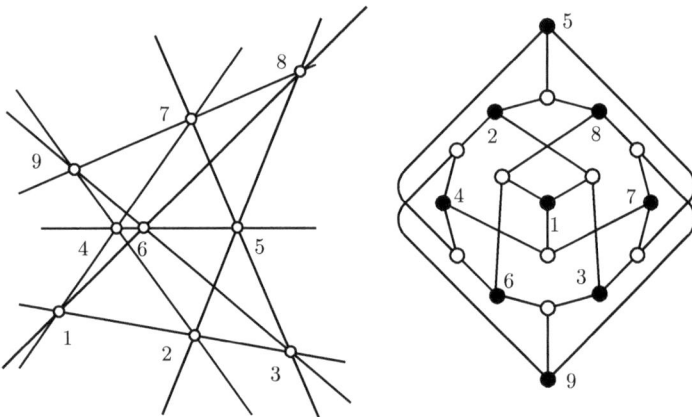

Fig. 5.23 The Pappus configuration and its Levi graph

5.3.3 The Pappus Configuration

The symmetric configuration on 9 points which is based on the theorem of Pappus of Alexandria (see Fig. 5.23) was introduced in Sect. 1.1.3. It is one of the three symmetric configurations found in Sect. 5.2 with its Levi graph drawn symmetrically on the torus in Fig. 5.12a. This embedding makes clear that the Levi graph is both vertex transitive and self-dual.

The Pappus theorem itself implies that the configuration which bears his name is vertex transitive since the theorem remains true if the labels of the points on a line are permuted, as well as the labels of the three lines. In particular, the permutations (123)(456)(789) and (147)(258)(369) permute the points and the lines respectively and commute, generating a subgroup of the automorphism group of order 9 and isomorphic to $\mathbb{Z}_3^2$.

To finish generating the automorphism group, the stabilizer of 1 permutes the vertices in the Levi graph at distance 2 and 3 from 1, which form a 12-cycle. Every element of the color-preserving automorphism group of this 12-cycle extends to an automorphism of the Levi graph. So the stabilizer of 1 is the dihedral group of order 12 generated by (28)(47)(63) and (27)(47)(59). This gives $9 \cdot 12 = 108$ automorphisms of the Pappus configuration.

5.3.4 The Desargues Configuration

As mentioned in Chap. 1, if two triangles are in perspective from a point, Desargues' theorem says that the corresponding sides are in perspective from a line. See Fig. 5.24. This theorem involves ten points, the six points of the two triangles, the

5.3 Classical Configurations

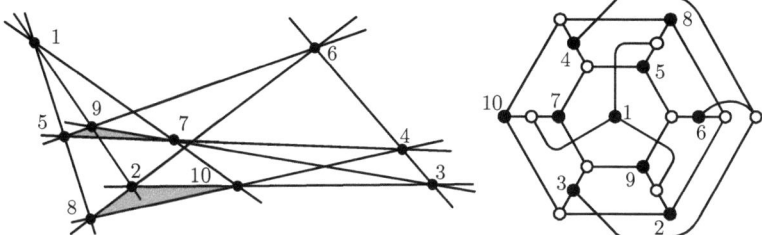

Fig. 5.24 The Desargues configuration and its Levi graph

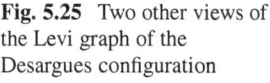

Fig. 5.25 Two other views of the Levi graph of the Desargues configuration

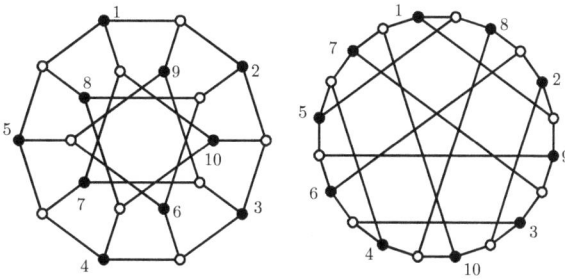

perspective point, and the three points of intersection on the perspective line, and ten lines, the six lines of the two triangles, the three lines to the perspective point, and the perspective line. These points and lines form the *Desargues configuration*. For the configuration derived from the Pappus theorem, a transitive automorphism group was expected from the nature of the theorem. For the Desargues configuration, while the nature of the theorem implies a self-dual configuration, which is obvious from the drawing of the Levi graph in Fig. 5.25, the fact that the graph is vertex transitive is not apparent until it is noticed that the Levi graph is in fact isomorphic to the generalized Petersen graph GP(10, 3); see Fig. 5.25. Like the Fano configuration and the Möbius–Kantor configuration, the Levi graph of the Desargues configuration is also Hamiltonian; however, $\mathbb{Z}_{20}$ does not act upon it. Fortunately, we already have a condition for deciding whether a generalized Petersen graph has a transitive automorphism group.

Since $3^2 \equiv -1 \pmod{10}$, we have seen that the two vertex permutations (12345) (67890) and (1930)(27)(5648) generate a group of automorphisms acting transitively on the generalized Petersen graph, the first preserving the outer ring and the second exchanging the inner and outer ring. The Levi graph as shown in Fig. 5.24 may be obtained from a hexagonal prism with the spoke edges subdivided and alternately attached to two additional vertices, one of which is the vertex 1 whose stabilizer is to be computed. Since the spoke edges are in one orbit of the hexagonal prism, the stabilizer of 1 is isomorphic to the automorphism group of the triangular prism, generated by the two involutions, (25)(34)(70)(98) and (27)(46)(58)(90), so the automorphism group of the Desargues configuration is of order $6 \cdot 20 = 120$, generated by (12345)(67890), (1930)(27)(5648) (25)(34)(70)(98), and (27)(46)(58)(90).

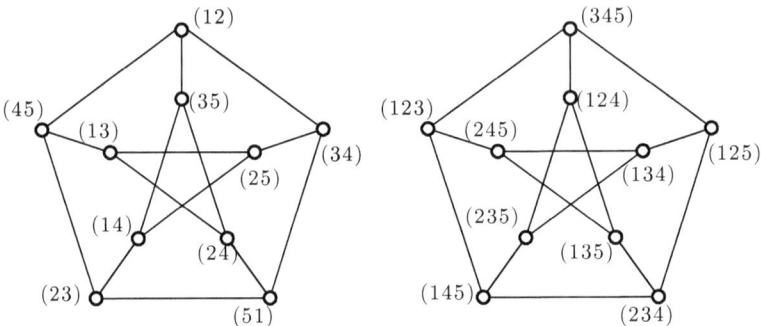

Fig. 5.26 Graphs on the 2-subsets and 3-subsets of a 5-set

Fig. 5.27 The Levi graph of the Desargues configuration as Kronecker double cover of the Petersen graph using the alternate labelings

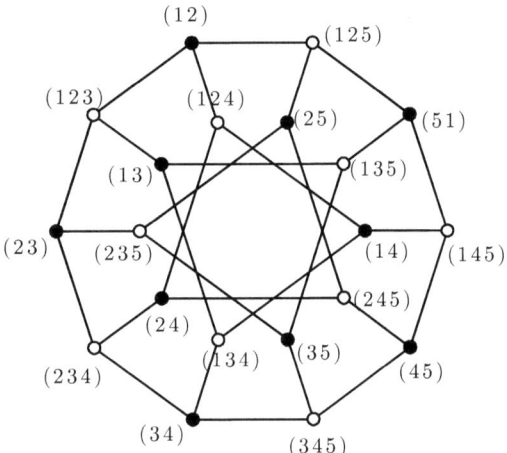

There is a second interpretation of the automorphism group of the Desargues configuration. We represent the vertices of the ordinary Petersen graph GP(5, 2) in two ways. First, the vertices are regarded as the 2-subsets of a 5-set, $\{1, 2, 3, 4, 5\}$, with adjacency between 2-subsets defined as having disjoint intersection; see Fig. 5.26. Using the set complement, there is an equivalent representation with the vertices regarded as 3-subsets, and adjacency between 3-subsets necessarily defined as having union $\{1, 2, 3, 4, 5\}$ or, equivalently, they must intersect in exactly one element.

Now, it has been remarked that the Levi graph of the Desargues configuration, GP(10, 3), is the Kronecker double cover (see Sect. 3.5.7) of the classical Petersen graph GP(5, 2), so let us use this representation with the alternate labels for the vertex covers; see Fig. 5.27. We can now succinctly describe the adjacency relation between the 20 sets making up the vertices of GP(10, 3) by requiring two sets to be adjacent if and only if they have different cardinalities and intersect in exactly 2 elements.

5.3 Classical Configurations 179

It now follows that the permutation group Sym(5) acts as a group of automorphisms on the Desargues graph, and it is easy to see that it is a faithful action. Since we already know there are exactly 120 automorphisms, the automorphism group is Sym(5).

The Desargues graph may also be regarded as the incidence graph of a rank 2 coset incidence structure.

Example 5.15. Consider the symmetric group Sym(5) generated by (12345), (12)(234) with two parabolic subgroups $\langle (12)(345), (34) \rangle$ and $\langle (45)(123), (12) \rangle$. Each is dihedral of order 12, and their intersection, the Borel subgroup, is isomorphic to the Klein four-group $\langle (12), (45) \rangle$. We leave it to the reader to verify that the incidence graph is indeed the Desargues graph; see Exercise 3.75.

5.3.5 The Cremona–Richmond Configuration

We can generalize the construction of the combinatorial Desargues configuration as subsets of a 5-set to subsets of a 6-set. There are $\binom{6}{2} = 15$ subsets of a 6-set $\{1, 2, 3, 4, 5, 6\}$ containing exactly 2 elements. There are also exactly $5 \cdot 3 = 15$ ways in which to partition 6 elements into 3 disjoint 2-subsets: The element 1 must be in some 2-subset, and there are five ways to choose its partner, then the least element not yet chosen is in some 2-subset, and there are 3 ways to choose its partner, and the last 2-subset is forced.

Now, we can define a bipartite graph on these 30 elements by setting a 2-subset adjacent to a 2-partition if the 2-subset is among the three 2-subsets of the partition. So, for example, (12) is adjacent to the three partitions (12)(34)(56), (12)(35)(46), and (12)(36)(45), while (12)(34)(56) is adjacent to the three 2-subsets (12), (34), and (56). There can be no 4-cycle since if two distinct 2-subsets are adjacent to the same partition, then they must be disjoint, say (ab) and (cd), and the 2-partition must be $(ab)(cd)(ef)$.

Moreover, the other 2-partitions adjacent to (ab) and (ef) are $(ab)(ce)(df)$ and $(ab)(cf)(de)$ for (ab) and $(ac)(bd)(ef)$ and $(ad)(bc)(ef)$ for (ef), and none of these have a 2-subset in common, so this graph has no 6-cycles either and must have girth at least 8. A quick inspection of Fig. 5.29 tells us that the girth is exactly 8, so as the smallest 3-regular girth 8 graph is an 8-cage, called Tutte 8-cage.

Since the graph has girth at least 6, it defines a combinatorial configuration known as the Cremona–Richmond configuration. The points are the 2-subsets, and the lines are the 2-partitions.

The *Cremona–Richmond configuration* with 15 points and lines is the smallest (n_3) configuration with no triangles [103]; see Fig. 5.28.

Similar to the case of Sym(5) acting on the Desargues graph, Sym(6) acts as a group of automorphisms on the Cremona–Richmond graph. Clearly, the action is transitive and also faithful since if a permutation fixes the vertices (ij) and (ik), then it must fix i.

Fig. 5.28 The Levi graph of the Cremona–Richmond configuration

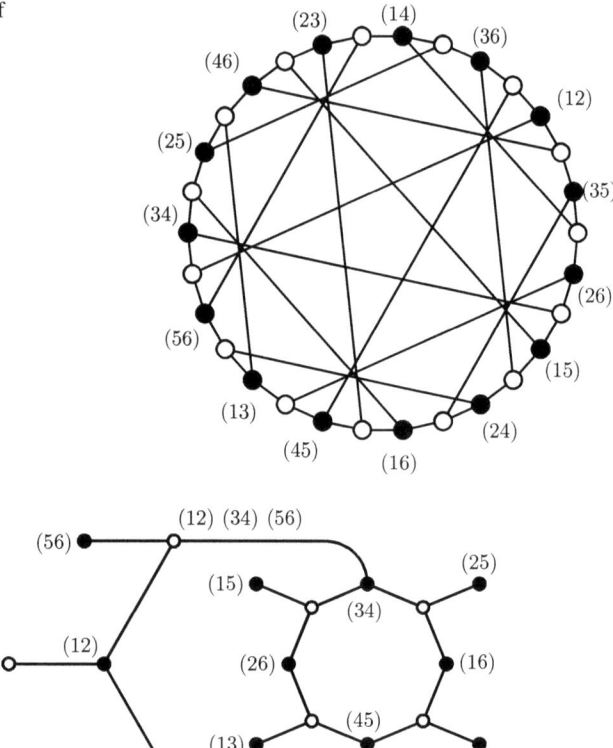

Fig. 5.29 A subgraph of the Cremona–Richmond graph focused on the 10-cycle {(12), (34), (51), (23), (45)}

We will show that every color-preserving automorphism of the Levi graph is induced by the action of the symmetric group. Let ϕ be any automorphism of the Cremona–Richmond graph. We will show that we can compose ϕ by a sequence of Sym(6)-induced automorphisms to yield the identity. Since Sym(6) acts transitively on the black vertices, there is a permutation σ_1 such that $\sigma_1\phi$ maps (12) to itself. The three lines incident to (12) are (12)(34)(56), (12)(35)(46), and (12)(45)(36), distinguished only by which element in $\{3, 4, 5\}$ is paired with 6, so there is a permutation σ_2 of $\{3, 4, 5\}$, which leaves (12) fixed and such that $\sigma_2\sigma_1\phi$ fixes (12) and all white vertices incident to it; see Fig. 5.29. Now the vertex 34 is either fixed or swapped with 56, and also 45 is either fixed or swapped with 36. The vertex (12) and its three white neighbors all are fixed by (34)(56), (35)(46), and (36)(45), and one of them will serve for σ_3 so that $\sigma_3\sigma_2\sigma_1\phi$ fixes (12) and all three lines incident to it, as well as (34) and (45). Now, the two paths of length four connecting 34 and 45 are either fixed or swapped, so either (15) is fixed or swapped with (25), with exactly the same situation with (23) and (13); so taking σ_4 to be either (12) or the identity, we have $\sigma_4\sigma_3\sigma_2\sigma_1\phi$ fix all five vertices $\{(12), (34), (51), (23), (45)\}$. Now, we can show $\sigma_4\sigma_3\sigma_2\sigma_1\phi$ fixes all black vertices and therefore all the vertices. We have that $\sigma_4\sigma_3\sigma_2\sigma_1\phi$ fixes all 2-subsets of $\{1, 2, 3, 4, 5\}$ of the form $(i, i + 1)$ with indices modulo 5. Since $(i6)$ is the third vertex of the white vertex $(i6)(i + 1, i + 2)(i + 3, i + 4)$, except for 6 all indices modulo 5, $(i6)$ must be fixed. Lastly, for any $i \in \{1, 2, 3, 4, 5\}$, the white vertex $(i, 1 + 2)(i + 1, 6)(i + 3, i + 4)$ is fixed since two of its black neighbors $(i + 1, 6)$ and $(i + 3, i + 4)$ are fixed, and hence, its third neighbor $(i, i + 2)$ is fixed. So all black vertices are fixed.

Therefore, the automorphism group is Sym(6), and it is not hard to see that the self-dualities of the configuration correspond to the outer automorphisms of Sym(6) (see Sect.3.6) so the group of color-preserving and reversing automorphisms of the Levi graph is Aut (Sym(6)).

5.3.6 The Reye Configuration

For a general construction, suppose we have an n-set and we consider its 2-subsets. Take two copies of each subset, calling one positive and the other negative. So we have $2\binom{n}{2}$ elements which will be the points, and for the blocks, we will consider those 3-subsets of pairs such that their union has cardinality 3 and their triple intersection is empty. Incidence will be containment. Note that the positive and negative version of the same couple cannot be completed to a block. The couples of each block must be supported by three elements from the n-set, and they may occur in any sign combination, so altogether $8\binom{n}{3}$ blocks. Each point is incident to $4\binom{n-1}{2}$ blocks, and each block is incident to exactly 3 points.

The incidence graph of this incidence structure has 4-cycles such as

$$\left((ij)^+, \{(ij)^+, (ik)^+, (jk)^+\}, (ik)^+, \{(ij)^+, (ik)^+, (jk)^-\} \right)$$

To avoid 4-cycles, we will discard the negative blocks, that is, those which have 1 or 3 negative couples.

Now, it is an easy exercise to show that the Levi graph of this incidence structure has girth at least 6. So for any $n \geq 3$, we have a combinatorial configuration.

For $n = 3$, the configuration has 6 points and 4 lines and is isomorphic to the complete quadrangle.

For $n = 4$, there are $2\binom{4}{2} = 12$ points and $4\binom{4}{3} = 16$ lines, and the resulting configuration is called the *Reye configuration* (see Fig. 5.30) after the German mathematician Theodor Reye. The automorphism group is clearly transitive on the black vertices.

There are three couples which do not belong to blocks incident with $(10)^+$, namely, $(10)^-$, $(23)^+$, and $(23)^-$. So the stabilizer of $(01)^+$ must permute these three elements. The transformation sending every negative couple $(ij)^-$ to its

Fig. 5.30 The Levi graph of the Reye configuration

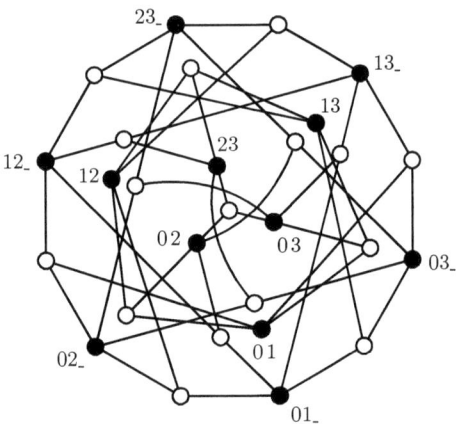

complementary negative couple $(kl)^-$ and fixing all the positive couples is an automorphism which fixes $(01)^+$ and $(23)^+$ and interchanges $(01)^-$ and $(23)^-$. We also have an automorphism which swaps every couple containing a fixed vertex i with its negative, leaving all the others fixed. So in the stabilizer of $(01)^+$, the orbit of $(01)^-$ has cardinality 3, and in the stabilizer of both $(01)^+$ and $(01)^-$, the orbit of $(23)^+$ has cardinality 2. Computing the 8 elements of the pointwise stabilizer of $\{(01)^+, (01)^-, (23)^+, (23)^-\}$ is left as an exercise. This gives a total of $24^2 = 576$ automorphisms of the Levi graph.

5.3.7 Möbius (8_4) Incidence Structures

In Sect. 2.4, we encountered the hypercube Q_n as the Cartesian product of n copies of K_2:
$$Q_n = K_2 \square K_2 \square \cdots \square K_2.$$

Since the Cartesian product of two bipartite graphs is again bipartite, the hypercubes are all examples of incidence graphs; however, none of the nontrivial hypercube graphs are Levi graphs of combinatorial configurations since they have girth 4. These graphs nevertheless encode an interesting class of classical combinatorial incidence structures of type (2_n^{n-1}).

Computing the automorphisms of an incidence structure based on the hypercube graph is straightforward. Realizing the hypercube graph with vertices placed at the n-tuples $\{(\pm 1, \pm 1, \ldots, \pm 1)\}$ in $\mathbb{R}^n$, the automorphism group of the hypercube is easily represented by the set of $n \times n$ matrices whose rows and columns each have exactly one nonzero entry of absolute value 1. There are $n!2^n$ such matrices, and there are accordingly $n!2^{n-1}$ color-preserving automorphisms of the incidence graph.

The ordinary cube graph Q_3, with a 2-coloring of the vertices, corresponds to a nonlineal incidence structure of type (4_3), in other words to a complete graph on

5.3 Classical Configurations

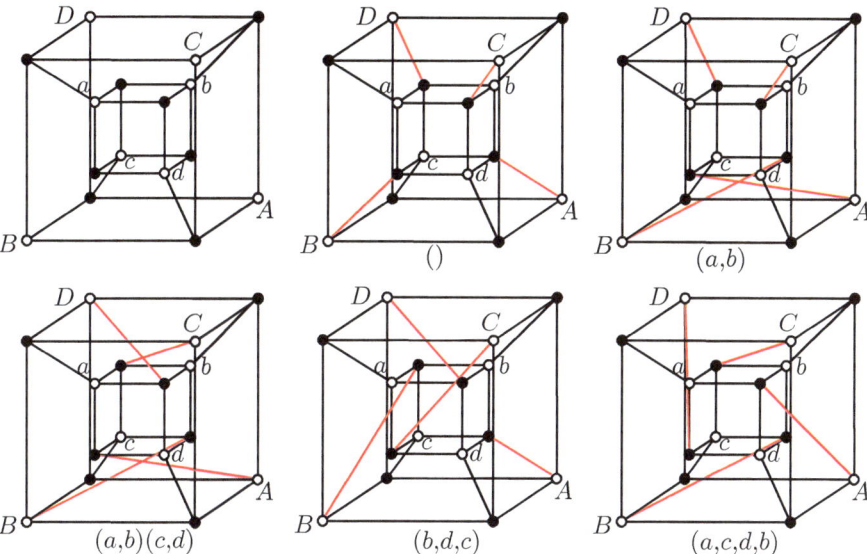

Fig. 5.31 The five four-dimensional Möbius hypercubes and their permutations. The *black* vertex labels are omitted

four vertices. This incidence structure has 24 automorphisms, comprising an index 2 subgroup of the full symmetry group of the cube. Note that this automorphism group is not the rotational group of the cube, which is also an index 2 subgroup of the full symmetry group of the cube, since some reflections fix the colors.

The hypercube graph Q_4 may be regarded as $K_2 \square Q_3$ and obtained from two cubes, an *inside* cube and *outside* cube, with corresponding vertices of the two cubes joined by a perfect matching of edges, such that the matched vertices on the inner and outer cubes are reversed in color. More generally, any two hypercubes Q_{n-1} with a perfect matching between them which respects the bipartitions give a bipartite graph which is the incidence graph of a (2_n^{n-1}) incidence structure, called a *Möbius* (2_n^{n-1}) *incidence structure of dimension n.*

It is not hard to see that there is, up to isomorphism, only one Möbius (4_3) incidence structure. Let us consider the Möbius (8_4) incidence structures.

Suppose we have a matching between two bicolored copies of Q_3 which reverses colors. Let us label the white vertices of the outer cube $\{A, B, C, D\}$. Observe that the antipodal vertices of the white vertices are black. Now label each white vertex of the inner cube with the lowercase letter corresponding to the white antipodal vertex of the black vertex on the outer cube to which it is attached; see the diagram on the top left of Fig. 5.31. With this labeling, to determine the matching, we need only specify how the uppercase outer white vertices are matched to the black inner vertices, the red edges in the figure. The black vertices of each cube will be labeled with the triple of white vertices to which they are incident in their cube.

To each such incidence graph, we can associate a permutation of the white vertices of the inner cube as follows: A white vertex x is adjacent to a black vertex on the outer cube PQR, whose antipodal white vertex S is attached to a black vertex uvw on the inner cube, having antipodal white vertex y; the permutation maps x to y. The hypercube corresponds to the identity permutation. We call such graphs *Möbius hypercubes*. They are the incidence graphs of the Möbius (8_4) incidence structures. Permuting the labels of a Möbius cube will alter the labels of its permutation, but not its cycle structure, in other words, not its conjugacy class. The five conjugacy classes of elements in Sym_4 give us five Möbius cubes in dimension 4.

5.4 Autopolar Combinatorial Configurations

Recall that a configuration is *autopolar* if it has a combinatorial self-duality of order two. Although, merely as a result of the rule of small numbers, most small examples of self-dual configurations are autopolar, it is not hard to construct self-dual configurations which have no involutory self-duality; see Sect. 4.5.3. In [7], it was shown that there is a unique smallest combinatorial (13_3) self-dual configuration that is not autopolar. In [63], self-dual but not autopolar (v_4) configurations are considered.

All configurations whose Levi graphs are generalized Petersen graphs are autopolar since bipartiteness requires that the outside cycle must be even, and the automorphism σ is an involutory self-duality; see Sect. 3.4.6. In particular, $G(8, 3)$, the Möbius–Kantor graph, defines an autopolar configuration.

Another autopolar configuration is the Fano configuration or Fano plane; see Fig. 5.32. In Sect. 1.1.1, we constructed the Fano configuration as the set of nonzero

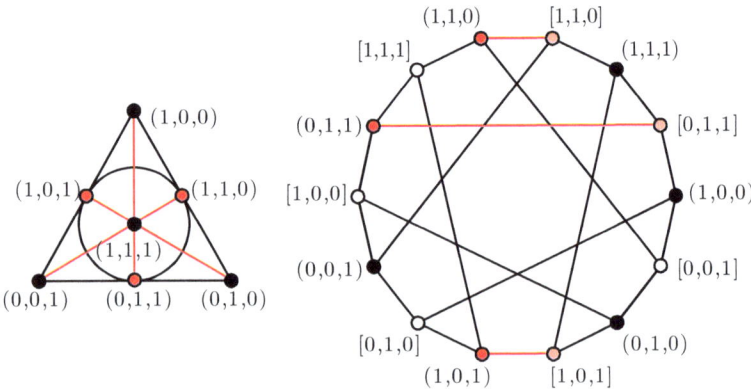

Fig. 5.32 The Levi graph of the Fano configuration, also known as the Heawood graph or the unique 6-cage. The left–right mirror symmetry gives an autopolarity

5.4 Autopolar Combinatorial Configurations

three-dimensional vectors over $\mathbb{Z}_2$, with seven column vectors representing points and seven row vectors representing the lines, and the incidence relation being that $\mathbf{l} \cdot \mathbf{r} = 0$, where the dot product is naturally taken over $\mathbb{Z}_2$. There are many autopolarities, among them is the involution which interchanges row and column vectors, realized in Fig. 5.32 by a mirror reflection.

Let G be an arbitrary bipartite graph with adjacency matrix B. Then by ordering the vertices of G so that each vertex in the first bipartition precedes each vertex of the second, the adjacency matrix B has the following block form:

$$B = \begin{bmatrix} 0 & A \\ A^T & 0 \end{bmatrix}$$

The matrix A is called *biadjacency matrix* of G. It is the incidence matrix of the configuration. If both bipartite sets have the same cardinality, which must be the case if G is the Levi graph of a self-dual configuration, then of course A is a square matrix. G admits a self-duality if and only if there are permutation matrices Π and Π' such that $\Pi^T A \Pi' = A^T$.

Given an autopolar configuration with involutory self-duality σ, we may choose any ordering $v_1, \ldots, v_k$ on the points and take ordering $\sigma(v_1), \ldots, \sigma(v_k)$ on the lines. Then the following result is clear.

Theorem 5.16. *A configuration is autopolar if and only if it admits a symmetric incidence matrix A.*

Here we follow Haemers et al. [49] and extend their definitions that were introduced for designs to general configurations or even more generally to incidence structures or equivalently to bipartite graphs.

A point (line) p is called *absolute* with respect to autopolarity π if it is incident with its image $\pi(p)$. Given a symmetric incidence matrix A, the corresponding polarity is the map that takes the ith point to the ith line, and absolute points correspond to the 1s on the diagonal. Without loss of generality, we may assume that the points and lines are enumerated in such a way that the matrix A has a block form:

$$A = \begin{bmatrix} A_1 & C \\ C^T & A_2 \end{bmatrix}$$

where A_1 and A_2 are symmetric matrices, A_1 has ones on the diagonal, and A_2 has zeros on the diagonal.

Thus, $A_1 - I$ and A_2 are adjacency matrices of graphs Γ_1 and Γ_2, say. Γ_1 and Γ_2 are called the graphs *induced* by the autopolarity. See Fig. 5.33.

The matrix C is called *the autopolar structure* of the configuration with respect to the given autopolarity. Of course, C represents an incidence structure but need not be a configuration. The Levi graph Γ_3 of the incidence structure defined by C is called the *third autopolar graph*. See Fig. 5.34. So the autopolar Fano configuration has been decomposed into $\Gamma_1 = 3K_1$, $\Gamma_2 = K_1 \cup C_3$, and $\Gamma_3 = K_{1,3}$. The Heawood graph admits 28 autopolarities, each of them includes the same autopolar graphs.

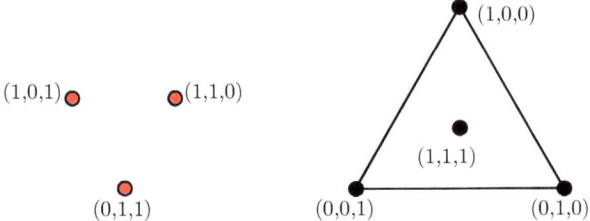

Fig. 5.33 The graph on the absolute points of the Fano configuration and for the nonabsolute points

Fig. 5.34 The third autopolar graph of the Fano configuration with respect to the autopolarity defined by the transpose

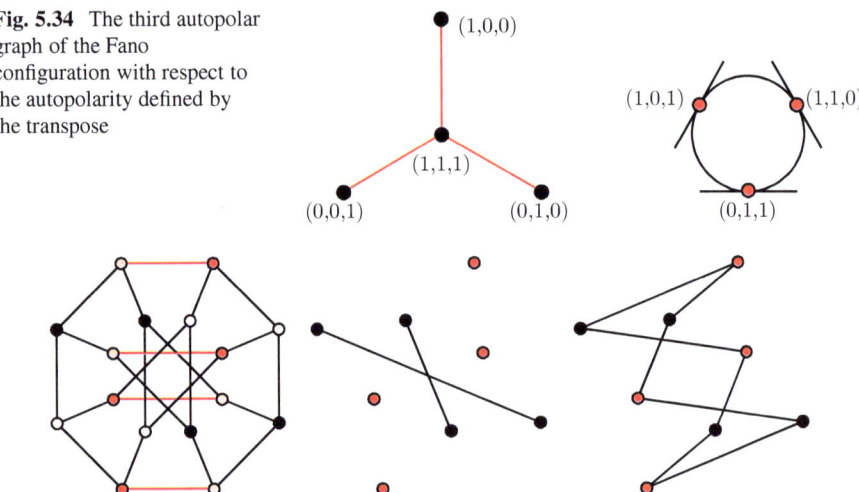

Fig. 5.35 The autopolar decomposition of the Möbius–Kantor graph induced by a mirror reflection in the diagram

The decomposition of an autopolar configuration into the absolute graph, the nonabsolute graph, and the absolute/nonabsolute Levi graph is called the *autopolar decomposition*. The autopolar decomposition is said to be *regular* if both graphs are regular and is called *strongly regular* if both graphs are strongly regular.

There are two special cases. If the absolute autopolar graph Γ_1 is empty, then there are no absolute points, and the Levi graph of the configuration is actually a cover over Γ_2; in fact, the Levi graph of the configuration is the Kronecker cover over Γ_2. Similarly, if the nonabsolute autopolar graph Γ_2 is empty, then the Levi graph is again a cover and corresponds to the lexicographic product of Γ_1 and K_2.

The generalized Petersen graph GP(8, 3), the Möbius–Kantor graph, is the Levi graph of an (8_3) combinatorial configuration. The Möbius–Kantor graph admits 18 autopolarities. Six of them (see Fig. 5.35) have $\Gamma_1 = 4K_1, \Gamma_2 = 2K_2$, and $\Gamma_3 = C_8$ and are therefore regular, while 12 of them (see Fig. 5.36) have $\Gamma_1 = 2K_1; \Gamma_2$ is composed of two triangles joined by an edge and $\Gamma_3 = 2(P_3 \cup K_1)$.

5.4 Autopolar Combinatorial Configurations

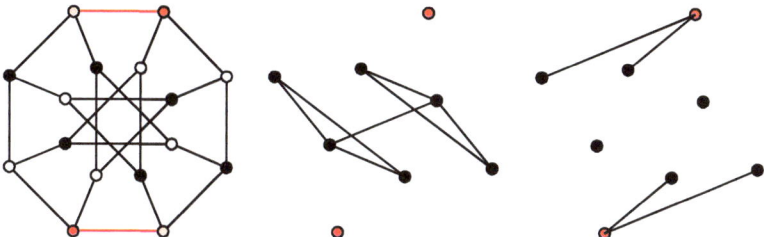

Fig. 5.36 A second autopolar decomposition of the Möbius–Kantor graph induced by an automorphism with two absolute vertices

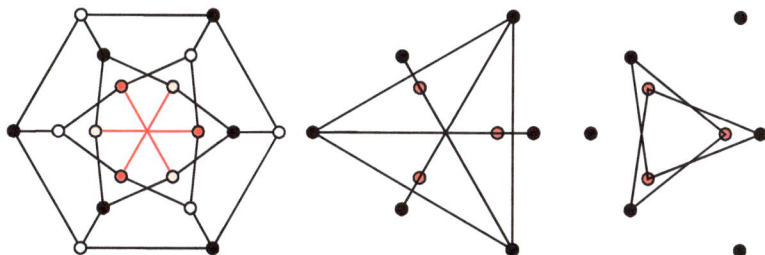

Fig. 5.37 The autopolar decomposition the Pappus graph with respect to a rotation of 180°

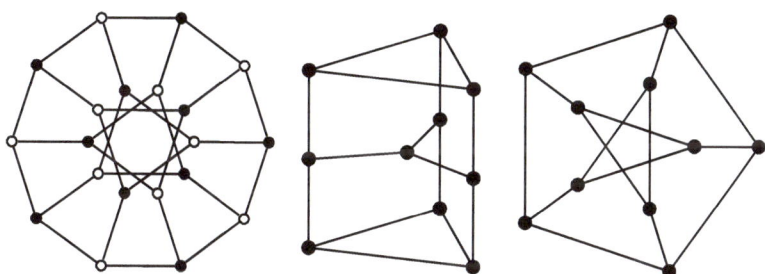

Fig. 5.38 The Desargues graph $G(10,3)$ expressed as a double cover of two nonisomorphic graphs

The Levi graph of the (9_3) Pappus configuration admits 18 autopolarities. All of them have $\Gamma_1 = 3K_1$; Γ_2 is composed of a triangle with three pending edges and $\Gamma_3 = 3K_1 \cup C_6$ (Fig. 5.37).

The other exceptional generalized Petersen graph GP(10, 3), the Desargues graph, admits 26 autopolarities. They fall in three classes. There are 15 autopolarities which induce a decomposition with $\Gamma_1 = 4K_1$; Γ_2 is a union of two paths P_3, joined by an edge and $\Gamma_3 = 2K_1 \cup C_8$. There are ten autopolarities which induce a decomposition with $\Gamma_1 = \emptyset$, Γ_2 a prism Π_3 with an additional vertex joined to the midpoints of all three vertical sides (see Fig. 5.38) and $\Gamma_3 = \emptyset$. Lastly, there is a single autopolarity which induces a decomposition with $\Gamma_1 = \emptyset$, $\Gamma_2 = G(5,2)$, and $\Gamma_3 = \emptyset$. Since for each of the second two autopolar decompositions the first autopolar graph Γ_1 is empty, as we noticed above, the Desargues graph is the canonical double cover of the second autopolar graph, Γ_2, so expressing $G(10,3)$ as the Kronecker cover of two nonisomorphic graphs, one of them being the Petersen graph.

We remark that the smallest pair of nonisomorphic generalized Petersen graphs with the same Kronecker cover consists of $G(12, 2)$ and $G(12, 4)$.

5.5 Cyclic Haar Graphs and Cyclic Configurations

Cyclic Haar graphs constitute an important family of bipartite regular graphs. Originally, the motivation for their introduction came from operator theory. Haar graphs were formally introduced in [52, 80] where the basic theory was presented and, among other things, their close connection with cyclic configurations was established.

Cyclic Haar graphs may be defined in a number of equivalent ways.

5.5.1 Cyclic Haar Graphs as Cyclic Covering Graphs Over a Dipole

We define a cyclic Haar graph to be a cyclic covering graph over a dipole. More precisely, let Θ_r be a dipole with r parallel edges. Let $S \subseteq \mathbb{Z}_n$ be a set of voltages $|S| = r$ which are assigned to the (parallel) arcs of Θ_r. This voltage assignment defines a regular n-fold covering graph with $\mathbb{Z}_n$ acting regularly on each fiber. We denote by $H(S, n)$ the corresponding cyclic Haar graph.

Example 5.17. Letting $S = \{0, 1, 3\}$ and $n = 8$ gives rise to the Möbius–Kantor graph $H(\{0, 1, 3\}, 8)$, alias generalized Petersen graph GP(8, 3).

Using this definition, it is clear that each cyclic Haar graph is bipartite since any covering graph over a bipartite graph is bipartite. In this case, the covering is regular. It follows that all black vertices are in the same orbit, and all white vertices are in the same orbit. However, cyclic Haar graphs are even more symmetric. They are vertex transitive. The easiest proof follows from the fact that each cyclic Haar graph is a Cayley graph.

5.5.2 Cyclic Haar Graphs as Certain Cayley Graphs for the Dihedral Group

We have seen that the dihedral group D_n of order $2n$ acts as the group of symmetries of a regular n-gon in the Euclidean plane and that those symmetries may be divided into two classes: There are n *rotations* (including the identity) and there are n *reflections*. If D_n is given by the group presentation

$$D_n = \langle r, s | r^n = s^2 = 1, rs = sr^{-1}\rangle,$$

then the powers of r are rotations of the n-gon. All other elements are reflections. Each reflection can be written in a unique way as sr^i for $i \in \mathbb{Z}_n$.

We note that a graph isomorphic to a Cayley graph of a dihedral group generated by reflections is a cyclic Haar graph. This statement may be used as alternate definition of a cyclic Haar graph.

Example 5.18. The Möbius–Kantor graph is isomorphic to the Cayley graph $\text{Cay}(D_n, \{s, sr, sr^3\})$.

Theorem 5.19. *Any cyclic Haar graph $H(S, n)$ is the Cayley graph of a dihedral group, and conversely, each Cayley graph $\text{Cay}(D_n, \{sr^i\} | i \in S)$ is isomorphic to the cyclic Haar graph $H(S, n)$.*

The proof of this theorem can be found in [52]. The following result follows readily.

Corollary 5.20. *Every cyclic Haar graph is vertex transitive.*

Although the previous result shows vertex transitivity of cyclic Haar graphs, we challenge the reader to find an explicit automorphism that interchanges black and white vertices of $H(S, n)$.

5.5.3 Associating a Cyclic Haar Graph to a Number

Let N be a natural number expressed in binary:

$$N = b_{n-1}2^{n-1} + b_{n-2}2^{n-2} + \cdots + b_1 2^1 + b_0 2^0.$$

Define a bipartite graph $\text{Ha}(N)$ on the vertex set

$$\{x_i, y_i \mid 0 \leq i \leq n-1\}$$

by setting x_i and y_{i+k} adjacent if and only if $b_k = 1$, with indices modulo n. This allows us to associate to natural number N a cyclic Haar graph we denote by $\text{Ha}(N)$.

Alternatively, we can encode the sequence b_i of binary digits in the coefficients of a polynomial $p_N(x) = b_{n-1}x^{n-1} + b_{n-2}x^{n-2} + \cdots + b_1x^1 + b_0x^0$ of degree N, so $p_N(2) = N$. On the other hand, given N, we may define the subset $B(N)$ of $\mathbb{Z}_n$ as follows:

$$B(N) = \{i \in \mathbb{Z}_n | b_{n-i-1} = 1 \text{ in the binary expansion of } N\}.$$

This gives the following:

Theorem 5.21. *For each cyclic Haar graph $H(S,n)$, there exists a natural number N such that $H(S,n)$ is isomorphic to $\mathrm{Ha}(N)$ where $S = B(N)$.*

All three viewpoints are equivalent and define the same class of graphs.

Example 5.22. The Möbius–Kantor graph is isomorphic to $\mathrm{Ha}(133)$. The natural number 133 is not unique with this property. There are other numbers N for which the graph $\mathrm{Ha}(N)$ is isomorphic to the Möbius–Kantor graph.

The example above raises the interesting question of which pairs of cyclic Haar graphs are isomorphic. We address this question in the next subsection.

5.5.4 Isomorphisms of Cyclic Haar Graphs

If we are given two cyclic Haar graphs $H(S,n)$ and $H(S',n')$, then they can be isomorphic only if $n = n'$. It is straightforward to show that for any nonzero $a \in \mathbb{Z}_n^*$ and any $b \in \mathbb{Z}_n$, the graphs $H(S,n)$ and $H(aS+b,n)$ are isomorphic. This leads us to define two equivalence relations among the subsets of $\mathbb{Z}_n$. We say that $S \sim T$ if $T = aS + b$ for some $a, b \in \mathbb{Z}_n$ with $a \neq 0$, and we say that $S \cong T$ if the Haar graphs $H(S,n)$ are isomorphic to $H(T,n)$. Clearly, the first equivalence implies the second; however, the converse is not always true. For a counterexample, consider $H(\{0,4,5,7\},8) \cong H(\{0,1,4,7\},8)$.

For cyclic Haar graphs of valence 2, it is trivially true that the equivalence relations $\sim$ and $\cong$ are the same. The same may be shown for cyclic Haar graphs of valence 3. Although we have a counterexample of for cyclic Haar graphs of degree 4, we can show that for those cyclic Haar graphs corresponding to cyclic configurations $\sim$ and $\cong$ are again the same; see [56].

Theorem 5.23 ([56]). *Let $H(A,n)$ and $H(B,n)$ be two connected cyclic Haar graphs of girth 6 which have valency at most 4. Then $H(A,n) \cong H(B,n)$ if and only if $B = aA + b$ for some $a \in \mathbb{Z}_n^*$ and $b \in \mathbb{Z}_n$.*

5.5.5 Cyclic Configurations and Cyclic Haar Graphs

A cyclic configuration (v_r) is a combinatorial configuration that admits an automorphism of order v that cyclically permutes the points and lines, respectively. There is a simple test to check whether a configuration is cyclic. Without loss of generality, we may assume that the points of the configuration are labeled with elements of $\mathbb{Z}_v$. A configuration is cyclic if and only if it admits a configuration table with the property that each column is obtained from the previous column by adding 1 mod v.

Example 5.24. The Möbius–Kantor configuration is cyclic. (See the configuration table.)

The Levi graph of a cyclic configuration is a cyclic Haar graph.

Theorem 5.25. *A graph is the Levi graph of a cyclic configuration if and only if it is a cyclic Haar graph of girth 6.*

Proof. Consider the Levi graph of a cyclic configuration. Take the quotient graph with respect to the cyclic automorphism. Since the graph is bipartite and the action is transitive on each color, the quotient must be a dipole. Hence, the Levi graph is a cyclic Haar graph. The converse follows directly from the definition of cyclic Haar graph.

Theorem 5.26. *A cyclic Haar graph of valence ≥ 3 has girth equal to either 4 or 6.*

Proof. If the valence is at least 3, then there are three parallel edges in the voltage graph carrying voltages, say, a, b, c. The closed walk $a - b + c - a + b - c$ has net voltage zero; hence, it lifts to a 6-cycle.

Table 5.2 presents an enumeration of cyclic Haar graphs on n vertices. The last column enumerates connected cyclic configurations on n points.

5.6 Exercises

Exercise 5.1. Draw the Levi graph of the incidence structure defined by the complete bipartite graph $K_{3,3}$.

Exercise 5.2. Draw the incidence graph of the incidence structure defined by the powerset $\mathcal{P}(\{a, b, c\})$.

Exercise 5.3. Draw incidence graphs of the incidence structure defined by all arrangements of lines with four lines.

Exercise 5.4. Show that any near pencil is an example of an autopolar linealbalanced incidence structure.

Exercise 5.5. Label the vertices of the two graphs in Fig. 5.5 by the cosets of appropriate subgroups of the quaternions.

Table 5.2 Cyclic Haar graphs on n vertices

n	All	Connected	Girth 6	Both	n	All	Connected	Girth 6	Both
1	0	0	0	0	26	10	7	7	5
2	0	0	0	0	27	8	6	6	5
3	0	0	0	0	28	16	9	10	7
4	1	1	0	0	29	5	5	4	4
5	1	1	0	0	30	24	13	16	11
6	2	2	0	0	31	6	6	5	5
7	2	2	1	1	32	18	9	11	7
8	4	3	1	1	33	11	9	9	8
9	2	2	1	1	34	12	9	9	7
10	4	3	1	1	35	12	9	9	8
11	2	2	1	1	36	29	13	19	11
12	8	5	3	3	37	7	7	6	6
13	3	3	2	2	38	14	10	11	8
14	6	4	3	2	39	14	11	12	10
15	6	5	4	4	40	30	15	20	13
16	9	5	4	3	41	7	7	6	6
17	3	3	2	2	42	32	17	24	15
18	10	6	5	4	43	8	8	7	7
19	4	4	3	3	44	22	13	16	11
20	12	7	6	5	45	21	13	17	12
21	9	7	7	6	46	16	12	13	10
22	8	6	5	4	47	8	8	7	7
23	4	4	3	3	48	46	19	33	17
24	22	11	13	9	49	12	10	10	9
25	6	5	4	4	50	24	15	18	13

Exercise 5.6. Let $\mathbf{A}$ be the matrix obtained from $\mathbf{N} \cdot \mathbf{N}^T$ by replacing the diagonal entries by zeros. Then $\mathbf{A}$ is the adjacency matrix of the Menger graph of the structure $\mathcal{C}$.

Exercise 5.7. Construct an incidence structure that is (a) simple but not cosimple, (b) cosimple but not simple, (c) simple and cosimple, and (d) neither simple nor cosimple.

Exercise 5.8. If graph X is viewed as an incidence structure, what is its incidence graph?

Exercise 5.9. Consider the Petersen graph PG$(5, 2)$ as an incidence structure.

1. What is its incidence graph?
2. What is its Menger graph?

Exercise 5.10. Recall that the dual Menger graph is the Menger graph of the dual incidence structure. Show that the Menger graph and the dual Menger graph do not together determine the incidence structure by finding two nonisomorphic incidence structures sharing both Menger and dual Menger graphs.

5.6 Exercises

Exercise 5.11. Show that the ij entry in the incidence matrix $N \cdot N^T$ gives the number of blocks common to points i and j and the $i-i$ entry gives the number of blocks incident with i.

Exercise 5.12. Show that the Levi graph of a *generalized digon* is a complete bipartite graph.

Exercise 5.13. Show that the smallest generalized quadrangle is the Cremona–Richmond configuration whose Levi graph is the Tutte 8-cage.

Exercise 5.14. Prove that the Desargues graph GP(10, 3) is an induced subgraph of the 5-cube Q_5.

Exercise 5.15. Determine the Levi graphs of rank 3 geometries determined by three embeddings of the cube Q_3: the first one in the sphere and two in torus.

Exercise 5.16. Each hypercube Q_d obviously defines a rank d incidence structure. What is its Levi graph?

Exercise 5.17. Each complete graph viewed as a simplex K_d obviously defines a rank d incidence structure. What is its incidence graph?

Exercise 5.18. Let

$$P = \{A(i, j), B(i, j), C(i, j); i, j \in \{0, 1\}\}$$

be the set of 12 points, and let $i, j, k, l, m, n \in \{0, 1\}$.

$$L = \{A(i, j), B(k, l), C(m, n)\} \mid i + k + m \equiv 0 \bmod 2, \, j + l + n \equiv 0 \bmod 2$$

be the set of 16 lines. Let incidence be defined by set membership. Show that this rank 2 incidence structure is isomorphic to the Reye configuration restricted to points and lines.

Exercise 5.19. Match the hexagonal surface tilings of Figs. 5.12 and 5.13 with the three combinatorial (9_3) configurations.

Exercise 5.20. Show that the Levi graphs of the (8_3) configuration and the (9_3) configurations are not planar.

Exercise 5.21. Either find a hexagonal tiling of the projective plane by the Levi graph of a (9_3) configuration or prove that none exists.

Exercise 5.22. Which generators of the automorphism group of the Fano configuration given in Sect. 5.3.1 are redundant? Find a nonredundant set of generators.

Exercise 5.23. Show that the representation of Sym(5) onto the automorphism group of the Levi graph of the Desargues configuration described in Sect. 5.3.4 is faithful.

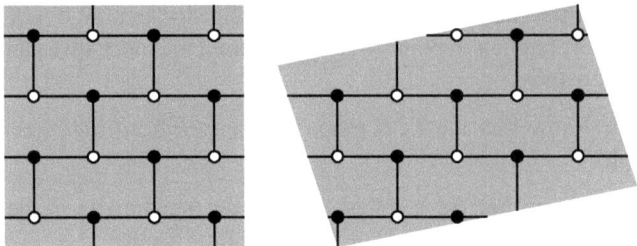

Fig. 5.39 Two cubic torus graphs

Exercise 5.24. Show that the Levi graph of the Reye configuration as defined in Sect. 5.3.6 is of girth 6.

Exercise 5.25. Find all automorphisms of the Reye configuration which fix $(01)^+$, $(01)^-$, $(23)^+$, and $(23)^-$.

Exercise 5.26. Consider the eight black vertices A_0, B_0, C_0, D_0, A_1, B_1, C_1, D_1 together with eight white vertices each of which is a 4-set of the form $\{A_i, B_j, C_k, D_l\}$ with $i + j + k + l$ odd. Let incidence between the black and white vertices be defined by set membership.

Show that this incidence structure is not a combinatorial configuration.

Decide whether or not it is one of the Möbius hypercubes.

Exercise 5.27. Figure 5.39 shows two cubic graphs embedded on the torus. Which of them defines a combinatorial configuration? Is the combinatorial configuration isomorphic to one of the classical configurations?

Exercise 5.28. The graph on the left of Fig. 5.39 also shows a cubic graph embedded on the Klein bottle. Does this graph define a combinatorial configuration? If so, try to provide an isomorphism to one of the classical configurations.

Exercise 5.29. Define a Levi graph by taking the black vertices to be the ordered pairs (i, j) and the white vertices $[i, j]$ with $i, j \in \mathbb{Z}_7$ and $i + j \in \{0, 1, 4\}$. We say that black vertex (i, j) is adjacent to $[k, l]$ if they agree in the first coordinate or the second coordinate, but not both.

Show that this describes a combinatorial configuration.

Show that it is self-dual and that the automorphism group has order 336.

(This is the (21_4) configuration of Felix Klein, [45].)

Exercise 5.30. Compute the automorphism group of the Möbius hypercube (ab).

Exercise 5.31. Compute the automorphism group of the Möbius hypercube $(ab)(cd)$.

Exercise 5.32. Compute the automorphism group of the Möbius hypercube (abc).

Fig. 5.40 A voltage graph over the Pappus graph

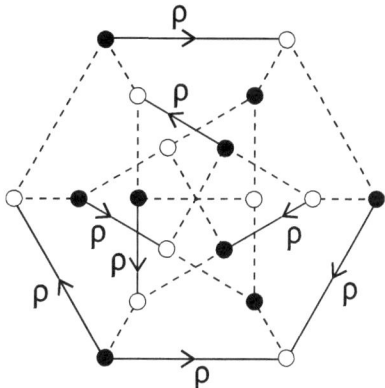

Exercise 5.33. Compute the automorphism group of the Möbius hypercube ($abcd$).

Exercise 5.34. Show that if the first autopolar graph is empty, then the Levi graph of the autopolar configuration is the Kronecker cover over Γ_2.

Exercise 5.35. Show that if the second autopolar graph is empty, then the Levi graph of the autopolar configuration corresponds to the lexicographic product of Γ_1 and K_2.

Exercise 5.36. Describe the automorphism of the Möbius–Kantor graph which induces its second autopolar decomposition presented in Sect. 5.4.

Exercise 5.37. Work out the details of the autopolar decompositions of the Desargues graph.

Exercise 5.38. Consider the $\mathbb{Z}_3$-voltage graph over the Pappus graph depicted in Fig. 5.40. The broken lines carry identity voltages 0, whereas $\rho = 1 \in \mathbb{Z}_3$. What is the covering graph?

Exercise 5.39. Find a suitable quotient for Balaban 10-cage and then express the cage as a $\mathbb{Z}_{10}$-covering graph over the quotient.

Exercise 5.40. Represent the Coxeter graph as a sevenfold covering graph over a cubic graph on 4 vertices with 3 loops.

Exercise 5.41. Show that a circulant is a cyclic Haar graph if and only if it is bipartite.

Exercise 5.42. Show that the Petersen graph is not a cyclic Haar graph.

Exercise 5.43. Show that the Möbius–Kantor graph $G(8, 3)$ [32,64] is a cyclic Haar graph.

Exercise 5.44. Show that the integers 143559 and 137331 are Haar equivalent, i.e. equivalent under $\cong$ but not cyclic equivalent, i.e. equivalent under $\sim$.

Exercise 5.45. Show that the numbers 69, 70, 81, 88, 98, and 104 form a cyclic and Haar equivalence class and show that their Haar graph is isomorphic to the Heawood graph.

Exercise 5.46. Show that there are exactly 12 numbers cyclically equivalent to 33478, and find the smallest number in its equivalence class.

Exercise 5.47. Show that the bipartite complement of a cyclic Haar graph is also a cyclic Haar graph. Determine n such that Ha(n) is a bipartite complement of Ha(75).

Exercise 5.48. Prove that if the number of binary digits of n, $k(n)$, is prime and $n \neq 2^{k(n)-1}$, then the Haar graph Ha(n) is connected.

Exercise 5.49. Prove that for odd n the Haar graph Ha(n) is Hamiltonian.

Exercise 5.50. Show that odd prisms and even Möbius ladders are not cyclic Haar graphs.

Exercise 5.51. Prove that the Haar graph Ha(2^k) is disconnected for all k.

Exercise 5.52. Show that the graphs Ha(34) and Ha(40) are both disconnected and isomorphic to $2C_6$.

Exercise 5.53. Classify those positive integers n such that Ha(n) is disconnected.

Exercise 5.54. Compile a list of all positive integers up to 100 that are disconnected.

Exercise 5.55. A graph is a cyclic Haar graph if and only if it admits an automorphism with precisely two orbits of equal size (and no other orbit) which are independent sets of vertices.

Chapter 6
Geometric Configurations

6.1 Geometric Planes

6.1.1 From Euclid to Descartes and Beyond

The oldest realm of geometric discourse is the Euclidean plane, E^2, which made its first formal appearance in the guise of the five postulates of Euclid:

E1 Given two points A and B, there exists a line segment $\overline{AB}$ containing them.
E2 A line segment can be extended indefinitely in both directions.
E3 Given any two points A and B, there exists a circle C having A as the center and segment $\overline{AB}$ as the radius.
E4 Right angles are congruent.
E5 Given two lines a and b which intersect a transverse line such that the sum of the interior angles on one side is less than two right angles, there exists a point of intersection of a and b on that side of the transversal.

After Descartes, there was a second, highly successful, competing point of view, namely, to regard E^2 as $\mathbb{R}^2$, the space of two-dimensional vectors over the real numbers.

We will consider several other sorts of planes.

Suppose an incidence structure $\mathcal{A}$ satisfies the following three axioms:

A1 Through any pair of distinct points, there passes exactly one line.
A2 Through any point P, there is exactly one line l' which is parallel to a given affine line l.
A3 There are at least three points not on the same line.

Then the incidence structure $\mathcal{A}$ is called an *affine plane*. Of course the Euclidean plane is an affine plane, and moreover, since that fact can be verified using only the vector space structure of $\mathbb{R}^2$, it is not surprising that for any field K, the two-dimensional vectors K^2 over K form an affine plane. In particular, the two-dimensional vector space over the finite field $\mathbb{F}(p^n)$ is an affine plane, denoted by AG$(p^n, 2)$. Here p is a prime, the *characteristic* of the field $\mathbb{F}(p^n)$, and the prime power p^n determines the field up to isomorphism; see Sect. 6.2.1.

For an incidence structure, we have defined two distinct blocks as being parallel if they do not intersect. If we further stipulate that a line is parallel to itself, then, in an affine plane $\mathcal{A}$, the relation $\parallel$ is an equivalence relation. An equivalence class of parallel lines is called a parallel class or a *parallel pencil* of lines. As an incidence structure, the pencil of Example 5.4 consists of a collection of lines with one point in common. The motivation for the consideration of pencils and the discrepancy of the terminology bring us to the next idea.

6.1.2 The Projective Plane

The next notion of the plane arises in the context of representing a three-dimensional scene on an ordinary two-dimensional plane. A painter drawing a picture of a horizontal landscape on a vertical canvas will observe that each parallel pencil of lines in the landscape, traditionally rendered as railroad tracks, will appear to merge at a particular point on the line of the horizon. The mathematical artist will certainly appreciate that neither the vanishing points nor the horizon line itself is actually parts of the landscape being depicted, although they are parts of *observation* of the scene, and they will occur on the plane canvas.

Taking this as motivation, given an arbitrary affine plane $\mathcal{A}$, we will extend it by adding for each parallel pencil of lines a *point at infinity* which will be said to be incident to each line in its associated pencil, and we will add as well to the collection of lines one additional *line at infinity* which will only be incident to the points at infinity. The new incidence structure, consisting of the points and lines of $\mathcal{A}$, together with points at infinity, sometimes called *ideal points*, and the line at infinity is called the *extended plane*.

In the extended plane, any two distinct points still determine a unique line, since a point and an ideal point are joined by the line through the ordinary point in the parallel class of that ideal point and two ideal points lie on the line at infinity. Moreover, any two distinct lines have a point of intersection, since two parallel lines now meet at the point of infinity of their pencil, and as for the line at infinity, it meets each line in the ideal point associated with the pencil of that line. Lastly, it is not hard to show that there exist at least four points such that no three of them are collinear; see Exercise 6.4.

The properties become the requirements for the *projective plane*, which is a structure satisfying the following three axioms:

P1 For any two distinct points P and Q, there exists a unique line l connecting them.

P2 For any two distinct lines l and m, there exists a unique point P in their intersection.

6.1 Geometric Planes

P3 There exist at least four points P, Q, R, S such that no three of them are collinear.

Note that the third axiom is desirable if for no other reason than to exclude the near pencil itself (see Sect. 5.4) which is one dimension too low and is in any case not the extended plane of any affine plane. In a projective plane, for any point P, the set of all lines incident to P is called a *pencil*. The dual statement of the third axiom,

P3* There exist at least four lines p, q, r, s such that no three of them are concurrent.

is a consequence of these axioms, so, since the first and the third axiom are dual to one another, the dual incidence structure is also a projective plane.

The extended Euclidean plane is not the only projective plane. For example, there is also the *rational projective plane*, PG $(2, \mathbb{Q})$, consisting of the points of the Euclidean plane with rational coordinates and the lines $Ax + By + C = 0$ with A, B, and C rational and the ideal points corresponding to rational slopes and the line at infinity. In general, given any field, the extended affine plane yields a projective plane PG $(2, \mathbb{F})$.

A more exotic example of a projective plane is the Moulton plane, [71], an incidence structure which in its typical description has points of two classes, ordinary points and ideal points, and lines of three classes, ordinary lines, bent lines, and a line at infinity. The ordinary points are points of the extended Euclidean plane. The ordinary lines are the lines of the Euclidean plane which have positive slope, and bent lines are angles with vertices on the y axis which have negative slope m to the left of the y axis and steeper negative slope $2m$ to the right of it. As with the real projective plane, parallelism is an equivalence relation on the ordinary and bent lines, and the equivalence classes are the ideal points. The ideal points are the points incident to the line at infinity. It is true but not obvious that the Moulton plane is not isomorphic to the real projective plane.

The Fano configuration, Sect. 1.1.1, satisfies all three axioms; therefore, not only are its points and lines realizable in an abstract projective plane, it *is* a projective plane, the Fano plane.

6.1.3 Homogeneous Coordinates

Cartesian coordinates greatly enhanced the ability of mathematicians to work with the Euclidean plane. Homogeneous coordinates have played the same role for the extended plane.

Let us first choose notation for the extended Euclidean plane. We already have ordinary Cartesian coordinates (x, y) for the ordinary points. For the ideal points, let us denote the pencil of parallel lines of the form $y = mx + b$ by (m) and lastly the pencil of vertical lines $x = a$ by simply (∞). For the lines, let $y = mx + n$ be denoted by $[-m, 1, -n]$, vertical lines $x = n$ by $[1, 0, -n]$, and the line at infinity by $[\infty]$.

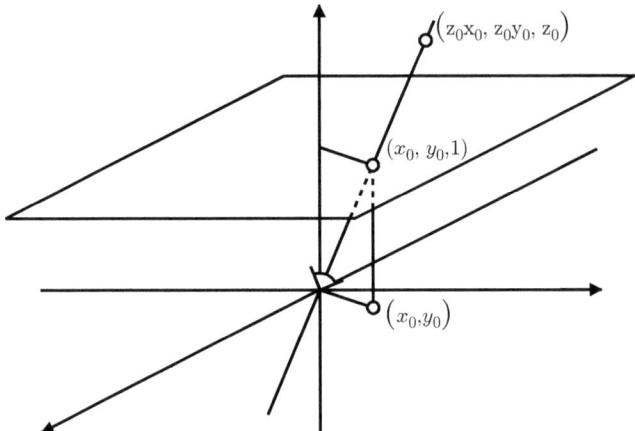

Fig. 6.1 Homogeneous coordinates. Extended plane is obtained from the affine plane $z = 1$

While these "coordinates" are perfectly serviceable as labels, they give us no algebraic facility whatever. For that, let us imagine ourselves in the situation of a mathematical painter, painting a three-dimensional scene with a plane canvas. As a mathematical painter, we will of course place ourselves at the origin, and to avoid negative numbers, we will not look down on the scene; we will look up at the plane $z = 1$ from below, see Fig. 6.1.

We might now coordinatize the extended plane like so:

$$(x, y) \mapsto (x, y, 1), \quad [-m, 1, -n] \mapsto [-m, 1, -n],$$
$$(k) \mapsto (1, k, 0), \quad [1, 0, -n] \mapsto [1, 0, -n],$$
$$(\infty) \mapsto (0, 1, 0) \quad [\infty] \mapsto [0, 0, 1]$$

The ordinary points correspond to points on the scene $z = 1$; the points at infinity correspond points on the *limiting line*, obtained by taking a line segment from the eye to a point on a fixed line in the scene and letting that point travel out to infinity. The limiting line lies on the plane $z = 0$. The choice of coordinates for the lines was made so that the points and the lines satisfy the crucial property that a point in the extended plane with the coordinates (x, y, z) is incident with a line with the coordinates $[a, b, c]$ if and only if $ax + by + cz = 0$. This coordinatizes the points of the extended plane with a subset of $\mathbb{R}^3$. It is not a subspace of $\mathbb{R}^3$; it does not even contain the zero vector. We will call it the $z = 1$ *model*.

Since our eye is at the origin, to pass from the points on the scene to the corresponding points on the canvas, in other words, to *raytrace* the scene, we need to scale the coordinates of each point on the scene, including the ideal points, by a nonzero scalar. The line coordinates may also be scaled without disturbing the crucial property. This suggests an equivalence relation on the points and lines of

6.1 Geometric Planes

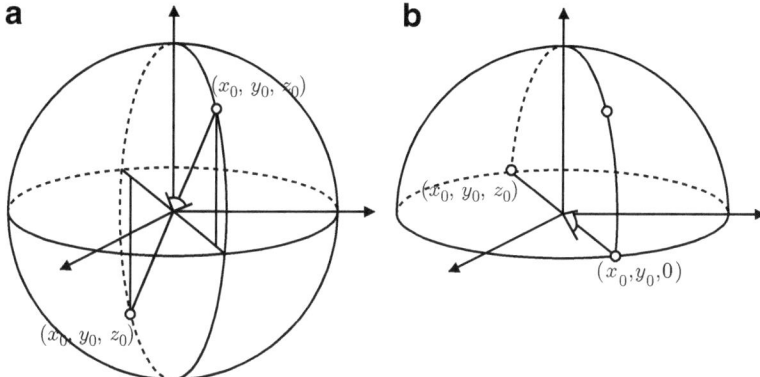

Fig. 6.2 Homogeneous coordinates. Extended plane is obtained from the unit sphere

($\mathbb{R}^3 - \mathbf{0}$), setting two points to be equivalent if they are raytraces of the same point in the $z = 1$ model and setting two lines equivalent if they span a plane through the origin:

$$(x, y, z) \simeq (kx, ky, kz), k \neq 0 \qquad [a, b, c] \simeq [ka, kb, kc], k \neq 0.$$

This specification by equivalence classes gives an alternate model for the extended plane. The points are equivalence classes of nonzero triples (x, y, z), and similarly, the lines are equivalence classes of triples $[a, b, c]$, and point (x, y, z) is incident to line $[a, b, c]$ if and only if $ax + by + cz = 0$. This model is called the *subspace model* since every equivalence class is a subspace of $\mathbb{R}^3$: Point equivalence classes are lines through the origin; line equivalence classes are planes through the origin. Incidence is orthogonality of the corresponding triples.

In the subspace model, each point equivalence class is a line through the origin of $\mathbb{R}^3$, and it intersects the unit sphere given by $x^2 + y^2 + z^2 = 1$ at exactly two antipodal points; see Fig. 6.2a. This gives us our final model, the *unit sphere model*, in which pairs of antipodal points on the unit sphere are the points and great circles are the lines. This makes it clear that topologically the projective plane is a surface (Fig. 6.2b). If we consider only the closed upper hemisphere, then there is one representative for each equivalence class except for the equator, which is the form in which we met the projective plane in Chap. 4.

This development of homogeneous coordinates has been done in the case of the real projective plane; however, it carries through with obvious modifications for the extended affine plane with respect to any field.

6.1.4 Calculations in the Real Projective Plane

Homogeneous coordinates give us a useful tool for calculations in the real projective plane.

Computing the line between two points: Let A be a point in the real projective plane with homogeneous coordinates (x, y, z) and A' be a point with homogeneous coordinates (x', y', z'). Then the homogeneous coordinates $[a, b, c]$ of the line (AA') passing through these two points are given by the vector product $[a, b, c] = (x, y, z) \times (x', y', z')$.

Computing the point of intersection of two lines: Let l be a line in the real projective plane with homogeneous coordinates $[a, b, c]$ and l' be a line with homogeneous coordinates $[a', b', c']$. Then the homogeneous coordinates (x, y, z) of the intersection (ll') of these two lines are given by the vector product $(x, y, z) = [a, b, c] \times [a', b', c']$.

A test for collinearity: Let A, A', and A'' be three points in the real projective plane with respective homogeneous coordinates (x, y, z), (x', y', z'), and (x'', y'', z''). These three points are collinear if and only if the corresponding points in $\mathbb{R}^3$ are in the same two-dimensional subspace, that is, the determinant

$$\det \begin{bmatrix} x & y & z \\ x' & y' & z' \\ x'' & y'' & z'' \end{bmatrix}$$

is zero.

Polarity: Consider the circle $x^2 + y^2 = 1$ in the plane $z = 1$. In the subspace model, the circle generates a "light cone" with coordinates satisfying $F(x, y, z) = x^2 + y^2 - z^2 = (x, y, z) \cdot (x, y, -z) = 0$. The polar of the projective plane with respect to the circle is obtained via the correspondence

$$(a, b, c) \longleftrightarrow [a, b, -c]$$

This correspondence is an involution interchanging points and lines which preserves incidences. It is easy to verify the following properties of the polar for some special points and lines:

- A point P on the circle is the polar of the tangent to the circle at P.
- The polar of the center of the circle is the line at infinity.
- The polar of a line through the origin is the point at infinity on a perpendicular line.

These properties suggest an elementary construction for the polars of the ordinary points with respect to the circle. Specifically, to construct the polar correspondence with respect to a given circle of radius r and center O, a point P on the circle and the tangent p to the circle at P are polar to one another; a point P exterior to the circle and the line p joining the points of tangency of the two tangents to the circle through P are polar to one another; and a point P interior to the circle and the line p consisting of those points whose tangents to the circle meet the circle in two points which are collinear with P are also polar to one another; see Fig. 6.3. In general, the polar of a point P distinct from O is a line p which is perpendicular to the ray OP which satisfies the condition that if Q is the intersection of p and OP, then the product of the lengths of the segments OP and OQ is r^2; see Fig. 6.4.

6.1 Geometric Planes

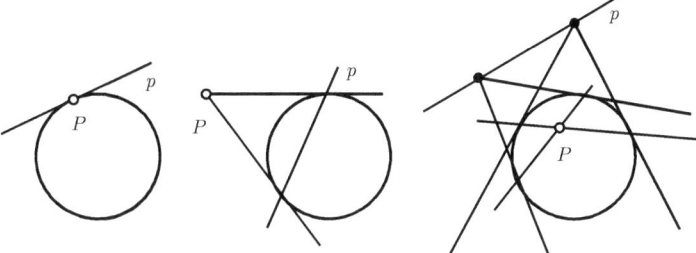

Fig. 6.3 Constructing polars of ordinary points with respect to a circle

Fig. 6.4 Constructing a point P in the exterior of a circle and its polar line p, together with a point Q in the interior of a circle and its polar line q

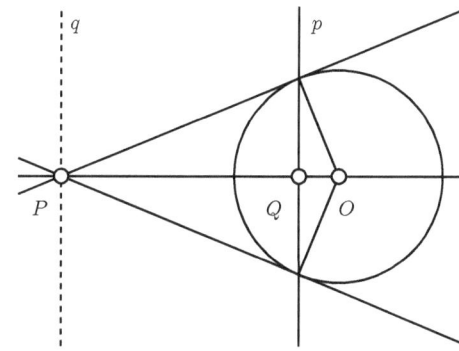

Proof of Pascal's Theorem

A conic is an equation of the form $ax^2 + bxy + cy^2 + dx + ef + g = 0$, which we scale to normalize with $g = 1$. So 5 distinct points give 5 linear equations in 5 unknowns.

Suppose we want an equation for the conic containing the points $A, B, C, D,$ and E, each given in homogeneous coordinates. Let the variable point be $X = (x, y, 1)$, and consider the three vectors

$$H = (A \times B) \times (D \times E)$$
$$I = (B \times C) \times (E \times X)$$
$$J = (C \times D) \times (X \times A)$$

and consider the determinant of the 3×3 matrix with H, I, and J as columns. Two of these columns, I and J, are linear in X, so setting the determinant equal to zero gives a quadratic equation in x and y and so describes a conic. Moreover, we verify that if $X \in \{A, B, C, D, E\}$, then the determinant is zero, so the determinant equation is in fact the conic through those 5 points:

If $X = A$ or $X = E$, then one of the column vectors, J or I, respectively, is zero; hence, the determinant is zero.

If $X = C$, then $I = (B \times C) \times (E \times C)$ is orthogonal to both $B \times C$ and $E \times C$, so I lies along the intersection of the planes spanned by B and C and E and C, so

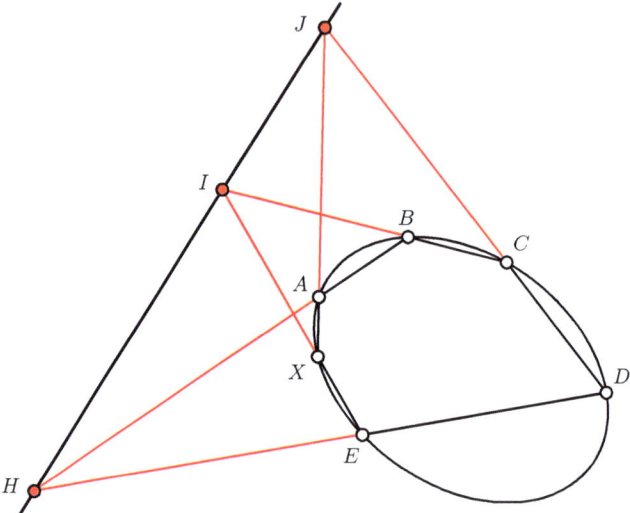

Fig. 6.5 Constructing a point X on the conic through A, B, C, D, and E: Construct H and choose any line through H, and construct I, its intersection with BC and J, and its intersection with CD. The corresponding point on the conic is the intersection of IE and JA. As the slope of the line through H varies, the point X will traverse the conic

I is a scalar multiple of C but, for the same reason, so is $J = (C \times D) \times (C \times A)$, so two columns are linearly dependant, and so the determinant is zero.

If $X = B$, then $I = (B \times C) \times (E \times B)$ is a scalar multiple of B. Moreover, $H = (A \times B) \times (D \times E)$ lies in the plane spanned by A and B and so does $K = (C \times D) \times (B \times A)$, so all three columns belong to the plane spanned by A and B, so the determinant is zero.

Similarly, if $X = D$, then $J = (C \times D) \times (D \times A)$ is a scalar multiple of D, and the other two columns are vectors in the plane spanned by D and E, so again the three columns are linearly dependent.

Thus, all five points A, B, C, D, and E lie on the conic defined by setting the determinant equal to zero. See Fig. 6.5. Moreover, we can interpret the columns of this matrix. If P and Q are plane points in homogeneous coordinates, then $P \times Q$ gives the normal vector of the plane they span, which represents the line between the projective points in homogeneous coordinates. Similarly, if P and Q represent projective lines, then $P \times Q$ gives homogeneous coordinates of their point of intersection. The three columns of the matrix above are therefore the homogeneous coordinates of the intersection points of the lines joining opposite sides of the hexagon (A, B, C, D, E, X), and to say that the determinant is zero is to say exactly that those three points are collinear; see Fig. 6.5.

So six points lie on a conic if and only if the three intersection points of the opposite sides of any hexagon having them as vertices are collinear.

6.1 Geometric Planes

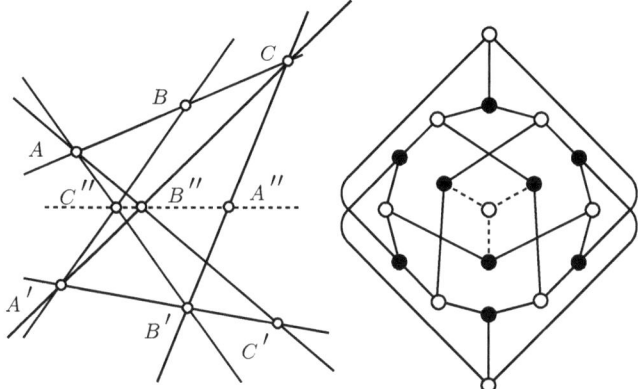

Fig. 6.6 Pappus theorem: In the configuration on the *left*, the *dotted line* is forced. In the Levi graph on the *right*, the existence of the *white vertex* and the *dotted edges* is implied

6.1.5 The Theorems of Pappus and Desargues

We conclude this section with two classical theorems.

The first one is known as the *Pappus theorem* (Fig. 6.6).

Theorem 6.1 (Pappus). *Let A, B, C be three collinear points, and let A', B', C' be another triple of collinear points. Let A'' be the intersection of BC' and $B'C$, B'' the intersection AC' and $A'C$, and C'' the intersection of AB' and $A'B$. Then the points A'', B'', and C'' are collinear.*

The other important theorem is the *theorem of Desargues*.

Theorem 6.2 (Desargues). *Let ABC and $A'B'C'$ be two triangles. Let A'' be the intersection of BC and $B'C'$; let B'' be the intersection of AC and $A'C'$ and C'' be the intersection of AB and $A'B'$. The lines AA', BB', and CC' intersect in a common point O if and only if A'', B'', and C'' are collinear.*

Both theorems hold in the extended Euclidean plane, i.e. in the real projective plane. Both theorems are expressed purely in terms of incidence. This means that for any incidence structure, whether or not the theorems of Pappus or Desargues are valid. For example, the Pappus theorem holds if the incidence structure is the projective plane over any field $\mathbb{F}$. These are not all the projective planes, however, and the theorem of Pappus will fail for the projective plane over any noncommutative division ring. The Desargues theorem is more general. It is valid in the projective plane constructed from any division ring, commutative or not, so in particular for any field. There do exist projective planes in which the Desargues theorem does not hold, which are called *non-Desarguesian*.

Theorem 6.3. *If the Pappus theorem holds for an incidence structure $\mathcal{C}$, then the Desargues theorem holds for $\mathcal{C}$ too.*

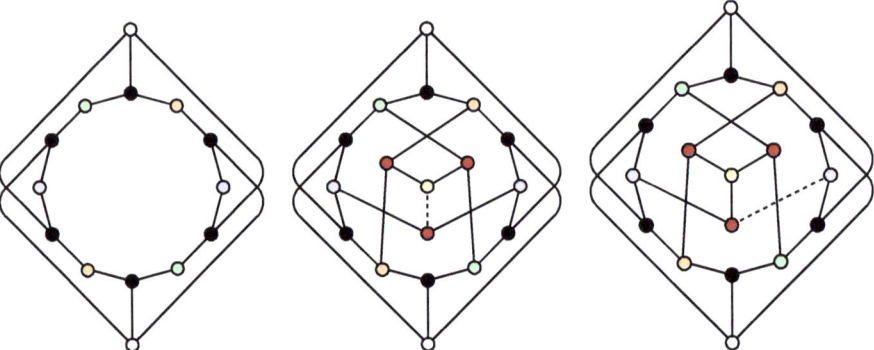

Fig. 6.7 A Pappus 12-cycle and two ways to complete a Pappus 12-cycle to a Pappus graph. The vertices have been colored to assist the visual identification of Pappus 12-cycles in the following proof. The *dark-red vertices* are "*black*," and the pairs of pastel shades are "*white*"

6.1.6 Proving Desargues from Pappus

Let us say that an incidence structure S is *Pappian* if for every subgraph of the Levi graph which is isomorphic to the Levi graph of the Pappus configuration minus an edge, the missing edge is also in the Levi graph of S. Similarly, we can define an incidence structure S to be Desarguesian if for every subgraph of the Levi graph which is isomorphic to the Levi graph of the Desargues configuration minus an edge, the missing edge is also in the Levi graph of S.

A *Pappus 12-cycle* is a 12-cycle in a Levi graph such that each of the two alternating triples of black vertices is adjacent to white vertex outside the 12-cycle; see the diagram on the left of Fig. 6.7.

The proof of the following lemma is left as an exercise.

Lemma 6.4. *Suppose C satisfies Axioms* **P1** *and* **P2**.
Then the incidence structure C is Pappian if and only if every Pappus 12-cycle in the Levi graph can be uniquely completed to a Pappus graph.

Theorem 6.5. *Suppose C satisfies Axioms* **P1** *and* **P2**.
If C is Pappian then it is Desarguesian.

In particular, a Pappian projective plane is Desarguesian. The argument of the following proof is freely adapted from Hilbert and Cohn-Vossen [51].

Proof. Suppose that we have an incidence structure C as in the hypotheses, whose incidence graph contains a subgraph isomorphic to the Desargues graph minus an edge (see Fig. 6.11) where the vertices are chosen using notation compatible with the depiction of the Desargues configuration in Fig. 6.8. Since the structure satisfies properties **P1** and **P2**, we will only label the points and refer to the unique block incident to points X and Y as $X \cdot Y$.

6.1 Geometric Planes 207

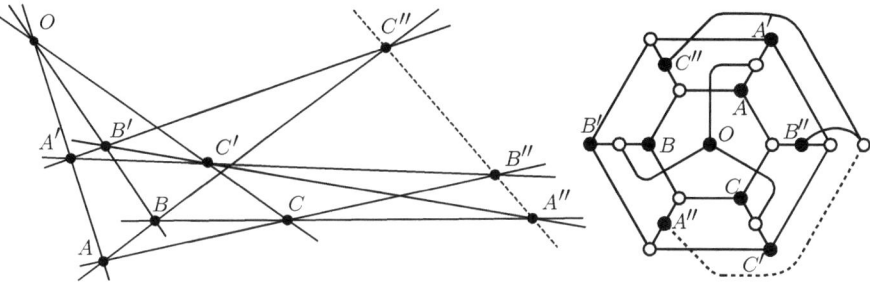

Fig. 6.8 Desargues theorem

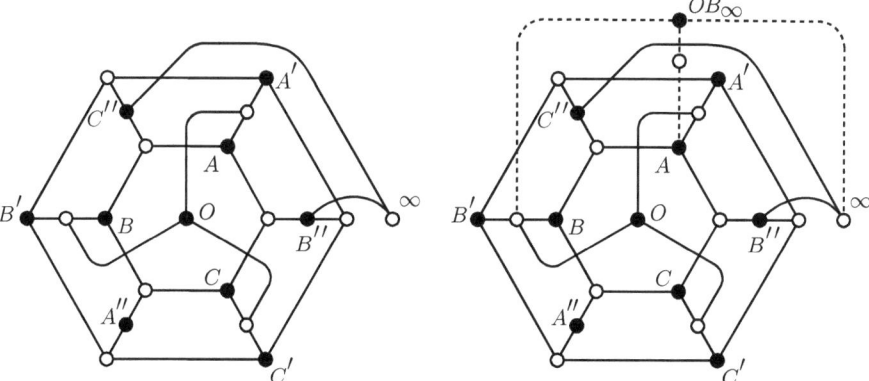

Fig. 6.9 A subgraph of a Levi graph containing a Desargues graph minus an edge. On the right vertices, OB_∞ and $A \cdot OB_\infty$ have been constructed

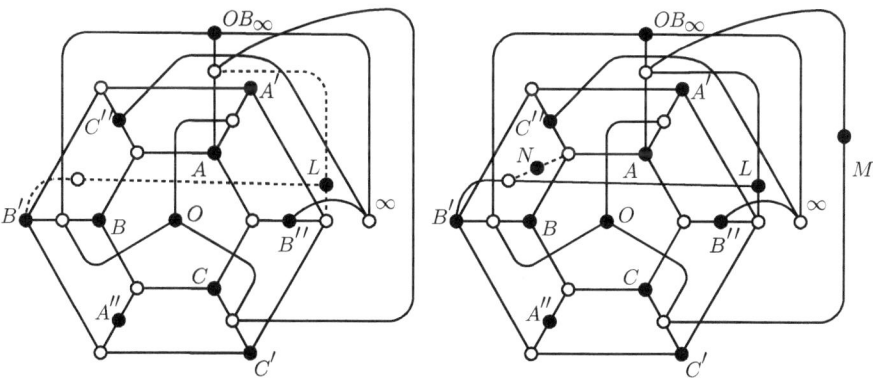

Fig. 6.10 On the *left*, L and $L \cdot B'$ have been constructed. On the *right* N and M

Using properties **P1** and **P2**, construct sequentially a point OB_∞ incident to the block ∞ and $O \cdot B$; the block $OB_\infty \cdot A$ (see Fig. 6.9); a point L incident to that block and the block $A' \cdot C'$; a block incident to L and B' (see Fig. 6.10); a point N

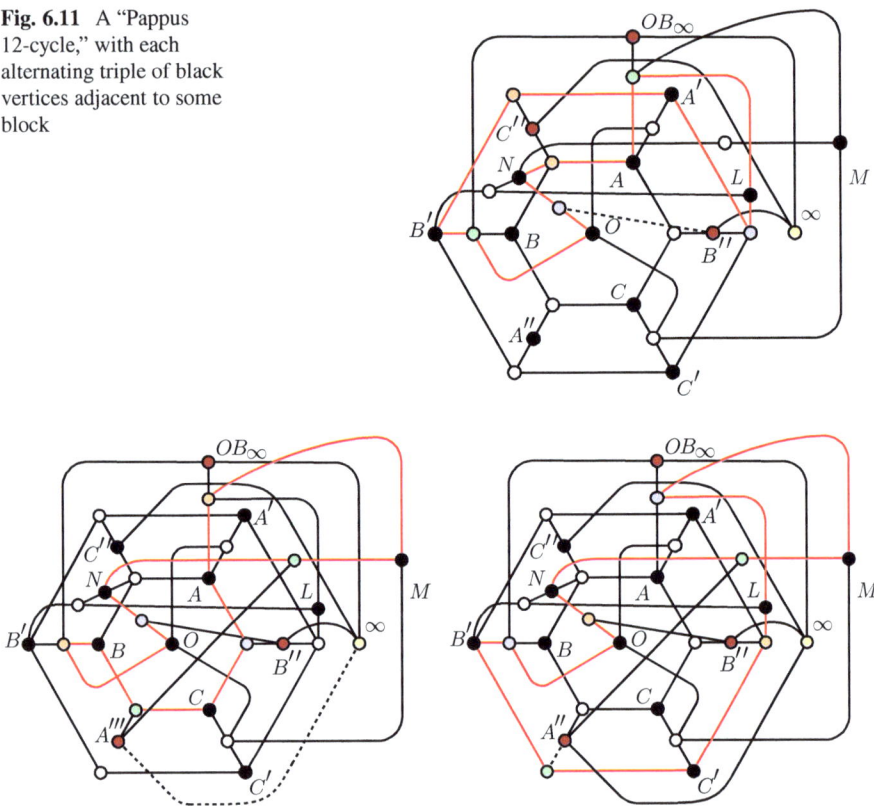

Fig. 6.11 A "Pappus 12-cycle," with each alternating triple of black vertices adjacent to some block

Fig. 6.12 Pappus 12-cycle $ONMACB$ on the left and Pappus 12-cycle $ONMLC'B'$ on the right

incident to that block $A \cdot B$; and lastly a point M incident to $OB_\infty \cdot A$ and $C \cdot C'$. This completes the construction of the diagram on the left of Fig. 6.11. With these new points, we find a 12-cycle with black vertices $ONALA'B'$ with alternate triples of back vertices, $\{O, A, A'\}$ and $\{N, L, B'\}$, each adjacent to some white vertex, in other words, a Pappus 12-cycle. Adding $N \cdot O$, we have a subgraph of the correct type to imply the incidence between B'' and $N \cdot O$; see Fig. 6.11.

Next, we consider the 12-cycle $ONMACB$; see the graph on the left of Fig. 6.12. This is again a Pappus 12-cycle, and it implies an adjacency between A''', here defined as the point incident to the blocks labeled $M \cdot N$ and $A \cdot A$, and the point ∞.

Lastly, we consider the 12-cycle $ONMLC'B'$ which again is a Pappus 12-cycle and implies an adjacency between A'' and $B' \cdot C'$ and moreover, by property **P2**, identifies A''' with A'' since they are both adjacent to $B \cdot C$ and $B' \cdot C'$.

Thus, the Desargues graph minus an edge has been completed, and the Desargues theorem holds.

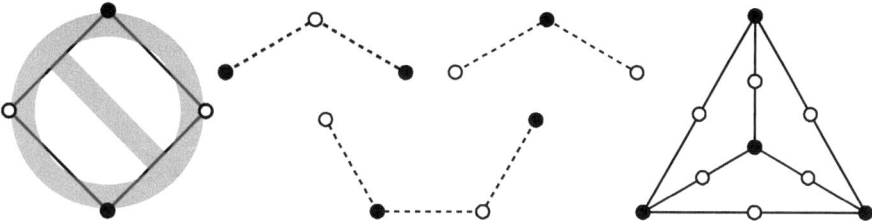

Fig. 6.13 Diagrammatic representation of the axioms for the incidence graph of a projective plane: bipartite, diameter three, girth 6, and containing an $S(K_4)$

6.2 Finite Projective Planes

As pointed out above, the Fano configuration not only has a realization in a projective plane, but it is the entire projective plane constructed as $PG(2,\mathbb{Z})$, the extended affine plane over the field $\mathbb{Z}_2$. We will show below that, although not directly stated in the axioms, the points and lines of every finite projective plane form a combinatorial configuration.

First, let us interpret the axioms in the language of the incidence graph. Axioms **P1** and **P2** imply that the distance between any two vertices of the same color is 2. Axiom **P3** says there exist four black vertices such that each of the $\binom{4}{2}$ pairs is adjacent to a white vertex, and since none of these white vertices is adjacent to more than two of the four black vertices, the six white vertices are distinct; in other words, the incidence graph contains the subdivision graph of the complete graph on four vertices, $S(K_4)$, as an induced subgraph; see Sect. 2.4. For vertices of distinct colors, their distance is either 1 or 3; if a and b are not adjacent and are both in $S(K_4)$, then their distance is 3, and if a, say, is not in $S(K_4)$, there is a path of length two to a vertex in $S(K_4)$ and a path from the middle vertex of that path to b. Since the diameter is three, the girth is at most six, and since there are no 4-cycles, the girth must be exactly six. So the incidence graph is bipartite of diameter 3 and girth 6. So a graph G is the incidence graph of a projective plane if and only if it is a bipartite graph of diameter 3, girth 6, and contains an induced subgraph isomorphic to $S(K_4)$; see Fig. 6.13.

Theorem 6.6. *Let $\mathcal{P}$ be a projective plane, and suppose there exists a line l with exactly $n + 1$ points incident to it.*

Then there exists at least one point P with exactly $n + 1$ lines incident with it.

Proof. Suppose the points incident to l are $P_1, P_2, \ldots, P_{n+1}$. Consider the incidence graph of $\mathcal{P}$. At most, two points of the induced $S(K_4)$ are among the P_i, so denote one of the other two by P. There is a path of length 2 from point P to each of the P_i (see Fig. 6.14a) and the white vertices of these paths must be distinct; otherwise, a 4-cycle would be formed with l.

So the point P has $n + 1$ neighbors. If P has another neighbor, l_0, then there would be a path of length 2 to l, and the interior vertex of this path cannot be P,

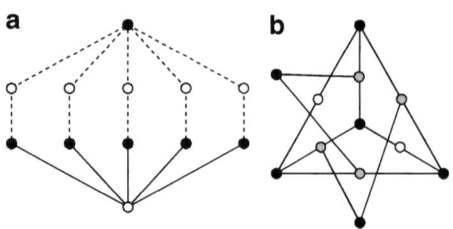

Fig. 6.14 A white vertex of degree n implies a *black vertex* of degree n and $S(K_4)$ with colors reversed

since it is not adjacent to l nor to the P_i, since that would form a 4-cycle, giving a new neighbor of l, a contradiction. □

Theorem 6.6 is an example of a statement implying its *dual statement*. In general, if any statement follows from the Axioms **P1**...**P3**, then its dual statement, in which the terms point and line are interchanged, must also follow from the axioms. This must always occur in a projective plane because Axiom **P1** is the dual of Axiom **P2** and the following theorem.

Theorem 6.7. *In any projective plane, there exist at least four lines k,l,m,n such that no three of them are concurrent.*

In other words, the incidence graph contains a copy of $S(K_4)$ as in Fig. 6.13 with the colors reversed.

Proof. By **P3**, there exists an induced subgraph isomorphic to $S(K_4)$. Choose a pair of white vertices subdividing a pair of independent edges in K_4, and consider the other four white vertices. Of the six pairs of these four vertices, four are adjacent to black vertices of the $S(K_4)$, and the other two pairs cannot be adjacent to any black vertices of $S(K_4)$ without creating a 4-cycle, and since the black vertices joining those two pairs cannot coincide without forming a 4-cycle, we have the required $S(K_4)$ with colors reversed. See Fig. 6.14b. □

Theorem 6.8. *Let $\mathcal{P}$ be a projective plane, and suppose there exists a line l with exactly $n+1$ points incident to it.*

Then every point P is incident with exactly $n+1$ lines, and every line l is incident with exactly $n+1$ points.

Proof. The argument in Theorem 6.6 and its dual imply that any pair of oppositely colored nonadjacent vertices in the incidence graph have the same valence. Considering the copy of $S(K_4)$ guaranteed by the axioms, since any two of its black vertices are nonadjacent to some white vertex, all the black vertices have the same valence, and since each white vertex is nonadjacent to some black vertex, all the vertices of the $S(K_4)$ have the same valence.

Now, any vertex not in the $S(K_4)$ cannot be incident to all the vertices of the $S(K_4)$ of the relevant color since some have distance in $S(K_4)$ which would create a 4-cycle; thus, that vertex and hence all vertices of the incidence graph have the same valence as those of the $S(K_4)$. □

6.2 Finite Projective Planes

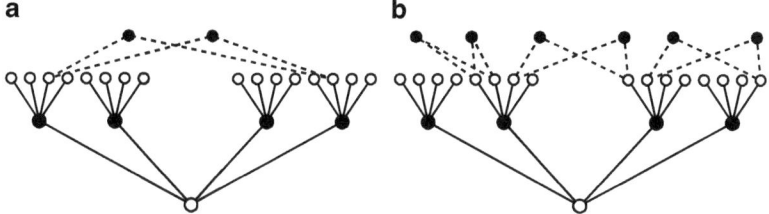

Fig. 6.15 To avoid 4-cycles (**a**) If two tuples correspond to distinct positions on the grid, (**b**) in each position the same number does not occur twice, two different positions do not agree in two coordinates, and each row and column has distinct entries

The number n in Theorem 6.8, one less than the valence of the incidence graph, is called the *order* of the projective plane. Since every finite projective plane has an order, it is possible to axiomatize the finite projective planes of a particular order by adding

P4 There exists a line l with exactly $n+1$ points incident to it.

If $\mathcal{P}$ is a finite projective plane of order n, any line l is incident to $n+1$ points, and each of these points is incident to n lines besides l, and these new lines must be distinct to avoid 4-cycles with l in the Levi graph. Moreover, since the diameter is 3, these $1 + (n+1)n$ lines exhaust all the lines of the projective plane $\mathcal{P}$. By duality, there must as well be $1 + (n+1)n$ points. Thus, every finite projective plane of order n forms a balanced combinatorial configuration of type $((n^2 + n + 1)_{n+1})$.

The construction of the previous paragraph can be used to establish a connection between finite projective planes and Latin squares. Consider the Levi graph of the finite projective plane. As before, start with a fixed line l; let its neighboring points be linearly ordered $\{0, 1, \ldots, n, n+1\}$ (left to right in the figure) on the next level, and then let the $n+1$ sets of second neighbors each be linearly ordered $\{1, 2, \ldots, n\}$ on the third level, with the remaining n^2 points comprising the fourth and final level with respect to l; see Fig. 6.15. Each vertex on the fourth level is adjacent to exactly one vertex in each of the sets of lines on the third level; all elements of a line set on the third level are adjacent to the same vertex on the second level. So we define for each of the n^2 black vertices on the fourth level an $n+1$-tuple in which the ith coordinate is the number of the white vertex on the third level which is the interior point on the path of length 2 to the i'th point on the second level. To avoid 4-cycles, no pair of vertices on the fourth level can have tuples which agree in two coordinates; see Fig. 6.15a.

We now use the tuples defined in the previous paragraph to create a system of $n-1$ Latin squares. First, take the 0th and nth coordinates of the tuples to associate each vertex on the fourth level with a unique position on an $n \times n$ grid. Now, enter the first coordinate of each fourth level vertex (i, j) into the ij position on the grid. If there are two grid entries with the same number in the same row, that would correspond to a 4-cycle in the Levi graph, and the same with the columns. So entering the first coordinates of the tuples into the grid gives us a *Latin square*. Of course so do all

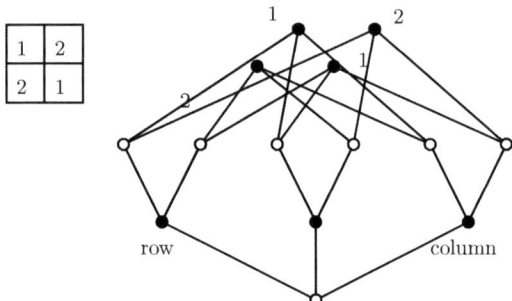

Fig. 6.16 The 1 in the 11 position of the grid indicates that a vertex labeled 1 on the fourth level is adjacent to the 1st of each of the three pairs of white vertices on the second level. The 1 in the 22 position indicates that a vertex labeled 1 on the fourth level is adjacent to, respectively, the 2nd, 1st, and 2nd vertex of the pairs of white vertices on the third level. The 2's in the 12 and 21 positions of the grid indicate two vertices labeled 2 on the fourth level adjacent to the 1st, 2nd, and 2nd (resp. 2nd, 2nd, and 1st) points of the pairs of white vertices on the third level. The resulting 14 vertex graph is another view of the Levi graph of the Fano plane

the coordinates $i = 1, \ldots, n - 1$. Lastly, there are no 4-cycles between the third and fourth levels with two white vertices associated with different points, which is exactly the condition that no two positions of the grid can have same pair of entries on two different Latin squares; in other words, the $n - 1$ Latin squares are *mutually orthogonal*.

This process can be reversed; see Exercise 6.10. In other words, given any collection of $n - 1$ mutually orthogonal $n \times n$ Latin squares, we can construct the Levi graph of a projective plane of order n. This process is carried out in Fig. 6.16 in which $n = 2$. In this case, we start with a 2×2 array of entries with only a single label, so orthogonality is not an issue, and the result is the Fano plane.

For order $n = 3$, it is not hard to construct two orthogonal 3×3 Latin squares. We remark that the name "Latin square" does not indicate that the concept was a favorite of the ancient Romans. It is an artifact of Euler's notation, in which he used the letters $a, b, c, \ldots$ instead of $1, 2, 3, \ldots$, and to distinguish different squares, Euler would use different alphabets, starting with the Latin, then the Greek, and so on. So for $n = 3$, he would consider a "Graeco-Latin" square such as

$$\begin{array}{|c|c|c|} \hline a\alpha & b\gamma & c\beta \\ \hline b\beta & c\alpha & a\gamma \\ \hline c\gamma & a\beta & b\alpha \\ \hline \end{array} ; \tag{6.1}$$

see Fig. 6.17. Notice that it is not significant in the construction of the Latin squares from the finite projective plane which two vertices are associated with the rows and columns of the grid. The conditions to be satisfied are uniform across the levels. If we switch, for example, the row numbers and the Latin labels, keeping the ordering the same, it will give a different "Graeco-Latin" square; in this case,

6.2 Finite Projective Planes 213

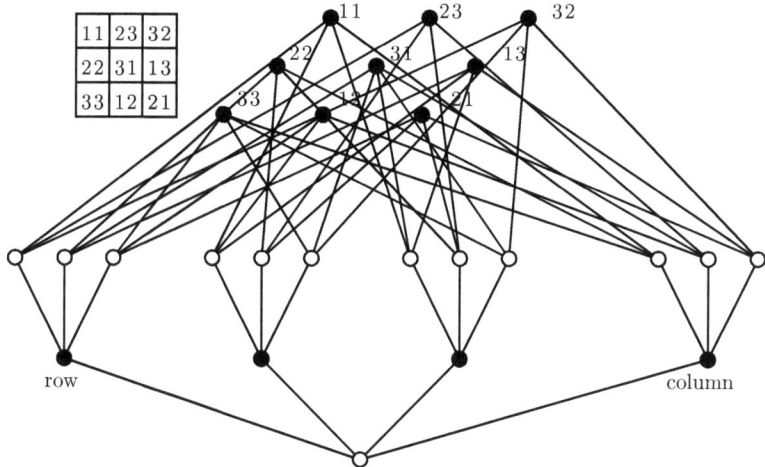

Fig. 6.17 A 3×3 Graeco-Latin square and the Levi graph of its associated order 3 projective plane

Fig. 6.18 The Levi graph of the projective plane constructed from the Graeco-Latin square (6.1)

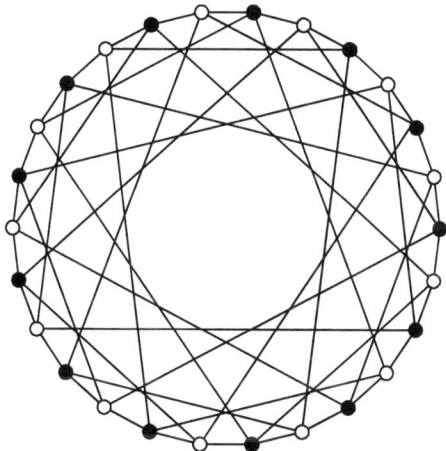

it will interchange the second two columns. The order 3 projective plane which results has $3^2 + 3 + 1 = 13$ points, a (13_4) configuration, whose Levi graph is more symmetrically drawn in Fig. 6.18.

Euler's search for mutually orthogonal Latin squares led him in 1782 (see [30]) to formulate his famous "36 Officers Problem," which starts with an army having 6 regiments and 6 ranks and asks for a method to march 36 officers in 6 rows and 6 columns so that no column contains two officers of the same rank or regiment. The fact that the problem could not be solved was not discovered for over 100 years, but even with a solution, a projective plane of order six would require the officers not just to have ranks and regiments but three other qualities as well. In particular, there is no projective plane of order six.

Verifying that the following

111	243	324	432
222	134	413	341
333	421	142	214
444	312	231	123

is a triple of mutually orthogonal Latin squares will establish the existence of the projective plane of order four and also give an immediate construction for its $2 \cdot 21$ vertex Levi graph. The same is true for the 5×5 array

0000	1234	2413	3142	4321
1111	2340	3024	4203	0432
2222	3401	4130	0314	1043
3333	4012	0241	1420	2104
4444	0123	1302	2031	3210

giving a quadruplet of mutually orthogonal Latin squares and establishing an order 5 projective plane with 31 points and 31 lines. This quadruplet is of a particularly regular form. In general, given a prime p, we may form a sequence of $p-1$ mutually orthogonal Latin squares by placing in the (i, j) square the $(n-1)$-tuple

$$(i+j, i+2j, i+3j, \ldots, i+(p-1)j)$$

with indices modulo p.

So we have two constructions for projective planes of order 5, each giving a balanced combinatorial configuration of type (31_6): one via the extended affine plane over the field $\mathbb{Z}_5$ and the second being via the four mutually orthogonal 5×5 Latin squares. It turns out that these two configurations are isomorphic and there is only one projective plane of order 5, PG(2,$\mathbb{Z}_5$), and its incidence table may be taken to be

```
0 0 0 0 0 0 1 1 1 1 1 2 3 4 5 2 2 2 2 4 5 3 5 4 3 5 3 4 5 4 3
1 6 b c d e 6 7 8 9 a 6 6 6 6 7 9 a 8 7 7 7 a 9 8 9 a a 8 8 9
2 7 i f g h b i g f h j h f g b d e c e d c b b b c d c e d e
3 8 j n q p c n j m k n i k m f i m k j k m n p q h f g f h g
4 9 k t s r d p t r s r t q p g u q p u r s v s r j j i i m k
5 a m u w v e q v w u s w v u h v t w w t v w t u q p r s n n
```

There are also unique projective planes of order 7 and 8 corresponding to the extended affine planes over $\mathbb{Z}_7$ and $\mathbb{F}_8$, respectively. For $9 = 3^2$, there is a finite field $\mathbb{F}_9$, but there are exactly three other projective planes of order 9 besides $GP(2,9)$; see [58]. None of them satisfy the theorem of Desargues. So far, although there exist finite projective planes not isomorphic to the extended affine plane over a finite field, there is no known example of a projective plane whose order does not agree with that of an extended affine plane over a finite field, in other words, of order a power of a prime.

6.2 Finite Projective Planes

It has been shown, by Bruck and Ryser (see [14, 15]) that if a finite projective plane of order n exists and $n \equiv 1$ or $2 \pmod 4$, then n is the sum of two squares. Since 14 is not the sum of two squares, it follows that there is no projective plane of order 14. On the other hand, an order 10 projective plane, while not forbidden by the Bruck–Ryser condition, has by exhaustive methods been shown not to exist; see [59]. No stronger general existence criterion is known.

6.2.1 Affine and Projective Realizations over Finite Fields

There are two kinds of finite fields. The *prime fields* are those of cardinality p, where p is a prime.
$$\mathbb{F}_p = \mathbb{Z}_p = \{0, 1, \ldots, p-1\},$$
with addition and multiplication modulo p. The additive group is cyclic of order p, with generators all nonzero elements, in particular the multiplicative unit 1. The multiplicative group is of order $p - 1$, and it is true but not obvious that the multiplicative group of $\mathbb{Z}_p$, and indeed of every finite field, is cyclic.

The other class of fields are of cardinality $q = p^n$, for $n > 1$. These fields may be regarded as being obtained by successively adjoining roots of irreducible polynomials. For example, $\mathbb{F}_4$, the field on $q = 4 = 2^2$ elements, has the elements $\{0, 1, \alpha, \alpha+1\}$ with the addition still modulo 2, and $\alpha^2 = \alpha + 1$. The elements α and $\alpha + 1$ are roots of $x^2 + x + 1$ which is not factorable over $\mathbb{Z}_2$. For more details on finite fields, their properties, and constructions, see [96]. For our purposes, it is enough to know that a finite field $\mathbb{F}_q = \mathbb{F}_{p^n}$ exists for all n and for every p prime.

Suppose $\mathbb{F}^d$ is a vector space of dimension d over a finite field $\mathbb{F} = \mathbb{F}_q$, which can be regarded as the set of d-tuples of elements of $\mathbb{F}$, $\mathbb{F}^d = \{f_1, \ldots, f_d\}$. This gives us a configuration of points and lines, since, just as in $\mathbb{R}^d$, two lines can only meet in at most one point. The vector space has $|\mathbb{F}|^d = q^d = p^{dn}$ points, and to determine the number of lines through any point, we count the number of lines through the origin. Each line in $\mathbb{F}^d$ contains d points. Each line through the origin is determined by one nontrivial point, of which there are $q^d - 1$ choices with each line being overcounted $q - 1$ times, so altogether there are
$$\frac{q^d - 1}{q - 1}$$
one-dimensional subspaces.

Note that each of these one-dimensional subspaces is an additive subgroup of $\mathbb{F}^d$ and the $q^d/q = q^{d-1}$ cosets of this subgroup are the lines in $\mathbb{F}^d$ which are parallel to the subspace, so there are
$$q^{d-1} \frac{q^d - 1}{q - 1}$$
lines in $\mathbb{F}^d$.

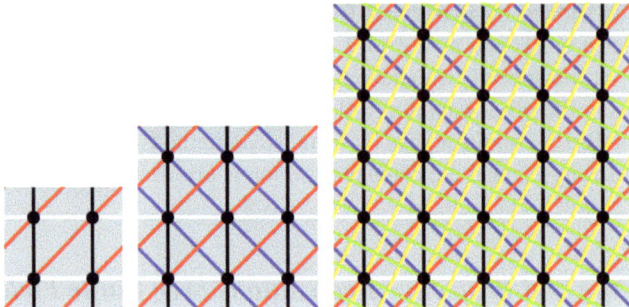

Fig. 6.19 Configurations of lines on the square torus give rise to doubly periodic configurations in the plane

We conclude that each vector space $\mathbb{F}^d$ contains a

$$\left(q^d_{\frac{q^d-1}{q-1}}, \left[q^{d-1}\frac{q^d-1}{q-1}\right]_q\right)$$

configuration of points and affine lines.

In the case of $\mathbb{F}^d = (\mathbb{Z}_2)^2$, we have a configuration of type $(4_3, 6_2)$, which also models the configuration of points and lines in a tetrahedron. For $(\mathbb{Z}_2)^3$, we have an $(8_7, 28_2)$ configuration.

For $\mathbb{F}^3 = (\mathbb{Z}_3)^2$, we have a $(9_4, 12_3)$ configuration, and for $(\mathbb{Z}_5)^2$, we have a $(25_6, 30_5)$ configuration, any of which can be regarded as configurations of straight lines on a square torus (see Fig. 6.19) since the action of $3\mathbb{Z} \times 3\mathbb{Z}$ on the square integer lattice has quotient $\mathbb{Z}_3 \times \mathbb{Z}_3$. Similarly, the vector space configurations of $\mathbb{Z}_p^2$ can be realized for all primes p as configurations of straight lines on a square torus.

The lines of any vector space configuration are naturally partitioned into parallel classes of q lines, each of which can be independently removed to form a smaller configuration. For example, the configurations of Fig. 6.19 can be altered to produce balanced configurations by removing the appropriate number of parallel classes (see Fig. 6.20) giving a balanced configuration of type (p_p^2). Removing any one of the parallel classes of the $(9_4, 12_3)$ configuration corresponding to $\mathbb{F}^2 = (\mathbb{Z}_3)^2$ yields the Pappus configuration (Fig. 6.21).

We can also consider projective spaces over finite fields. For $d = 3$, these will be the projective plane in the abstract sense. In the subspace model, the projective points are the lines through the origin, of which there are $(q^d - 1)/(q - 1)$ in $\mathbb{F}^d$. The projective lines are the planes through the origin.

To count the number of projective lines, we count the unordered pairs of linearly independent points. There are $q^d - 1$ choices for one nontrivial vector and $q^d - d$ vectors not linearly independent with the first, and since the order does not matter, we have $(q^d - 1)(q^d - q)/2$ unordered pairs of linearly independent vectors. Since each plane has been counted $(q^2 - 1)(q - q)/2$ times, the number of planes through

6.2 Finite Projective Planes

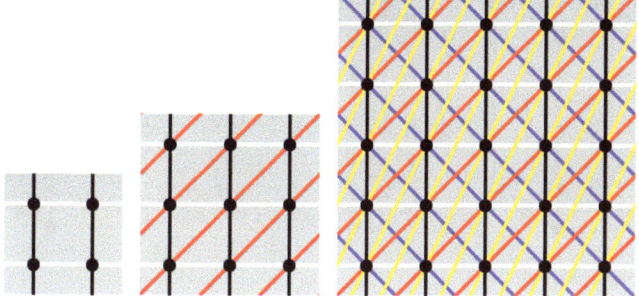

Fig. 6.20 Balanced configurations (4_2), (9_3), and (25_5)

Fig. 6.21 The Pappus configuration in $\mathbb{Z}_3^2$

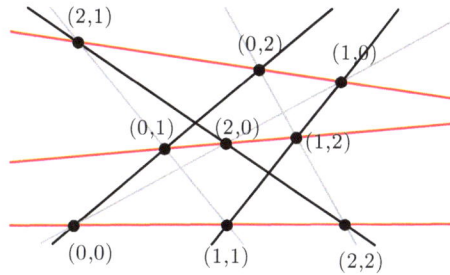

the origin is

$$\frac{(q^d-1)(q^d-q)}{(q^2-1)(q^2-q)} = \frac{(q^d-1)(q^{d-1}-1)}{(q^2-1)(q-1)}.$$

There are $(q^2-1)/(q-1) = (q+1)$ projective points on each projective line, and each projective point is contained on

$$\frac{(q^d-q)}{(q^2-q)} = \frac{(q^{d-1}-1)}{(q-1)}$$

projective lines. Thus, $\mathbb{F}^d$ yields a projective

$$\left(\left(\frac{q^d-1}{q-1} \right)_{\frac{(q^{d-1}-1)}{(q-1)}}, \left(\frac{(q^d-1)(q^{d-1}-1)}{(q^2-1)(q-1)} \right)_{q+1} \right)$$

configuration.

As expected for $(\mathbb{Z}_2)^3$, we get the Fano plane, (7_3), and in general for $d = 3$, we get a balanced configuration of the form

$$\left(\left(\frac{q^3-1}{q-1} \right)_{q+1} \right) = \left((q^2+q+1)_{q+1} \right),$$

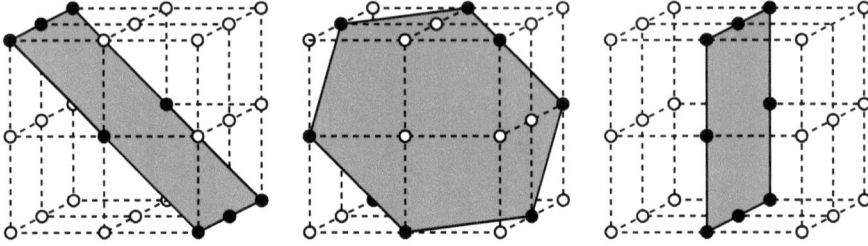

Fig. 6.22 Subspaces of $\mathbb{Z}_3^3$

so $|\mathbb{F}| = p^n = q$ is the order of the projective plane, and it follows that projective planes of order p^n exist for all primes p and exponents n. So the projective planes (13_4), (21_5), and (31_6) of orders 3, 4, and 5 all exist and are realizable as geometric projective planes.

The automorphism group of any projective plane over a field $\mathbb{F}$ of order $q = p^n$ will be the quotient of the group of invertible 3×3 matrices over $\mathbb{F}$, the general linear group $GL(3, \mathbb{F})$, or simply $GL(3, q)$, by the group of invertible diagonal matrices, i.e. the *projective linear group*, $PGL(3, q)$. The cardinality of $PGL(3, q)$ is therefore

$$[q^3 - 1] \cdot [q^3 - q] \cdot [q^3 - q^2]/(q - 1)^3.$$

For the Fano plane the group of invertible diagonal matrices is trivial, so this gives an alternative representation for the 168 automorphisms we found in Sect. 5.3.1 as a group of matrices.

Just as the vector space configurations could be realized as configurations of points and lines in the square torus, the projective planes which are constructed from extended affine planes can be represented as points and planes in a cubical three torus. So the 13 points in the configuration of type (13_4), the projective plane of order 3 constructed from the vector space $\mathbb{Z}_3^2$, correspond to the 13 pairs of antipodal integer points on the unit cube in $\mathbb{R}^3$. The 13 lines, corresponding to 2-dimensional subspaces in $\mathbb{Z}_3^3$, may be associated with certain subsets of 8 vertices. Nine of these subsets correspond to planes through the origin of the unit cube, and four correspond to hexagons in the cube together with the vertices of the antipodal pair on the normal line to the hexagon; see Fig. 6.22.

Generalizing the projective configurations, we can consider *Grassmannians*, the sets of subspaces of rank k in $\mathbb{F}^d$. Proceeding as before, there are

$$\frac{(q^d - 1)(q^{d-1} - 1) \cdots (q^{d-k+1} - 1)}{(q^k - 1)(q^{k-1} - 1) \cdots (q - 1)}.$$

subspaces of rank k in $\mathbb{F}^d$. Moreover, two subspaces of rank $k + 1$ intersect in at most one subspace of rank k. So we have a configuration of type

$$\left(\frac{(q^d - 1) \cdots (q^{d-k+1} - 1)}{(q^k - 1) \cdots (q - 1)}_{\frac{q^{d-k}-1}{q-1}}, \frac{(q^d - 1) \cdots (q^{d-k} - 1)}{(q^{k+1} - 1) \cdots (q - 1)}_{\frac{q^{k+1}-1}{q-1}} \right)$$

6.3 Realization of Classical Configurations 219

For example, $(\mathbb{Z}_2)^4$ gives rise to three Grassmannian configurations of type $(31_{15}, 155_3)$ and $(155_3, 31_{15})$ which are dual and a balanced configuration of type (155_7).

For affine configurations, we note that each subspace of rank k belongs to a parallel class of q^{d-k} affine planes, so there are affine configurations of type (v_r, b_k) with

$$v = q^{d-k}\frac{(q^d-1)\cdots(q^{d-k+1}-1)}{(q^k-1)\cdots(q-1)},$$

$$b = q^{d-k-1}\frac{(q^d-1)\cdots(q^{d-k}-1)}{(q^{k+1}-1)\cdots(q-1)},$$

$$r = \frac{q^{d-k}-1}{q-1}$$

$$k = q\frac{q^{k+1}-1}{q-1}$$

For $\mathbb{F}_8 = (\mathbb{Z}_2)^3$, this yields configurations of type $(8_7, 28_2)$ for points and lines and $(28_3, 14_6)$ for lines and planes.

6.3 Realization of Classical Configurations

6.3.1 Fano Plane

The Fano configuration is realized as the Fano plane, the extended affine plane over $\mathbb{Z}_2$. We have already computed the automorphism group of the Fano configuration in Sect. 5.3.1, represented as a permutation group. The geometric realization of the Fano configuration as the projective plane over $\mathbb{F}_2^3$ yielded a representation of the automorphism group as a matrix group in Sect. 6.2.1. We could look for other representations by considering realizations of the combinatorial configuration over other projective planes.

Let $\mathbb{F}$ be a field with additive unit 0 and multiplicative unit 1. If there is a realization of the Fano configuration in the projective plane over $\mathbb{F}$, then we can choose a basis in $\mathbb{F}^3$ so that the three noncollinear points a, b, and c have homogeneous coordinates $(1, 0, 0)$, $(0, 1, 0)$, and $(0, 0, 1)$, respectively; see Fig. 6.23. The points d and e, the third points on lines $[1, 0, 0]$ and $[0, 1, 0]$, must have two nonzero homogeneous coordinates to be distinct from the standard basis elements already chosen, $(0, y, 1)$ and $(x, 0, 1)$, respectively, with $x, y \neq 0$; see Fig. 6.24. The line between these points and $e = (x, 0, 1)$ and $b = (0, 1, 0)$ is given by $e \times b = [-1, 0, x] \sim [-1/x, 0, 1]$ since x is invertible in $\mathbb{F}$. Similarly $d \times a \sim [0, -1/y, 1]$.

Fig. 6.23 Plana Gino Fano

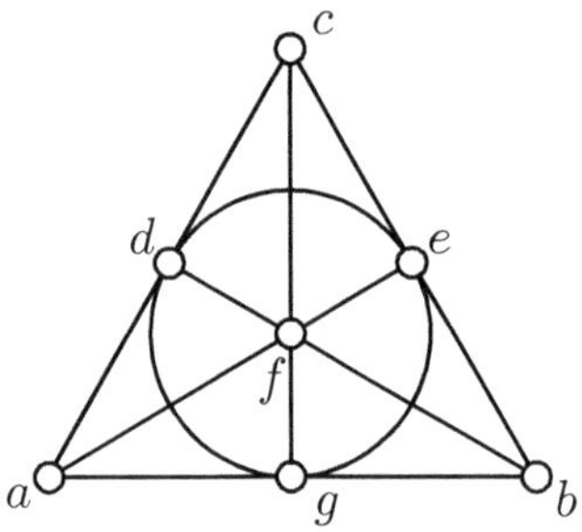

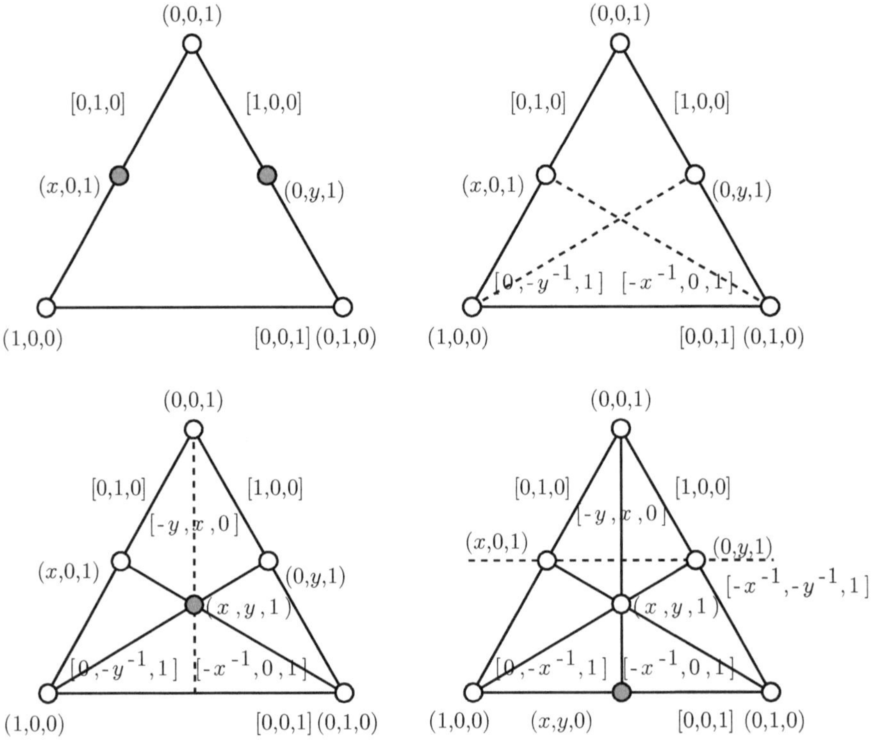

Fig. 6.24 Realizing the Fano plane over a projective plane over a field

Lines $[-1/x, 0, 1]$ and $[0, -1/y, 1]$ intersect at $f = (x, y, 1)$, and the line between c and f is given by $[-y, x, 0]$, which intersects the line $[0, 0, 1]$ at $g = (x, y, 0)$. But the line between d and e is $[-1/x, -1/y, 1]$, and $[-1/x, -1/y, 1] \cdot (x, y, 0) = -2$, so g is only incident with the line between d and e if $-2 \equiv 0$. So we have

Theorem 6.9. *The Fano plane is realizable in a projective space over a field if and only if the field is of characteristic* 2.

6.3 Realization of Classical Configurations

The method of proving this theorem suggests a procedure to generate conditions that a field must satisfy to support an (n_3) configuration.

Another direction suggested by this approach is the theorem of Steinitz which states that any combinatorial (n_3) configuration has a Euclidean representation with at most one curved line. The argument is as follows: Consider the Levi graph, and delete any edge. We now have a subgraph of the Levi graph with some vertex of valence less than 3. Remove that vertex, and the result, if not empty, still must have a vertex of valence less than 3; otherwise, the remainder of the Levi graph is connected and 3-valent and hence a complete connected component of the original. But the original Levi graph was connected by definition. Continue in this way until left with an empty graph. Reversing these moves, we have a sequence of instructions for the construction of the configuration in which at each stage one is adding the intersection of two lines, adding the line joining two existing points, placing a point on a line, drawing a line through a point, placing an isolated point, or placing an isolated line. Have at hand a set of algebraically independent transcendentals, and use them for any choices implied by the instructions to avoid degeneracies. The last incidence is between a point in the figure and a line which joins two points in the figure. Since the point and the line are already constructed, it is unlikely that this last requirement can be met. However, if the point does not lie on the line, the line may be deformed into a circle if necessary.

Theorem 6.10 (Steinitz). *Every (n_3) configuration has a Euclidean realization having at most one curved line.*

The proof of the Steinitz theorem successfully produces a Euclidean realization with all but one line straight; however, the method employed to avoid degeneracies cannot be augmented to actually prevent them. We will see in Sect. 6.4 that there are examples where, long before encountering the final incidence when the last line might have to be bent into compliance, undesired incidences intrude. Extra incidences, not required by the configuration which is to be realized, may begin to appear; a constructed point may be placed on an existing line; a constructed line might pass through an existing point, or, even worse, a constructed point might be forced to occupy the same position as a previously existing point. We call this type of realization a *weak* realization, and that is the type of realization guaranteed by the Steinitz theorem. It is clear that this is not the type of realization either intended by Steinitz or those who have afterward cited his result, since a much simpler weak realization can be had by simply mapping any configuration to the trivial configuration consisting of a single point and a single line passing through it. Without a specified bound on the number, type, or distribution of the degeneracies introduced in the construction, the main content of this theorem seems to be to provide a heuristic for producing a realization with one defective line and having "few" degeneracies.

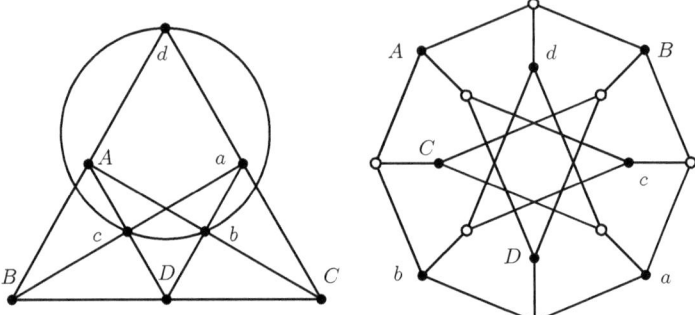

Fig. 6.25 Realizing the Möbius–Kantor configuration, $\omega = (1 - i\sqrt{3})/2$

Fig. 6.26 The Möbius–Kantor, $\omega = (1 - i\sqrt{3})/2$

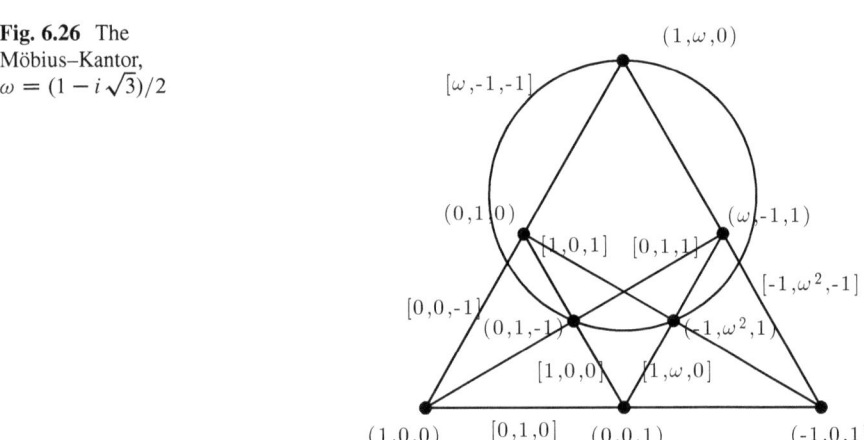

6.3.2 The Möbius–Kantor Configuration

We computed the automorphism group of the Möbius–Kantor configuration in Sect. 5.3.2. Unlike the Fano configuration, it does not come equipped with a natural geometric realization which immediately reveals all its structure and automorphisms. As such, it may be regarded as the smallest abstract configuration, the smallest purely combinatorial configuration, or, for that matter, the smallest interesting configuration (Fig. 6.25).

Like the Fano configuration, the (8_3) configuration also cannot be realized in the real projective plane. However, unlike the Fano configuration, that failure is not the result of a restriction of the field characteristic, since the Möbius–Kantor configuration can be constructed in complex projective space; see Fig. 6.26 in which ω is a primitive 3rd root of unity in $\mathbb{C}$, $\omega^3 = 1$.

The configuration does have realizations over finite fields. Take, for example, the 8 nonzero vectors in $\mathbb{Z}_3^2$, where the lines are the triples of distinct vectors $\{\mathbf{v}_1, \mathbf{v}_2, \mathbf{v}_3\}$ such that

6.3 Realization of Classical Configurations

Fig. 6.27 The octahedron decomposed into two complete quadrangles of triangles, one visible and the other invisible

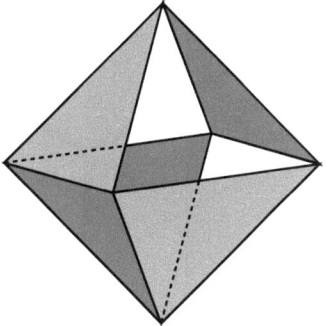

$$\mathbf{v}_1 + \mathbf{v}_2 + \mathbf{v}_3 = \mathbf{0}$$

For a higher dimensional geometric realization, we consider first the regular octahedron whose vertices are embedded in $\mathbb{R}^3$ as $\{\pm \mathbf{e}_i\}$, $i = 1, 2, 3$, in which the triangular faces are the convex hulls of triples of vertices for which $\mathbf{v} \cdot \mathbf{f} = 1$ for one of the eight vectors $\mathbf{f}_i = (\pm 1, \pm 1, \pm 1)$. These faces fall into two color classes, depending on whether $\mathbf{f}_i \cdot (1, 1, 1)$ is even or odd. See Fig. 6.27. Two triangles of any one of these color classes intersect in at most one vertex, so the vertices and color classes of triangles form an incidence structure consisting of 6 points and 4 "lines," a $(6_2, 4_3)$ configuration, in other words, a complete quadrangle.

Suppose we carry out a similar construction one dimension higher, decomposing the hyperplanes of the polyhedron in $\mathbb{R}^4$ with eight vertices $\{\pm \mathbf{e}_i\}$, $i = 1, \ldots, 4$. This is the *cross polytope*. The hyperplanes are sixteen regular tetrahedra. The tetrahedra come in two color classes, as before, with each tetrahedron sharing a face with four tetrahedra of the other color. Since these tetrahedra are indexed by the vertices of the hypercube, they form an (8_4) configuration from which one parallel class of incidences can be ignored to produce the (8_3) configuration; see Fig. 5.21 of Chap. 5.

The (8_3) configuration occurs in a two-colored tiling of the torus by 16 triangles with 8 points. The "lines" are one color class of triangles; see Fig. 6.28. While an interesting realization, it does not capture the full symmetry of the Möbius–Kantor configuration. For this, we consider a tessellation of the two-holed torus by octagons. The two-holed torus is framed as an octagon with opposite sides identified in pairs; see Fig. 6.29. The advantage of this map of the Levi graph of the (8_3) is that each of the automorphisms of the graph is realized as an automorphism of the map. The reflections indicated by the red, blue, and black lines correspond to the generators $(Ab)(aB)(Dd)$, $(Bc)(bc)(dD)$, and $(Bb)(Cd)(cD)$, all represented by reflections of the map and giving rise to a triangular fundamental region on which the group acts freely and transitively. So, as described in Chap. 4, the dual of the map defined by the reflection lines gives the Cayley graph of the automorphism group.

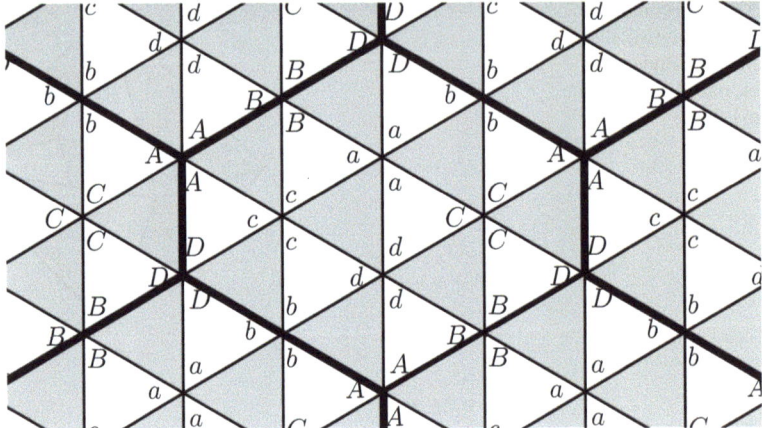

Fig. 6.28 A more exotic realization of the (8_3) configuration

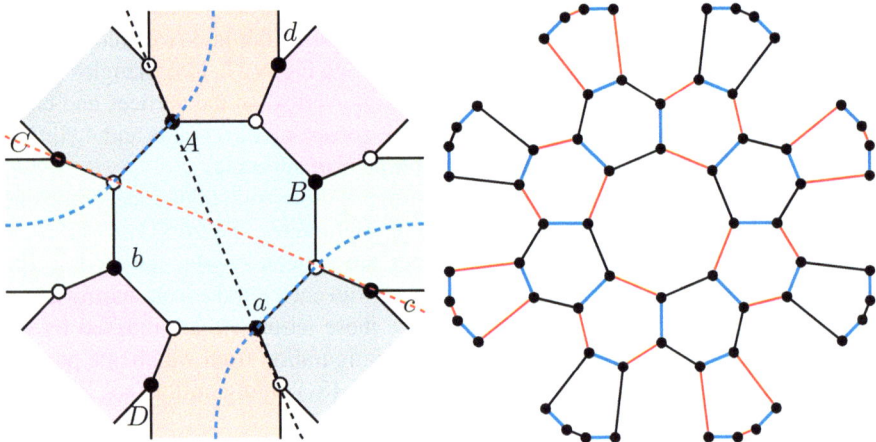

Fig. 6.29 Constructing the Cayley graph of the automorphism group of the Möbius–Kantor configuration. The graphs are drawn on an octagon with opposite sides identified. The labeling of the graph is consistent with Figs. 6.24 and 6.28. The octagons are each colored differently, and there is one seemingly tiny yellow octagon made up of the eight corners of the framing octagon. If the tessellation is straightened as described in Chap. 4, the octagons will all be equal-sized regular hyperbolic octagons

6.3.3 The Pappus Configuration

In constructing the combinatorial (9_3) configurations, we came across quite naturally the hexagonal surface embeddings of the Levi graph. In the case of the Pappus configuration, the automorphism group of the map is an index two subgroup of the automorphism group of the Levi graph, since in the Pappus configuration, there is a nontrivial automorphism fixing a line and all points on that line, while for

6.3 Realization of Classical Configurations

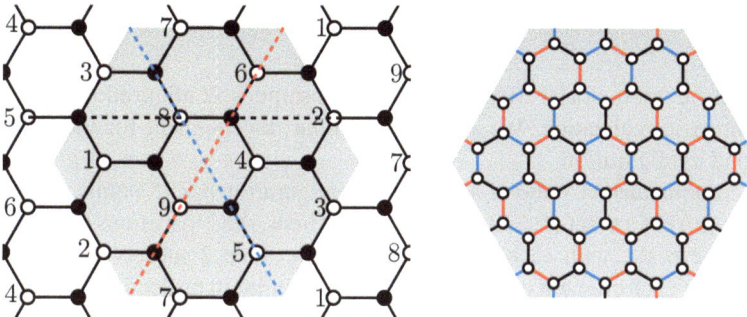

Fig. 6.30 Constructing the Cayley graph of the automorphism group of the Pappus configuration as a regular map of the torus

any connected map, such an automorphism must be trivial. So the Cayley graph indicated in Fig. 6.30 is for an index 2 subgroup of the automorphism group of the Pappus configuration.

6.3.4 The Desargues Configuration

We saw that the Desargues configuration is combinatorially realized as a Levi graph by the Kronecker double cover of the Petersen graph; see Sect. 5.3.4. The vertices of the double cover are partitioned into the ten black vertices, the 2-subsets of a five set, and ten white vertices, the 3-subsets of a 5-set. Adjacency between a 2-subset and a 3-subset is simple containment. It is possible to use this combinatorial representation to construct a geometric realization of the Desargues configuration. Suppose the numbers 1, 2, 3, 4, and 5 refer to five planes in a three-dimensional affine or projective space over a field. Assume that the planes are generically chosen, so that there is no parallelism or degeneracy. Then each triple (ijk) can be associated with the point of intersection of the three planes i, j, and k, and the pairs (ji) can be regarded as the intersection of the two planes i and j, that is, a line.

Following the combinatorial description, each point (ijk) lies on three lines (ij), (ik), and (jk), and every line contains 3 points. Thus, any generic choice of five planes leads to a realization of the (10_3) Desargues configuration in three-dimensional space. Projecting down one dimension gives a realization in the plane.

The three-dimensional structure, with 10 points, 10 lines, and 5 planes, is interesting in its own right. Each plane, say 1, contains 6 points, (123), (124), (125), (234), (235), and (245), and 4 lines 12, 13, 14, and 15, hence contains a complete quadrangle. Every point lies on 3 planes.

Dually we could let 1, 2, 3, 4, and 5 refer to five points. Two points determine a line, and three points determine a plane, giving a configuration of lines and planes which is the dual of the previous construction, again the Desargues configuration.

6.3.5 The Cremona–Richmond Configuration

The purely combinatorial definition of the Desargues configuration led directly to a geometric realization. We will see a similar development for the Cremona–Richmond configuration.

The combinatorial Cremona–Richmond configuration has 15 points, namely, the 2-subsets (ab) of a 6-set $\{1, 2, 3, 4, 5, 6\}$, and 15 lines, the 2-partitions $(ab)(cd)(ef)$ of the same 6-set, with each line incident to the three 2-subsets comprising the partition. Let each element of $\{1, 2, 3, 4, 5, 6\}$ be a point in projective 4-dimensional space such that the set of points $\{1, 2, 3, 4, 5, 6\}$ is in general position, that is, no three are collinear, no four are coplanar, and no five are cospatial.

Each pair (ij) of two points determines a line which intersects the hyperplane determined by the four points $\{1, 2, 3, 4, 5, 6\} - \{i, j\}$ of the complement of (ij) in a point we denote by $P_{(ij)}$.

Consider three pairs which partition $\{1, 2, 3, 4, 5, 6\}$, say (12), (34), and (56). They correspond to the three points $P_{(1,2)}$, $P_{(3,4)}$, $P_{(5,6)}$ which all lie on the hyperplane determined by $\{3, 4, 5, 6\}$: The whole lines determined by (34) and (56) are contained in that hyperplane, hence, in particular, $P_{(3,4)}$ and $P_{(5,6)}$, and the line determined by (12) intersects $\{3, 4, 5, 6\}$ exactly in the point $P_{(1,2)}$. In the same way, the three points are contained in the two hyperplanes determined by the sets $\{1, 2, 5, 6\}$ and $\{1, 2, 3, 4\}$. Since the points $\{1, 2, 3, 4, 5, 6\}$ are in general position, these three hyperplanes intersect in a line, so $P_{(1,2)}$, $P_{(3,4)}$, and $P_{(5,6)}$ are in fact collinear, and we naturally associate that line with the partition (12)(34)(56), giving exactly the incidences required. Thus, the construction of the points $P_{(i,j)}$ defines a four-dimensional realization of the Cremona–Richmond configuration.

As with the previous construction of the Desargues configuration, once the configuration is realized as a collection of points and lines in higher dimensions, lower dimensional configurations may be obtained by projection, although losing perhaps some additional geometrical features, in this case, the hyperplanes. See Fig. 6.31 in which the configuration has been projected with fivefold rotational symmetry.

6.3.6 The Reye Configuration

In Sect. 5.3.6, we defined the Reye configuration abstractly in terms of signed 2-subsets. For a realization, therefore, we ought to consider a geometric situation in which 2-subsets have some handedness. The classical realization is in terms of the internal and external points of similitude of two circles. Given two circles, a point of similitude between them is a point at which the two circles appear identical from central projection, that is, to an observer at that point. A point of similitude can be constructed as the intersection of the common tangents of the circles; see Fig. 6.32 in which the circles are not congruent and the sum of the radii is less than the distance

6.3 Realization of Classical Configurations

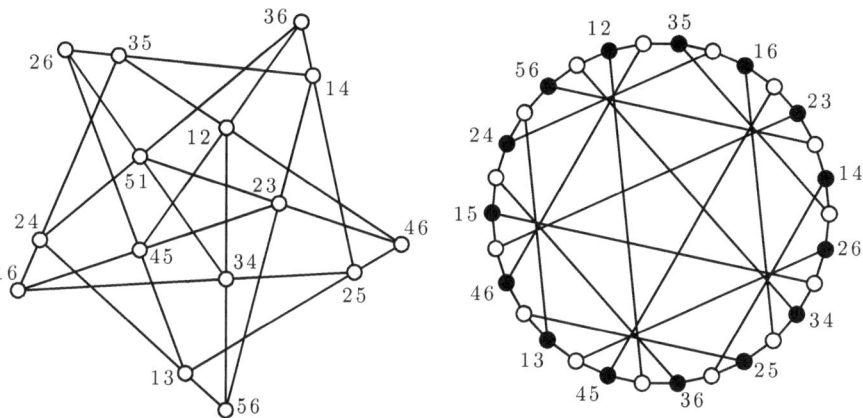

Fig. 6.31 The Cremona–Richmond configuration and its Levi graph

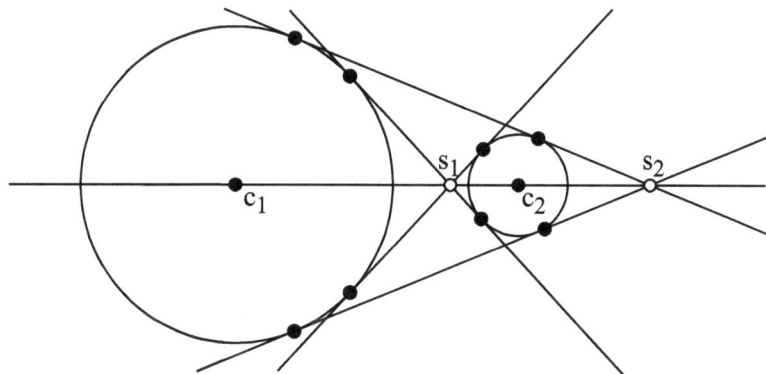

Fig. 6.32 Constructing the points of similitude of two circles

between the centers. In this case, there are two points of similitude, one *internal*, along the segment $[c_1, c_2]$, and one *external*. If we regard the centers of the circles as points on the real line, for instance, if we regard the plane to be the complex plane, the coordinates are easily seen to satisfy the equation

$$\frac{r_2}{r_1} = \frac{c_2 - s_1}{s_1 - c_1} = \frac{s_2 - c_2}{s_2 - c_1}$$

which gives

$$\frac{(c_1 - s_1)(c_2 - s_1)}{(c_2 - s_2)(c_1 - s_2)} = -1.$$

We note that the expression on the left is the *cross ratio*.

If the circles intersect or if one is contained in the interior of the other, then there is no internal point where the common tangents cross. In this case, an alternate

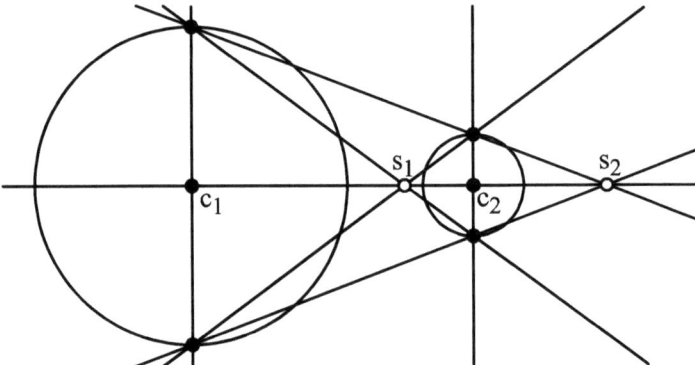

Fig. 6.33 An alternative construction of the points of similitude of two circles via perpendiculars through the centers

Fig. 6.34 The centers of similitude form a complete quadrangle

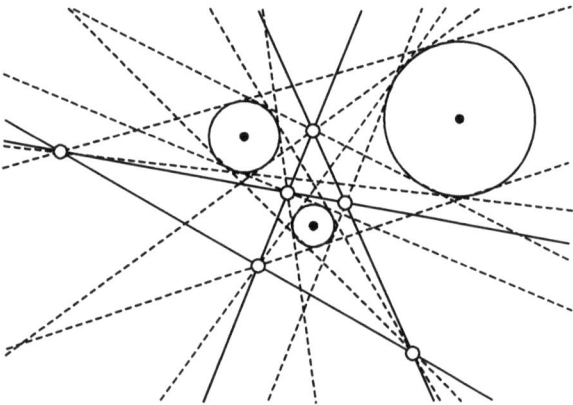

construction may be used which gives the same condition; see Fig. 6.33. If the radii are equal, then the internal point of similitude is at the midpoint of the segment connecting the centers of the circles, and the external point of similitude is the point at infinity of the line between the centers.

In either case, if the radii are given, we can find both s_1 and s_2.

$$s_1 = \frac{r_2}{r_1 + r_2}c_1 + \frac{r_1}{r_1 + r_2}c_2$$

$$s_2 = \frac{r_2}{r_2 - r_1}c_1 + \frac{r_1}{r_1 - r_2}c_2$$

For 3 circles in the plane, it is straightforward to show (see Exercise 6.12) that the six points of similitude will be arranged in a complete quadrangle; see Fig. 6.34. It is important to note the interaction between the internal and external centers. One of the lines of the quadrangle will contain all three external centers, while the

6.3 Realization of Classical Configurations

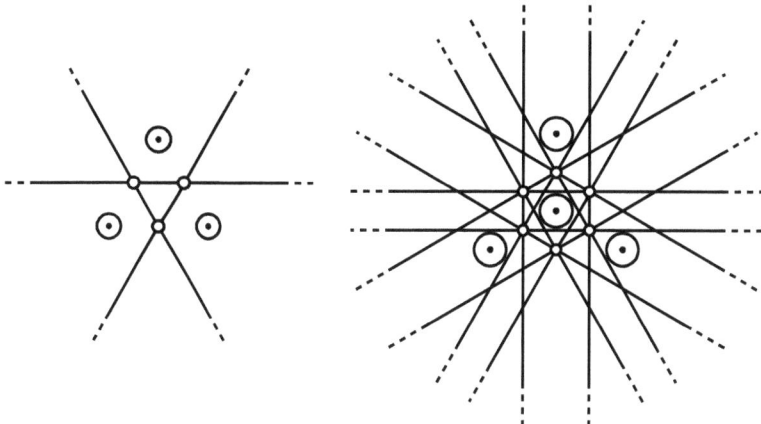

Fig. 6.35 Three and four *circles* and the configurations of their centers of similitude

other three lines will each contain two internal centers, in other words, exactly those triples of points for which the number of external centers is odd. The situation of a symmetric placement of circles is particularly easy to visualize and analyze; see Fig. 6.35 in which three congruent circles are placed at the vertices of an equilateral triangle. In this case, the quadrangle is realized as an equilateral triangle whose sides join two internal centers and three external centers lie along the line at infinity.

Moving on to three dimensions and the Reye configuration, consider four small congruent spheres centered at the vertices of a regular tetrahedron in three-dimensional projective space. The internal centers of similitude will be at the midpoints of the 6 edges, which are the vertices of a regular octahedron; see the diagram on the right of Fig. 6.35 which is viewed from above. The six external points are all on the sphere at infinity and are joined by four lines in a complete quadrangle, with each line of the quadrangle corresponding to one of the faces of the original tetrahedron of spheres. The condition on the internal and external centers for each line of each quadrangle, that the number of external centers is always odd, is exactly the condition we saw in Sect. 5.3.6 between signed 2-subsets to form the Reye configuration.

Geometrically our realization of the Reye configuration supports twelve planes: eight planes which contain one of the faces of the octahedron, with each of these eight planes containing four lines, including one line on the plane at infinity; three additional planes bisecting the octahedron, each containing four lines; and the final plane at infinity, itself with its four lines. These 12 planes give a plane-line configuration which is also a realization of the Reye configuration. It may, in fact, be constructed as the polar of the original Reye configuration, with the octahedron transformed to a cube and with the sphere at infinity transformed to its centroid; see Fig. 6.36.

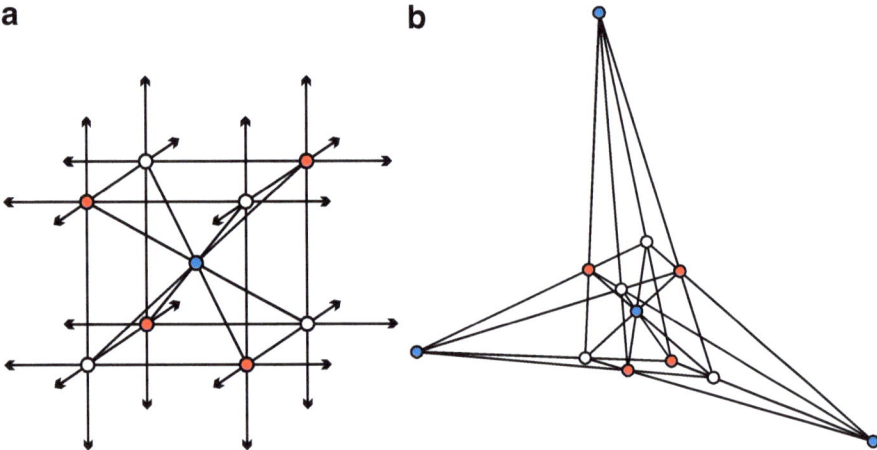

Fig. 6.36 Two views of the Reye configuration. The figure on the *left* shows the eight *red and white "cube"* vertices and the *blue "centroid"* vertex, with the three points at infinity indicated by the *arrows*. The figure on the *right* is foreshortened with all finite points, and projectively, vertices of any two colors could play the role of the cube vertices, and any vertex of the remaining color could play the role of the centroid

For convenience, we have considered the realization in which the spheres are congruent and symmetrically placed. Applying a projective transformation will destroy the symmetry but leave the incidences intact, just as for the plane. For a recent quantum mechanical application of the Reye configuration, see [2].

The Möbius (8_4) Spatial Configuration

The Möbius hypercubes of Sect. 5.3.7 may be realized as a collection of points and planes in $\mathbb{R}^3$. In any point-plane realization, the outer cube and the inner cube of the Möbius hypercube will correspond to a spatial tetrahedron, and the two spatial tetrahedra must be placed so that each vertex of one tetrahedron lies on a distinct plane of the other tetrahedron. This can clearly not be done if one requires that the vertices all lie in the interior of a face of the other tetrahedron.

The simplest construction is to consider two tetrahedra with equilateral bases, equal heights, each of whose fourth vertex lies on the perpendicular line through the base triangle which passes through the centroid of its three vertices. It is not necessary that the two base triangles be congruent. Place the tetrahedra such that the apex of one lies at the centroid of the base of the other and vice versa. The base triangles must then lie in parallel planes; see Fig. 6.37. Finally, we turn one tetrahedron along the centroid axis so that it is 90° out of phase with the other tetrahedron. Projecting along that axis, the base triangles will be such that the

6.3 Realization of Classical Configurations

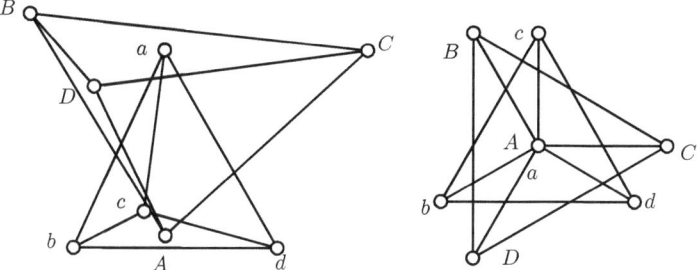

Fig. 6.37 Constructing a Möbius structure from two tetrahedra. The planes bcd and BCD are parallel. The view on the *right* is along the centroid axis

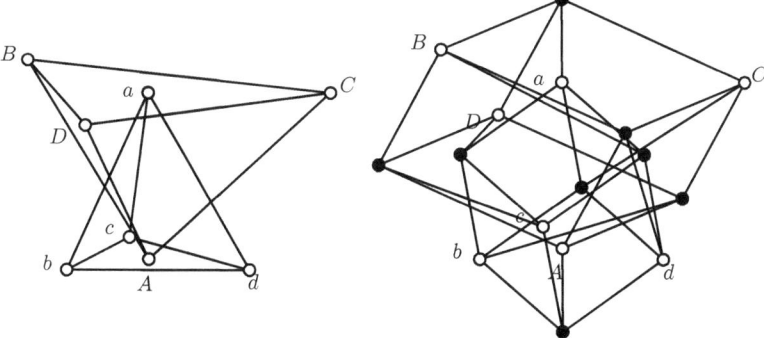

Fig. 6.38 The hypercube Q_4, the incidence graph of the Möbius structure of two twisted tetrahedra

altitudes of one base triangle will be parallel to the sides of the other. So, for example, side bc in one base triangle will be parallel to the altitude through D in the other base triangle and that altitude also passes through a. It follows that the points $\{a,b,c,D\}$ will be coplanar. Altogether we have 8 planes each incident with four points,

$$abcD \quad abCd \quad aBcd \quad Abcd$$
$$ABCd \quad ABcD \quad AbCD \quad aBCD$$

the six planes corresponding to the base sides of one base triangle and the altitude of the other, together with the planes of the two base triangles. The incidence graph of this (8_4) point-plane incidence structure must be one of the Möbius hypercubes; see section "Möbius (8_4) Incidence Structures". It is in fact the Möbius hypercube corresponding to the identity permutation, the true hypercube; see Fig. 6.38.

The incidence graph is the standard 3-dimensional hypercube, having 16 vertices, 32 edges, 24 faces, and 8 cells. The 24 faces belong to 6 parallel classes of four faces each. Each edge belongs to 3 faces, all of which are 4-cycles, so it is possible to delete a set of faces such that each edge belongs to exactly two faces, forming a polygonal surface. This yields a map with 16 faces, all of which are 4-cycles,

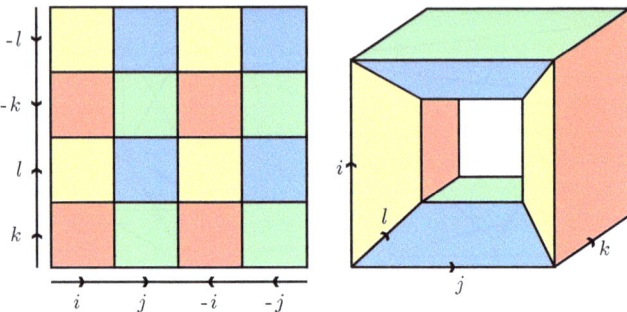

Fig. 6.39 One of the three regular maps of the hypercube graph on the torus with all faces 4-cycles

Fig. 6.40 The Cayley graph of the group of the $\{4,4\}_{4,4}$ regular map on the hypercube, an index 3 subgroup of the group of the hypercube

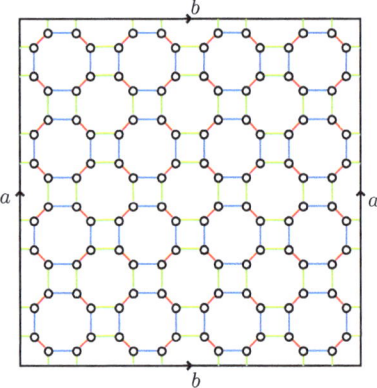

so the map must have Euler characteristic 0. Taking the standard representation of the hypercube, each edge is parallel to one of the four coordinate directions $\mathbf{i}$, $\mathbf{j}$, $\mathbf{k}$, or $\mathbf{l}$ in $\mathbb{R}^4$, and each face is spanned by a choice of any two of those four directions. Suppose we delete the face spanned by $\mathbf{i}$ and $\mathbf{j}$, say containing the points $\{\mathbf{x}, \mathbf{x}+\mathbf{i}, \mathbf{x}+\mathbf{k}, \mathbf{x}+\mathbf{i}+\mathbf{k}\}$. Then the 4-cycles $\{\mathbf{x}, \mathbf{x}+\mathbf{j}, \mathbf{x}+\mathbf{k}, \mathbf{x}+\mathbf{j}+\mathbf{k}\}$ share an edge with the deleted 4-cycles and so must remain and share the edge $\{\mathbf{x}, \mathbf{x}+\mathbf{k}\}$, so the face $\{\mathbf{x}, \mathbf{x}+\mathbf{k}, \mathbf{x}+\mathbf{l}, \mathbf{x}+\mathbf{k}+\mathbf{k}\}$ at $\mathbf{x}$ must also be deleted. By the connectedness of the complex, it follows that all the faces which are parallel to either $\{\mathbf{x}, \mathbf{x}+\mathbf{i}, \mathbf{x}+\mathbf{k}, \mathbf{x}+\mathbf{i}+\mathbf{k}\}$ or $\{\mathbf{x}, \mathbf{x}+\mathbf{k}, \mathbf{x}+\mathbf{l}, \mathbf{x}+\mathbf{k}+\mathbf{l}\}$ must also be deleted. Therefore, there are exactly three maps with 4-cycle faces possible: One with all faces spanned by $\mathbf{i}$ and $\mathbf{j}$ and those spanned by $\mathbf{k}$ and $\mathbf{l}$ are deleted; one with all faces spanned by $\mathbf{i}$ and $\mathbf{k}$ and those spanned by $\mathbf{k}$ and $\mathbf{l}$ are deleted; and one with all faces spanned by $\mathbf{i}$ and $\mathbf{l}$ and those spanned by $\mathbf{j}$ and $\mathbf{k}$ are deleted; see Fig. 6.39.

Since the group of the hypercube acts on the sets of deleted faces, the map is a regular map, and the Cayley graph of its automorphism group (see Fig. 6.40) is given by its barycentric subdivision, as we have seen in Chap. 4. The group of the map has 128 elements and is an index 3 subgroup of the full automorphism group of the hypercube, corresponding to the three ways of embedding the hypercube graph in

6.3 Realization of Classical Configurations

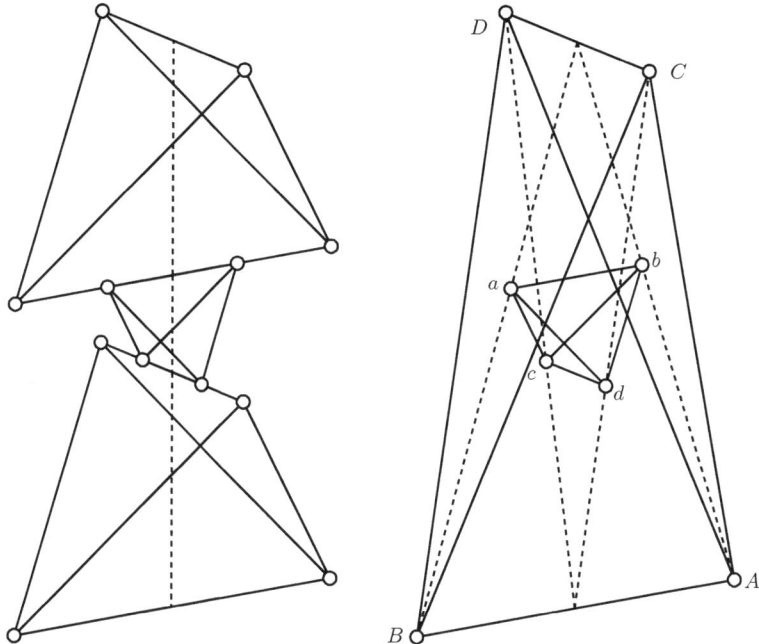

Fig. 6.41 Realizing a Möbius hypercube. The *dotted lines* are altitudes of the face triangles of the exterior tetrahedron which have been drawn in to aid the eye in determining that the points of the interior tetrahedron lie on the faces of the exterior tetrahedron

the torus as a regular map with 4-cycle faces. The three regular maps also correspond to the six ways to consistently cyclically order the four edges around a vertex of the hypercube, that is, to cyclicly order $\{\mathbf{i},\mathbf{j},\mathbf{k},\mathbf{l}\}$.

The pair of twisted tetrahedra in Fig. 6.38 is not the only realization of a Möbius hypercube as a point-plane incidence structure, that is, as a pair of tetrahedra in space with incidences matching the vertices of one to the planes of the other. If we start with a regular tetrahedron in $\mathbb{R}^3$ and consider a line passing through the centers of two antipodal edges, we may string two other regular tetrahedra along this line, one on top and one on the bottom; see Fig. 6.41. Since the tetrahedra are regular and the dihedral angles are all equal, the faces of the middle and top tetrahedron which share a line are coplanar in pairs, and so each of the top vertices of the top tetrahedron lies on a plane though a face of the middle tetrahedron. The analogous statement is true for the bottom tetrahedron. To realize the Möbius hypercube as a point-plane incidence structure, we require two tetrahedra; the first will be the middle tetrahedron in the stack of three, which we will call the *internal* tetrahedron, and the second tetrahedron will be assembled from the two top vertices of the top tetrahedron and the two bottom vertices of the bottom tetrahedron and will be called the exterior tetrahedron. Now, simultaneously scale the top and bottom tetrahedra so that the points of the interior tetrahedron lie on the faces of the exterior tetrahedron;

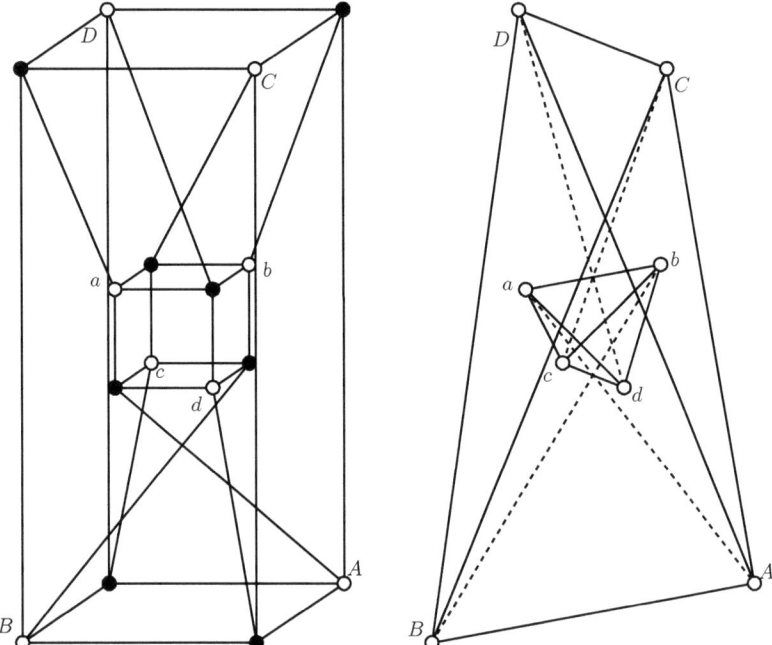

Fig. 6.42 The Möbius hypercube $(ab)(cd)$ realized as a pair of mutually inscribed tetrahedra. The *dotted lines* are altitudes of the face triangles of the interior tetrahedron drawn to illustrate that the points of the exterior tetrahedron lie on the faces of the interior tetrahedron

see Fig. 6.42. This incidence structure of two mutually inscribed tetrahedra realizes the Möbius hypercube $(ab)(cd)$.

6.4 Representations and Realizations

Let us here make precise the ideas of representation and realization of incidence structures considered earlier.

Let $\mathcal{C}$ and $\mathcal{C}'$ be two incidence structures with incidence graphs G and G'. A structure homomorphism from $\mathcal{C}$ to $\mathcal{C}'$ is a graph homomorphism

$$\phi : G \to G'$$

which respects the bipartitions (colors) of the incidence graphs. A structure homomorphism is said to be *direct* if it preserves the bipartitions, so it sends points to points and blocks to blocks and indirect otherwise. Homomorphisms are presumed to be direct unless specified otherwise.

6.4 Representations and Realizations

A structure homomorphism which is one to one on the vertices corresponding to points is said to be *point faithful* and, similarly, *line faithful* if it is one to one on the vertices corresponding to blocks. A homomorphism which is both point and line faithful is said to be *flag faithful*. A point and line faithful homomorphism is said to be a *representation* if the image of G is an induced subgraph of G'.

If the points and blocks of the image structure G' of a representation are actual points and lines in an affine or projective space, the representation is called a *realization*. For contrast, a structure homomorphism into real or projective space is called a *weak realization*. Sometimes a realization will be called a *strong realization* for emphasis.

The Euclidean plane $\mathbb{E}$ and the real projective plane PG$(2, \mathbb{R})$ are both infinite incidence structures. Given any finite incidence structure $\mathcal{C}$, *the (weak) realization defect* is the least number of lines that have to be removed from $\mathcal{C}$ in order to obtain an incidence structure which can be (weakly) realized in $\mathbb{E}$ or PG$(2, \mathbb{R})$.

The ambiguity implied by this definition is resolved by the following:

Proposition 6.11. *For any finite incidence structure $\mathcal{C}$, the defects with respect to $\mathbb{E}$ and* PG$(2, \mathbb{R})$ *are the same.*

Proof. Since the Euclidean plane embeds into the real projective plane, any structure that embeds in the Euclidean plane embeds also in the projective plane. On the other hand, if $\mathcal{C}$ can be realized in the projective plane PG$(2, \mathbf{R})$, then choose any point $p_0 \in$ PG$(2, \mathbf{R})$ which is not in the image of $\mathcal{C}$. There are finitely many lines among the infinite pencil of lines through p_0 which pass through points in the realization of $\mathcal{C}$, so there exists a line L through p_0 which intersects no point in the realization. Regarding L as the line at infinity, the realization into PG$(2, \mathbf{R})$ is then a Euclidean realization.

We have seen in Sect. 6.3.1 that the realization defect of the Fano plane is exactly 1 and the Möbius–Kantor configuration has defect at most 1. We may also restate Theorem 6.10.

Theorem 6.12 (Steinitz). *Every* (n_3) *configuration has weak realization defect 1.*

That Steinitz's argument can only guarantee that a weak realization with defect 1 is clear from the consideration of the combinatorial configuration $\mathcal{C}$ corresponding to the Levi graph of Fig. 6.43 in which two Pappus graphs have been joined together by deleting two edges, one in each copy of the Pappus graph, and their four endpoints rejoined as indicated by the bold edges. This move is called an incidence switch; see Sect. 6.5. Any two-dimensional Euclidean realization of $\mathcal{C}$ will have to manifest the deleted incidences by the theorem of Pappus. Restoring the dotted edges will result in a 4-cycle in the incidence graph, so a Euclidean representation must at least have two of the vertices on this 4-cycles represented by the same geometric object. That is, either the representation is not point faithful or not line faithful. For the line-deleted realization indicated by Steinitz's argument, one of the dotted incidences still must be manifest, so the realization must perforce be weak.

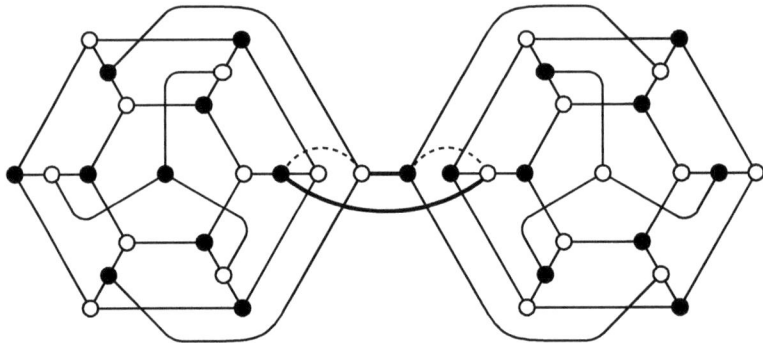

Fig. 6.43 Even after the deletion of any line, a Euclidean realization will be forced to include one of two dotted incidences by the Pappus theorem

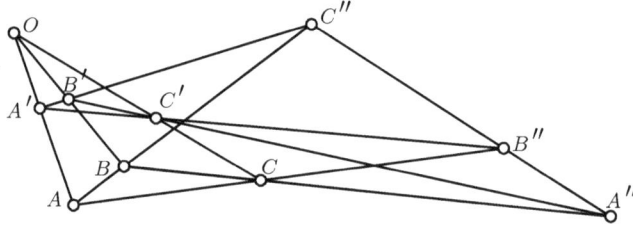

Fig. 6.44 The Desargues configuration is of dimension 3

6.4.1 The Dimension of a Realization

Given a realization $\mathcal{C} \to \mathbb{R}^n$, we say that the realization has *dimension* k if the images of the points of $\mathcal{C}$ span a k-dimensional affine subspace of $\mathbb{R}^n$. If a finite incidence structure $\mathcal{C}$ has a realization of dimension k, then it has a realization of dimension k' for all $2 \leq k' \leq k$. This is seen by projecting the k-dimensional subspace on any k'-dimensional subspace which contains none of the lines through any pair of points of $\mathcal{C}$. For any finite incidence structure, it is not possible to have any realization whose dimension is higher than the affine span of all of its points, so there is a well-defined maximum dimension for the set of realizations of any fixed incidence structure $\mathcal{C}$, which is defined to be the *dimension* of $\mathcal{C}$. With essentially the same proof as that of Proposition 6.11, we have

Proposition 6.13. *For any finite incidence structure $\mathcal{C}$, the dimensions of its realizations into Euclidean and projective space are the same.*

It is not hard to show, for example, that any realization of the Pappus configuration is of dimension 2. On the other hand, we have seen above that the Desargues configuration has a realization of dimension 3. Moreover, given any realization of the Desargues configuration (see Fig. 6.44), the affine space spanned

6.4 Representations and Realizations

by $\{O, A, B, C\}$ contains the third points of each of the six lines through these four points, which comprises all the remaining six points of the configuration. So the dimension of any realization of the Desargues configuration is at most the dimension spanned by these four points, that is, at most three. So the Desargues configuration has dimension 3.

A configuration also has a *dual dimension,* a useful concept since the dimension of the dual configuration need not agree with the dimension of the original configuration. Extreme examples of this are the $(n_{n-1}, \binom{n}{2}_2)$ configurations K_n whose points and lines are the vertices and edges of the complete graph. K_n clearly has dimension $n - 1$. Consider the points $(1, 2)$, $(2, 3)$, and $(3, 1)$ in the dual of K_n, whose span contains the lines 1, 2, and 3. The line 1 contains the point $(1, i)$, and the line 2 contains the point $(2, i)$; thus, the span of $(1, 2)$, $(2, 3)$, and $(3, 1)$ contains the line i which joins $(1, i)$ and $(2, i)$ and so contains all the lines of the configuration, so K_n has dual dimension at most 2. It is not difficult to see that the dual dimension is exactly 2. In particular, the dimension of the complete quadrangle is 2.

6.4.2 Lifting Configurations

The general problem of determining whether a finite incidence structure $\mathcal{C}$ is realizable is intrinsically difficult. A projective realization in $\mathrm{PG}(n, \mathbb{R})$ may be regarded as an $(n + 1) \times v$ *coordinatization matrix* of homogeneous coordinates which, without loss of generality, has the $n + 1$st row entirely of 1's;

$$M = \begin{bmatrix} m_{1,1} & m_{1,2} & \cdots & m_{1,v} \\ m_{2,1} & m_{2,2} & \cdots & m_{2,v} \\ \vdots & \vdots & \cdots & \vdots \\ m_{n,1} & m_{n,2} & \cdots & m_{n,v} \\ 1 & 1 & \cdots & 1 \end{bmatrix}.$$

Requiring a set of points to be collinear is to require that the relevant columns of this matrix to be a rank 2 submatrix. In particular, for points $\{i, j, k\}$, to be collinear in the projective plane requires that the ith, jth, and kth columns of the $3 \times v$ matrix

$$\begin{bmatrix} m_{1,1} & m_{1,2} & \cdots & m_{1,v} \\ m_{2,1} & m_{2,2} & \cdots & m_{2,v} \\ 1 & 1 & \cdots & 1 \end{bmatrix}.$$

to be of rank 2; in particular, the determinant must be zero,

$$\begin{vmatrix} m_{1,i} & m_{1,j} & m_{1,k} \\ m_{2,i} & m_{2,j} & m_{2,k} \\ 1 & 1 & 1 \end{vmatrix} = 0.$$

Fig. 6.45 The Fano-minus-one structure point faithfully represented on the line and then lifted into the plane

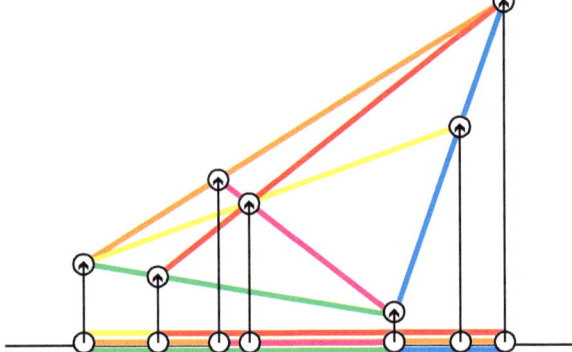

If we are searching for a realization, this requirement is a quadratic equation in the variables $m_{i,j}$, and the complete realization problem would require a solution of that system of quadratic equations.

One approach to overcoming this intractability in realizing configurations, as indicated by Steinitz's argument, is to sequentially introduce the points of the configuration into the plane until an obstruction intervenes and then, if possible, to backtrack. For the Pappus configuration, if each point is sequentially introduced and satisfies the constraints required among the previously introduced points, there will always be, by the Pappus theorem, at least one position to place the last point to realize the configuration. For the Desargues configuration, if one is unlucky enough to place the vertices $\{A, B, C, A', B', C'\}$ first, there is zero probability of being able to place point O on line $A''B''C''$. For the Fano configuration, of course, no amount of backtracking will result in a realization.

A different, more holistic approach is to place all the x-coordinates of the points first, without regard to the constraints, and then to use those constraints to select the corresponding y-coordinates. The initial placement can be regarded as a rather trivial representation onto a line, and the corresponding solution a *lift* into the plane; see Fig. 6.45. If we are lifting with respect to a single coordinate, then the coordinatization matrix has n rows of constants, including the rows of 1's, and one row of variables, so each determinant comprising a lifting constraint is a linear homogeneous equation in the lifting variables. For a (v_3) configuration, there are v variables, one for each point, and v linear constraints, one for each line. If the constraints are all independent, there will be the unique trivial solution, the trivial lift with "height" zero for every point. This never occurs, however, since there are always *trivial lifts*, that is, lifts of the $(n-1)$-dimensional affine space into another $(n-1)$-dimensional affine space. There is always an n-dimensional space of such trivial lifts. In order to have independent constraints with a nontrivial lift, therefore, it is necessary that the number of constraints not exceed the number of points minus 3. If we want not simply a nontrivial lift but a line faithful representation, it would seem that we would need much more, but fortunately there is a result that guarantees that this condition is sufficient, provided it applies not merely to the combinatorial incidence structure as a whole but as well to any substructure, [105].

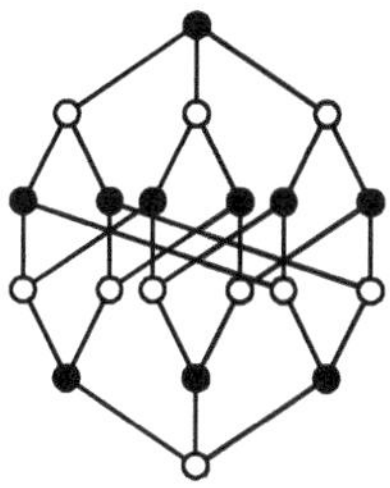

6.4 Representations and Realizations

$i = 3$	$a = 1$	$b = 6$	$l + 2p - 2 = 8$
$i = 9$	$a = 8$	$b = 6$	$l + 2p - 2 = 20$
$i = 21$	$a = 8$	$b = 9$	$l + 2p - 2 = 23$
$i = 27$	$a = 9$	$b = 9$	$l + 2p - 2 = 25$
$i = 30$	$a = 9$	$b = 9$	$l + 2p - 2 = 25$

Fig. 6.46 The incidence numbers associated with subconfigurations of the Pappus configuration

Theorem 6.14 (Whiteley). *Given an incidence structure $\mathcal{C}$ and a generic choice of x coordinates $\{x_1, \ldots, x_v\}$, there exists a lift $\{y_1, \ldots, y_v\}$ which is a strong realization of the incidence structure if and only if*

$$i \leq a + 2b - 3$$

is satisfied for all substructures with at least two blocks, where the substructure has a points, b blocks, and i incidences.

A realization is *generic* if any incidences it exhibits must be exhibited by all realizations of the structure. For a set of x coordinates to be generic, it is sufficient to take a set of algebraically independent transcendentals.

It seems that we have just the tools here to get a strong realization of a (v_3) configuration with some defects. If we delete a line and its three points, that eliminates 9 incidences and 3 points and 1 line, and we have for the subconfiguration $i \leq a + 2b - 4$. Then, given a strong realization of the subconfiguration, we have at most but a few undesired incidences, all clustered about the deleted line, which we can curve if necessary; see Fig. 6.46. However, to apply Whiteley's theorem, the count condition must also be satisfied by all subconfigurations of the configuration, and that will depend on the particular configuration. In the configuration of Fig. 6.45, each of the Pappus-minus-one graphs corresponds to an incidence structure which violates the condition.

6.4.3 General Lifting

There is a more general version of Whiteley's theorem which applies, for example, to realizations of point-plane incidence structures. A *picture* of an incidence structure $\mathcal{C}$ is an embedding π of the points of $\mathcal{C}$ into $(k-1)$-dimensional projective space PG $(k-1, \mathbb{R})$. A *lift* of the picture π is an embedding σ into PG $(k-1, \mathbb{R})$ which projects (vertically) onto π such that the set of vertices of any block span a $k-1$ dimensional subspace. A lift is called *sharp* if distinct blocks lift to distinct affine subspaces.

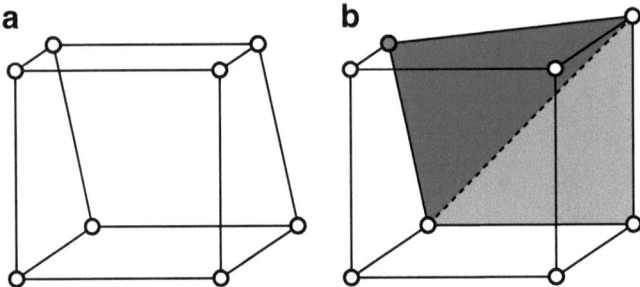

Fig. 6.47 The picture of the cube graph on the *left* is liftable to a trapezoidal prism. By contrast, the picture on the *right* has no lift, and the *dotted line* indicates a forced crease

Theorem 6.15 (Whiteley). *A generic picture in* $(k-1)$-*space of an incidence structure has a sharp lift if and only if*

$$i \leq a + kb - (k+1)$$

for all subincidence structures having at least two blocks.

It is possible to use this higher dimensional theorem to study the structure of point-line configurations of dimensionality greater than 2. For instance, let us consider the Reye incidence structure; see Fig. 6.36. A realization of the point-line Reye configuration in the projective plane may be obtained from the projection of the vertices of a three-dimensional cube, or more generally a parallelepiped, together with the centroid of the parallelepiped, which lies at the point of intersection of the four main diagonals, and three points at infinity corresponding to three parallel classes of lines. Conversely, given a plane drawing of the points and lines of the 1-skeleton of a parallelepiped, we can ask if it must necessarily be the projection of a three-dimensional parallelepiped. Clearly the main diagonals must be concurrent. Moreover, the four lines containing the edges corresponding to each perfect matching must be concurrent. In other words, it is a necessary condition for the existence of a lift that the plane picture of the cube graph be extendable to a Reye configuration. Is this condition sufficient?

If a plane picture $\pi(C)$ may not be completed to a Reye configuration because of the failure of some required concurrence, $\pi(C)$ will not be the projection of any parallelepiped, but it may still have a sharp lift; see Fig. 6.47a. On the other hand, it may happen that $\pi(V)$ has no sharp lift (see Fig. 6.47b) where the dotted line indicates a crease that would be forced when a sharp lift of the five white vertices and the faces they span is extended to the last gray vertex of C. We would like to determine geometrically when a plane picture $\pi(C)$ has a sharp lift.

A generic picture of the cube with 8 vertices and 6 faces violates the 2-dimensional lifting condition and so does not have a sharp lift. If we remove one vertex, say H, from the vertex set $\{A, B, \ldots, H\}$ as well as the three facial incidences containing H from the point-plane incidence structure of the cube, the

6.4 Representations and Realizations

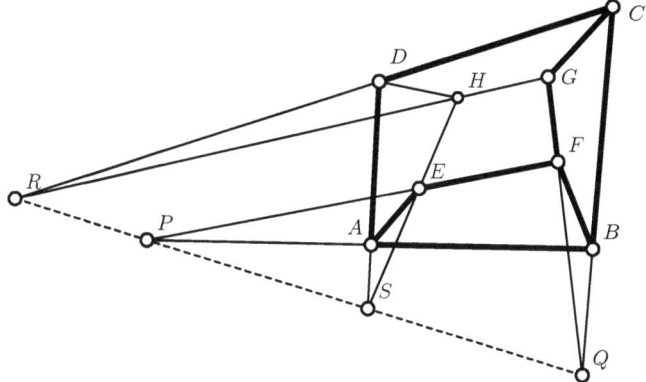

Fig. 6.48 A purely planar construction of $\pi(H)$ from $\pi(\mathcal{C} - H)$: If there is a lift, then the line joining $P = (EF \wedge AB)$ and $Q = (BC \wedge FG)$ must contain the points $S = (AD \wedge EH)$ and $R = (CD \wedge GH)$

resulting incidence structure does satisfy Whiteley's condition, and so do all its subincidence structures. Hence, a generic picture of $\mathcal{C} - H$ has a sharp lift into 3-space. Moreover, since the picture is generic, the three planes corresponding to the three faces incident to H intersect in one point, call it $\sigma(H)$, in space, which projects back to a point $\pi(H)$ in the plane, extending $\pi(\mathcal{C} - H)$ nongenerically to a liftable picture $\pi(\mathcal{C})$. Note that the same conclusion holds if we assume only that the three quadrilaterals of $\pi(\mathcal{C} - H)$ are each in general position.

Given a sharp lift of $\pi(\mathcal{C} - H)$, the planes through $\{\sigma(A), \sigma(B), \sigma(C), \sigma(D)\}$, and $\{\sigma(E), \sigma(F), \sigma(G)\}$ intersect in a line in space which projects to a line in the plane determined by the intersection point of lines $\pi(A)\pi(B)$ and $\pi(E)\pi(F)$, call it $\pi(P)$, as well as the intersection point of lines $\pi(B)\pi(C)$ and $\pi(F)\pi(G)$, call it $\pi(Q)$; see Fig. 6.48. Let $\pi(A)\pi(D)$ intersect $\pi(P)\pi(Q)$ at $\pi(S)$, so $\pi(H)$ must lie on $\pi(S)\pi(E)$. By the same token, $\pi(H)$ must lie on $\pi(R)\pi(G)$. Note that sharp lifts of $\pi(\mathcal{C} - H)$ give rise to precisely the same point $\pi(H)$.

Theorem 6.16. *Given the cube graph $\mathcal{C}$ with vertex set $\mathcal{V} = \{A, \ldots, H\}$ and given seven generic points $\{\pi(A), \ldots, \pi(G)\}$ in the plane, there is a unique eighth point $\pi(H)$ so that $\pi(\mathcal{C})$ is a plane picture of $\mathcal{C}$ with a sharp lift.*

If the picture of $\mathcal{C}$ has a sharp lift, then the quadrilaterals corresponding to a pair of opposite faces in $\mathcal{C}$ are perspective from a line. On the other hand, we have just seen that if the quadrilaterals corresponding to any pair of opposite sides in $\mathcal{C}$ are perspective from a line, then that picture is the unique picture corresponding to the sharp lift guaranteed by the incidences. So we have the following:

Theorem 6.17. *Given a picture $\pi(\mathcal{C})$ of $\mathcal{C} = (\mathcal{V}, \mathcal{E}, \mathcal{F})$ such that the points $\pi(f)$ are in general position for each $f \in \mathcal{F}$. If two quadrilaterals corresponding to a pair of opposite faces in $\mathcal{C}$ are perspective from a line, then all pairs of quadrilaterals in $\pi(\mathcal{C})$ corresponding to a pair of opposite faces in $\mathcal{C}$ are perspective from a line.*

Fig. 6.49 Corresponding pairs of points belonging to opposite quadrilaterals $ABCD$ and $EFGH$ induce lines intersecting at the four *gray* vertices

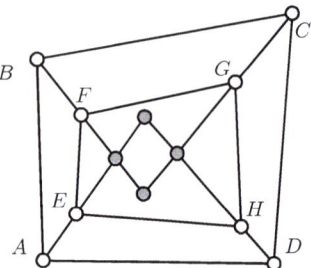

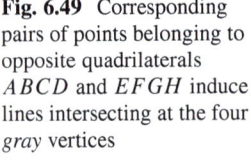

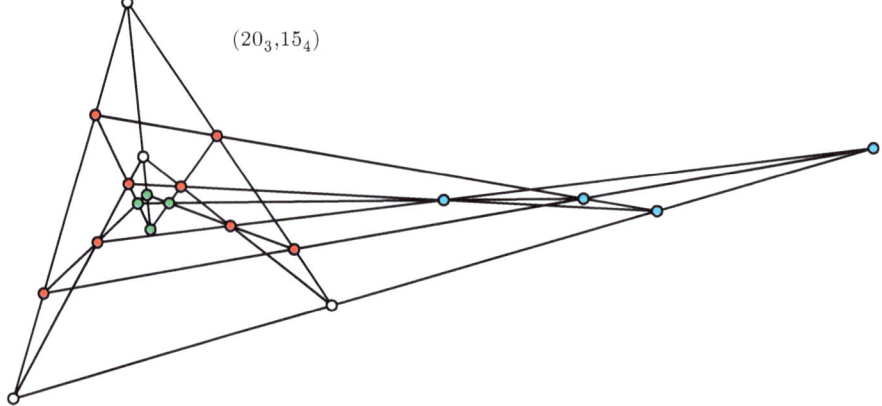

Fig. 6.50 The generalized Reye configuration in which the vertices of the cube graph are *red* and the 4-sets of intersection points of the lines connecting corresponding points of opposite quadrilaterals are colored *white*, *blue*, and *blue*

We call a picture $\pi(\mathcal{C})$ of $\mathcal{C} = (\mathcal{V}, \mathcal{E}, \mathcal{F})$ a *general picture* if the points $\pi(f)$ are in general position for each $f \in \mathcal{F}$. Suppose that we have a general picture $\pi(\mathcal{C})$ of $\mathcal{C}$ in which some, and hence all, pairs of opposite quadrilaterals are perspective from a line. So the picture lifts to a spatial picture $\sigma(\mathcal{C})$. If each of the three pairs of opposite quadrilaterals is also perspective from a point, perhaps at infinity, then the spatial and plane pictures can be completed to Reye configurations. If this is not the case, then the pairs of opposite quadrilaterals each give rise to up to four intersection points; see Fig. 6.49.

Note that the lines $\sigma(A)\sigma(E)$ and $\sigma(C)\sigma(G)$ will in general be skew in space and so need not intersect, similarly for the lines $\sigma(B)\sigma(F)$ and $\sigma(D)\sigma(H)$. The line of intersection of the opposite quadrilaterals $(\sigma(A)\sigma(D)\sigma(H)\sigma(E))$ and $(\sigma(B)\sigma(C)\sigma(G)\sigma(F))$ contains the point $\sigma(A)\sigma(E) \wedge \sigma(B)\sigma(F)$, as well as $\sigma(C)\sigma(G) \wedge \sigma(D)\sigma(H)$, $\sigma(B)\sigma(C) \wedge \sigma(A)\sigma(D)$, and $\sigma(F)\sigma(G) \wedge \sigma(E)\sigma(H)$. So we may add the three sets of four intersection points corresponding to the opposite quadrilaterals, and this set of twelve new points is partitioned by the three lines of perspectivity of the opposite quadrilaterals. Adding these three lines as well gives us a configuration of twenty points and fifteen lines of type $(20_3, 15_4)$ which we call the *generalized Reye configuration*; see Fig. 6.50.

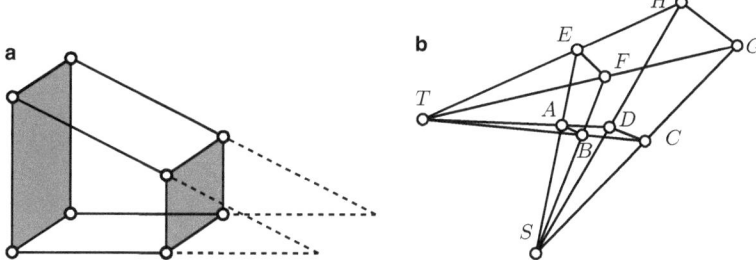

Fig. 6.51 On the *left*, only two of the three pairs of opposite quadrilaterals are perspective from a point. On the *right*, two pairs of opposite quadrilaterals perspective from a point imply the third pair is perspective from the line joining them

In analogy with the relationship between the Reye configuration and the parallelepiped, we may say the following:

Theorem 6.18. *A general picture $\pi(\mathcal{C})$ of the cube graph has a sharp lift if and only if it extends in the plane to a generalized Reye configuration.*

For a plane picture of a triangular prism, the theorem of Desargues provides an equivalence between the triangular faces being perspective from a point and perspective from a line. There is no such symmetry for the plane picture of $\mathcal{C}$. It is not true that if some pair of opposite quadrilaterals is perspective from a point, then all three pairs are. In Fig. 6.51a, we have a configuration in which all three pairs of opposite quadrilaterals are perspective from a line, so it has a lift, and the vertices of two pairs of opposite quadrilaterals are perspective from points, but the third, shaded pair is not in perspective from a point.

However, we can say the following:

Theorem 6.19. *If two pairs of opposite quadrilaterals in a general picture $\pi(\mathcal{C})$ of $\mathcal{C}$ are in perspective from points, then $\pi(\mathcal{C})$ has a lift.*

Proof. If two pairs of opposite quadrilaterals in a plane picture $\pi(\mathcal{C})$ of $\mathcal{C}$ are in perspective from two points S and T, then the third pair of opposite quadrilaterals has only S and T as points of perspectivity (see Fig. 6.51b) so the third pair of opposite quadrilaterals is perspective from the line ST; hence, $\pi(\mathcal{C})$ has a lift. □

6.5 The Grünbaum Incidence Calculus

In this section, we consider several techniques for creating new configurations from known ones. Combinatorially, the easiest thing to do to alter a configuration without changing its topological type is perform an *incidence switch*. Take two edges in the Levi graph such that the distance between their black and white endpoints is at least 5. Delete those two edges from the Levi graph, and connect the newly created

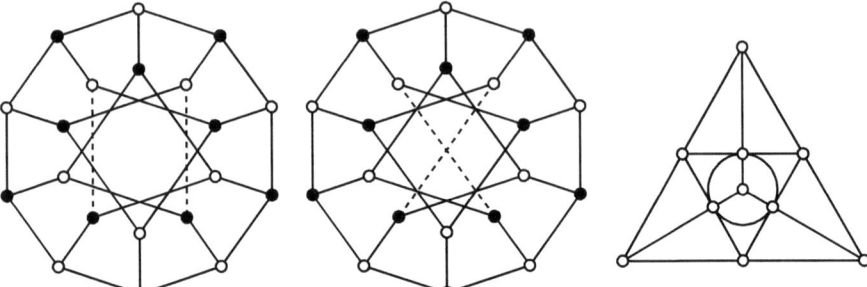

Fig. 6.52 Incidence switching on the Levi graph of the Desargues graph may yield the Levi graph of the anti-Desargues configuration, a non-Euclidean configuration

Fig. 6.53 The polar of the Pappus configuration is another differently embedded Pappus configuration

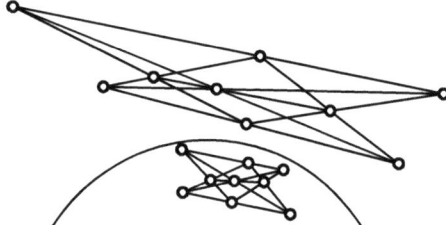

four vertices of valence 2 with the other two possible bipartition respecting edges among those four points. This operation is illustrated in Fig. 6.52 in which the Levi graph of the Desargues configuration, the generalized Petersen graph GP(10, 3), is altered to produce the Levi graph of another combinatorial (10_3) configuration, the *anti-Desargues configuration* which may be shown to be not realizable in Euclidean space. For realizability, simple incidence switching is too violent. We would like to consider a collection of moves which preserve realizability of an incidence structures and, ideally, preserve the regularity.

Polar Duality

We have already considered the polar correspondence in Sect. 6.1.4 and observed that it reverses the roles of the points and the lines of an incidence structure $\mathcal{C}$ while preserving incidence; see Fig. 6.53. The polar produces a new incidence structure, the dual of $\mathcal{C}$. So switching the colors of an incidence graph does not affect its realizability as a collection of points and lines in the Euclidean plane. So (v_n, b_m) configuration is realizable if and only if its dual (b_m, v_k) is realizable.

6.5 The Grünbaum Incidence Calculus

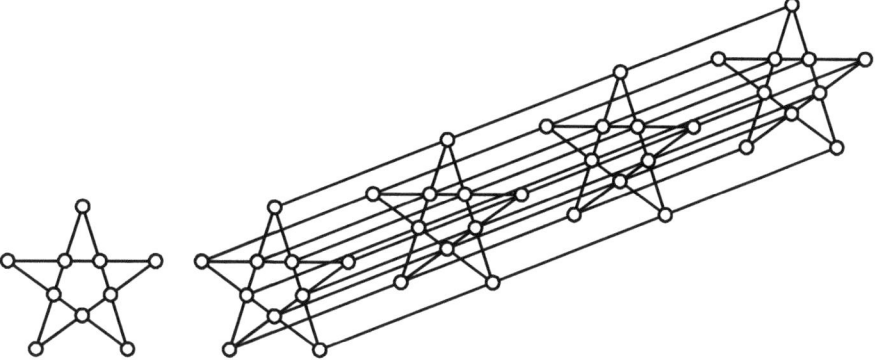

Fig. 6.54 The $(10_2, 5_4)$ pentagram transformed to a $(40_3, 30_4)$ configuration

Projective Transformation

While the polar duality is a completely safe operation in terms of 2-dimensional realizability, it does not produce much variety combinatorially. Even safer is to apply any projective transformation s to the realization, producing a new realization which is combinatorially identical. Nevertheless, translating, dilating, scaling, and rotating realizations are useful when combined with other operations, and so are included in the list for completeness.

Disjoint Union

Clearly the disjoint union of two incidence structures $\mathcal{C} \cup \mathcal{C}'$ is realizable if and only if each of $\mathcal{C}$ and $\mathcal{C}'$ is realizable. Using the simple operations so far, we can, given a few structures, produce copies of them and their duals, projectively transformed if desired.

n-Fold Parallel Replication

Given an incidence structure $\mathcal{C}$ and an integer n, the *n-fold parallel replication*, $n^*(\mathcal{C})$, is obtained by taking n copies $\mathcal{C}$ translated by various amounts in a fixed direction and adding a parallel class of lines, one for each point of $\mathcal{C}$ in the translation direction. See Fig. 6.54 in which a pentagram of type $(10_2, 5_4)$ has been replicated. In order for the resulting incidence structure to still be regular, if this is to be the only operation, it is necessary to perform a fourfold replication, so the type of $4^*\mathcal{C}$ is $(40_3, 30_4)$. Although the structure of $4^*\mathcal{C}$ is necessarily mundane, the fourfold

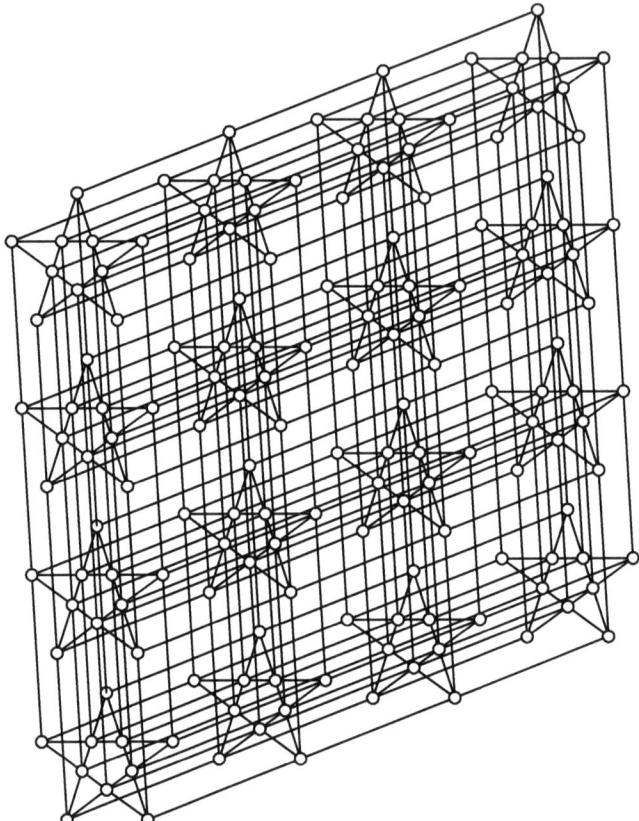

Fig. 6.55 The $(10_2, 5_4)$ pentagram trice replicated to form a balanced (160_4) configuration

replication of the pentagram does have the interesting effect of increasing the point valence while leaving the block valence fixed. So if we then form $4 * (4^*\mathcal{C})$, we obtain a balanced configuration; see Fig. 6.55. This works in general. If the block valence is less than the point valence, a sequence of replications is all that is required. If the block valence is greater than the point valence, then one simply performs the replications on the polar and repolarizes the balanced result.

Theorem 6.20. *Every configuration is a subconfiguration of a balanced configuration.*

If one starts with a balanced (v_n) configuration, an n-fold parallel replication will result in a nonbalanced $(nv_{n+1}, (n+1)v_n)$ configuration (see Fig. 6.56) which, if desired, can be rebalanced with a n-fold replication on the polar to produce a $(n(n+1)v_{n+1})$ configuration.

6.5 The Grünbaum Incidence Calculus

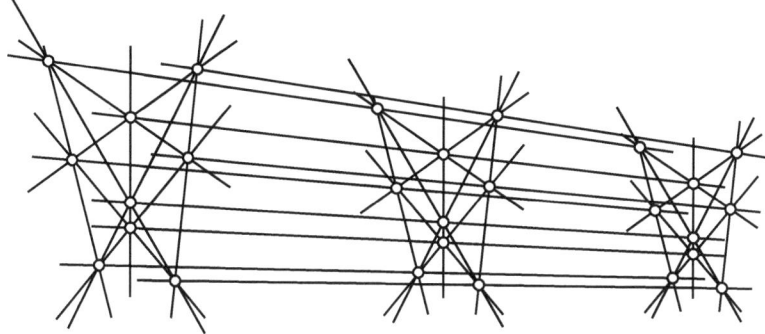

Fig. 6.56 Creating a $(27_4, 36_3)$ configuration from the (9_3) Pappus configuration. The parallel replication may of course be with respect to a finite vanishing point. If rebalanced with the polar replication, this yields a balanced (108_4) configuration

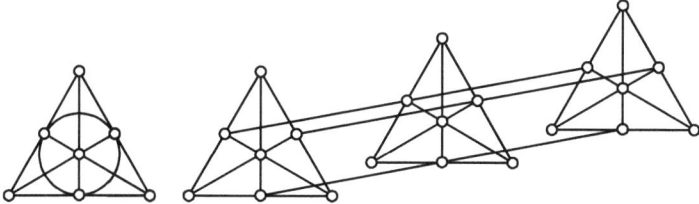

Fig. 6.57 A parallel switch on a Fano configuration

Parallel Switch

Given an incidence structure $\mathcal{C}$ and a block L, the *parallel switch*, $\sigma(\mathcal{C}, L, n)$, of $\mathcal{C}$ with respect to L is the structure obtained from n translated copies of $\mathcal{C}$, each with line L deleted, and, for those points incident to L in each copy, a line joining them parallel to the translation; see Fig. 6.57. If $\mathcal{C}$ is a (v_n, b_m) configuration, then the Levi graph of $\sigma(\mathcal{C}, L, m)$ also has girth at least 6 and so also is a configuration of type (mv_n, nb_m) and is simply denoted by $\sigma(\mathcal{C}, L)$. In particular, the parallel switch of a balanced configuration is balanced. In a case like the Fano plane, which is not realizable but is realizable with a line deleted, the parallel switch is realizable even though the Fano plane is not.

Incidence switching performed on a disconnected Levi graph may connect it, and the result will be realizable; however, the new realization may be weak.

Cartesian Product

Given incidence structures $\mathcal{C} = (V, B)$ and $\mathcal{D} = (W, A)$, we can form an incidence structure $\mathcal{C} \times \mathcal{D}$ on the point set $V \times W$ with blocks $(V \times A) \cup (W \times B)$ by setting (v, w)

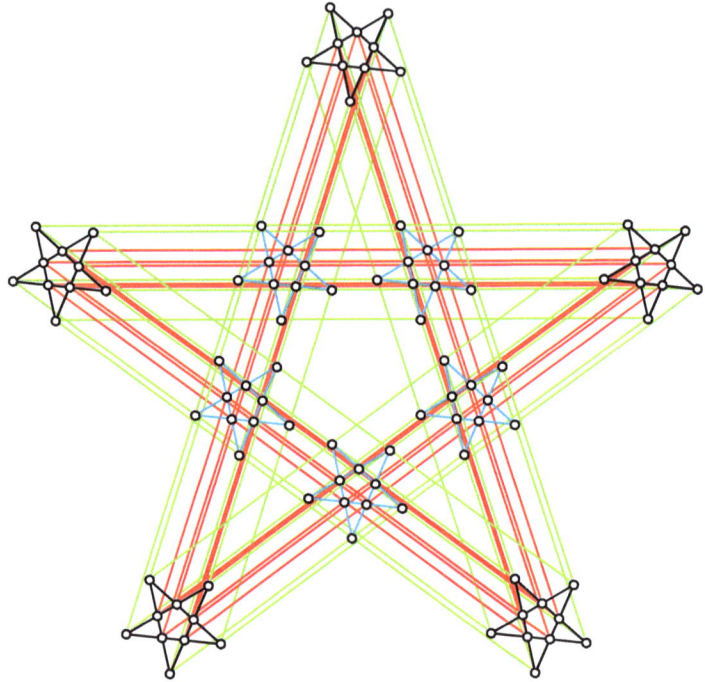

Fig. 6.58 The Cartesian product of an incidence structure with itself

incident to (v, a) if w is incident to a and (v, w) incident to (b, w) if v is incident to b. To see that the Cartesian product of realizable structures is realizable, one can first directly realize the product realization in $\mathbb{R}^4$ and then project back to the Euclidean plane. See Fig. 6.58. If the incidence structures are configurations of different block valence, then the result of the Cartesian product is not a configuration, since the two block classes will be of different types in the product. The Cartesian product of configurations of type (v_n, b_k) and $(v'_{n'}, b'_k)$ is of type $(vv'_{n+n'}, (vb' + v'b)_k))$.

For example, the $(10_2, 5_4)$ pentalateral configuration is realizable in the plane as a simple pentagram, and its Cartesian product with itself is a $(100_4, 100_4)$ configuration realizable in $\mathbb{R}^4$. The four-dimensional realization may be projected back onto the plane (see, for example, Fig. 6.58) in which the projection is given by $p(x_1, y_1, x_2, y_2) = (x_1 + x_2, y_1 + y_2)$. The colors in that figure arise from the two types of points on the pentalateral, the *exterior points*, which are not incident to the pentagonal face, and the *interior points* which are. The black and green lines are products with exterior points of one pentalateral or the other, and the blue and red lines are products with interior points. In Fig. 6.59, we see an ingenious projection of this same (100_4) product pentalateral configuration which is due to Gábor Gévay, in which he has first represented the product configuration with the symmetry of a right pentagonal prism in $\mathbb{R}^3$, before projecting orthogonally into $\mathbb{R}^2$. He makes also the following conjecture:

6.5 The Grünbaum Incidence Calculus

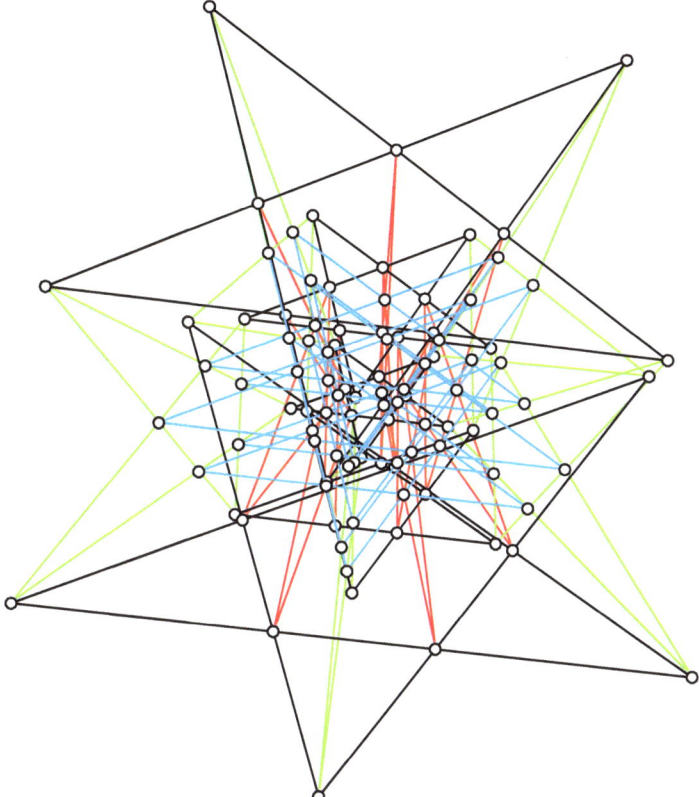

Fig. 6.59 Gábor Gévay's (100_4) configuration

Conjecture 6.21 (Gévay). Given any 3-dimensional realization of the (100_4) configuration of Fig. 6.58 in which points satisfy the collinearities of three of the four color classes of lines, then the collinearities of the fourth color class must also be satisfied.

The operations introduced so far are combinatorial operations which preserve realizability. They are perfectly well defined for any combinatorial configuration and can be completely specified as an operation on the incidence graph. The remaining operations are geometric and not combinatorial in nature and are defined in terms of a realized incidence structure.

Adding Realized k-Fold Incidences

Given a realization ρ of an incidence structure $\mathcal{C}$, particularly one with symmetry, there may be many collections of k geometrically coincident lines which are not

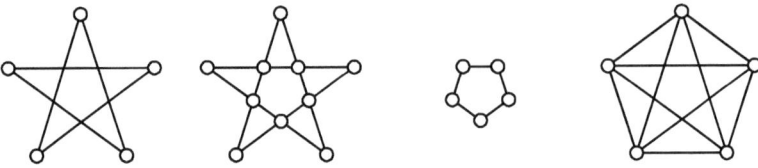

Fig. 6.60 The $(10_2, 5_4)$ configuration is $\kappa_2^p(\mathcal{P})$ for either the pentagon or the pentagram. For both, $\kappa_2^l(\mathcal{P})$ gives the complete 5-point

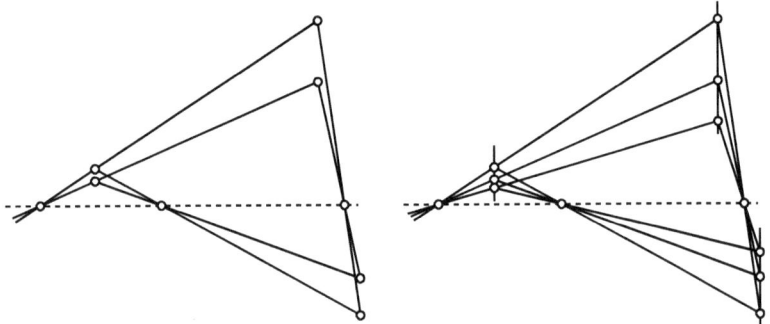

Fig. 6.61 Transforming a (3_2) configuration, that is, a triangle, into a $(9_2, 6_3)$ configuration as well as a (12_3) configuration

explicitly stated in the incidence structure. Adding them to the structure yields $\kappa_k^p(\mathcal{C}, \rho)$. The dual procedure is to add all lines which pass through k points in the structure, $\kappa_k^l(\mathcal{C}, \rho)$. The regular pentagon and the regular pentagram are both (5_2) incidence structures, $\mathcal{C}_5$. In their symmetric realizations ρ and ρ', respectively, $\kappa_2^p(\mathcal{C}, \rho) = \kappa_2^p(\mathcal{C}, \rho')$; see Fig. 6.60. It is also true that $\kappa_2^l(\mathcal{C}, \rho) = \kappa_2^l(\mathcal{C}, \rho')$.

Since the combinatorial operations yield not only a combinatorial configuration but also a realization with some parallelism, there will often be an opportunity to combine them with the geometric operations. For example, see Fig. 6.61 in which, for variety, the polar form of a parallel replication of a (3_2) triangle gives the $(9_2, 6_3)$ configuration on the left, as we saw before. (The polar form uses dilations of the configuration with respect to a fixed line, adding a point on that line incident to each line which intersects it there.) However, with a threefold polar parallel replication, the result is not a configuration until the κ_3^l operation yields a third line at each of the replicated triangle points, giving a (12_3) configuration. In general, this construction takes any realizable (v_k) configuration into a realizable $((k+2)v_{k+1})$ configuration.

Adding Extended Objects

If a Euclidean realization ρ of $\mathcal{C}$ has k-fold parallel pencils of lines, then their projective points at infinity may be included to form $\pi_k(\mathcal{C})$. Adding the line at

Fig. 6.62 Adding the three points at infinity joining the three parallel classes, this $(6_3, 9_2)$ configuration becomes the Pappus configuration

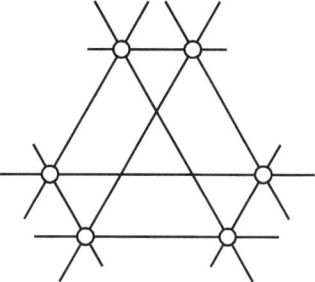

infinity is denoted by $\infty(\mathcal{C}, \rho)$. Lastly, for a rotationally symmetric configuration, there are often lines passing through the origin, and adding them is denoted by $o(\mathcal{C})$.

Altogether, we call this collection of operations, some combinatorial and some geometric, the *Grünbaum calculus*. For example, starting with a single point $\mathcal{P}$, two threefold parallel replications give $3 * (3 * (\mathcal{P}))$, a $(9_2, 6_3)$ configuration whose dual $(3 * (3 * (\mathcal{P})))^d$ has the symmetric realization ρ of Fig. 6.62, and applying π_3, we get $\pi_3((3 * (3 * (\mathcal{P})))^d, \rho)$, a (9_3) configuration, a realization of the Pappus configuration.

The Grünbaum calculus forms a powerful tool to study realizable configurations. In particular, using them, Grünbaum was able to show the following:

Theorem 6.22. *For each k, there exists a number $V(k)$ such that for each $v \geq V(k)$, there exists a geometric (v_k) configuration.*

For many more results of this character and using these techniques, the reader is recommended to consult the book by Grünbaum, [44].

6.6 Constructing Treelike Configurations

Suppose we wish to create a large realizable configuration with three vertices on every line and three lines through each point. We might start with a single point and choose generically any three lines through that point and then, continuing, for each of those lines, choose generically two additional points incident with them and, for each of those new points generically, choose three incident lines. Continuing in this way, we get an incidence structure in the plane whose Levi graph is a trivalent tree except for its pendant nodes, whose number is divisible by three. The task then is to join up the pendant nodes to complete the incidence structure to a configuration. Consider any triple of them, say three pendant nodes of the Levi graph corresponding to lines. Since the three pendant lines form a triangle, we may attach a complete quadrangle as shown in Fig. 6.63, where the red triangle represents the free lines to which the quadrangle is attached. Since all triangles are projectively equivalent, such a quadrangle always exists. If the pendent nodes are partitioned into triples and quadrangles are attached, then each of the free lines now has three

Fig. 6.63 Adding a quadrangle cap on a *triangle*

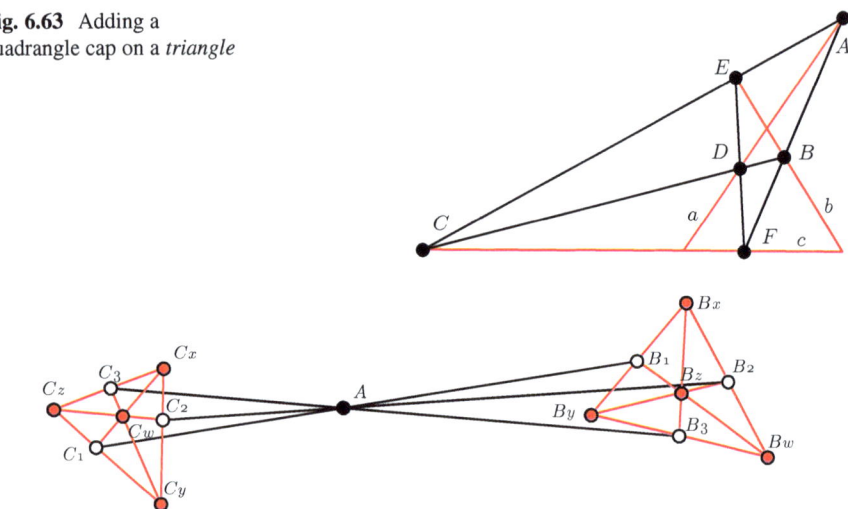

Fig. 6.64 Construction to level 1 with 2 dual caps attached at $3 \cdot 2^1$ vertices

vertices incident with it, the original one and two from the quadrangle. The lines of the quadrangles are each incident with three quadrangular points, and the vertices of the quadrangle are incident with two lines of the quadrangle and one of the three free lines. So the Levi graph is trivalent, and the incidence structure is a configuration.

If the construction ends with the pendant nodes in the Levi graph associated with points in the incidence structure, then these pendant vertices may be capped in groups of three with the dual of the quadrangular cap, K_4. In either case, if the full tree is used to level n, then the final configuration, completed with the revelent caps, will be a balanced $((2^{n+3} - 1)_3)$ configuration (Figs. 6.64 and 6.65).

6.7 Realizing the Gray Graph and Bouwer Graph

The Gray graph was introduced in Sect. 3.4.1 and was constructed by subdividing three copies of the complete bipartite graph $K_{3,3}$ and attaching each triple of vertices of valence 2 subdividing corresponding edges in the three copies of $K_{3,3}$ to a new vertex of valence three. With the Grünbaum calculus, we can construct a realization of the configuration associated with the Gray graph. As noted in Example 5.5, each subdivided $K_{3,3}$ is realized by a configuration consisting of two parallel classes of three lines, with nine points of intersection, the Cartesian product of two dual pencils. The required joining of three $S(K_{3,3})$'s is a threefold parallel replication or, equivalently, the Cartesian product with another dual pencil. The dual pencil itself is a threefold parallel replication of a single point P. So the graph may be succinctly expressed as

$$(3 * P) \times (3 * P) \times (3 * P) = (3 * P)^3.$$

6.7 Realizing the Gray Graph and Bouwer Graph

Fig. 6.65 Construction to level 3, with 2^3 dual caps attached at $3 \cdot 2^3$ vertices

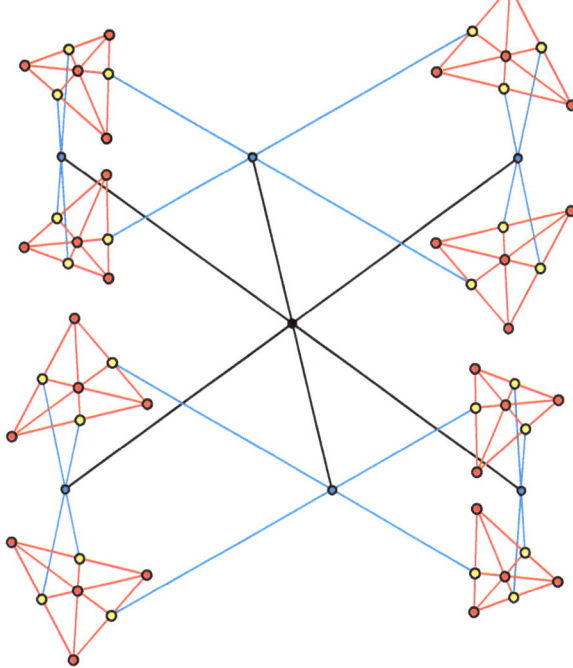

Fig. 6.66 Spatial version of the Gray configuration consisting of 27 points and 27 lines grouped in three classes, each containing 9 parallel lines

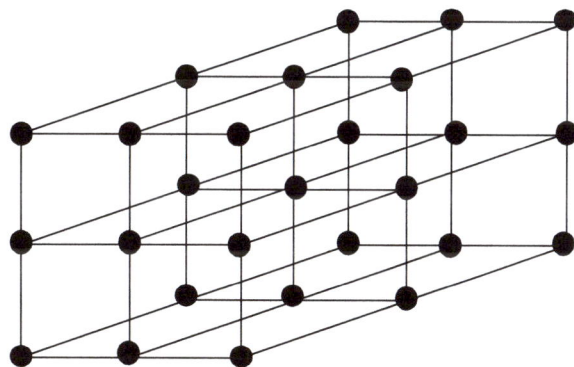

This expression immediately suggests a spatial realization of the Gray configuration by taking this third parallel class to be orthogonal to the two orthogonal classes of the Euclidean realization of $S(K_{3,3})$; see Fig. 6.66.

The Gray graph was introduced and studied because of its extremal properties and high degree of symmetry, both of which were obscured in its original construction. Indeed, it is remarkable that its true simplicity has taken so long to be revealed in the guise of the Levi graph of a simple product configuration.

The Gray graph is the unique smallest trivalent semisymmetric graph. In particular, this means that any trivalent edge but not vertex-transitive graph on 54 vertices has to be isomorphic to the Gray graph, [62].

The Gray graph has surprisingly many appearances in constructions. It can be verified, for instance, that it may be obtained from the generalized quadrangle $W(3)$ as defined in [35], p. 84. This is a semisymmetric tetravalent graph on 80 vertices of girth 8. In other words, $W(3)$ is a bipartite, regular, edge-transitive but not vertex-transitive graph. There are 54 vertices at distance 4 from each edge, and the graph induced on these 54 vertices is the Gray graph. Let $H_1 \subset H_2 \subset G$ be graphs. Define $G(H_1, H_2)$ to be a bipartite graph defined on all occurrences of H_1 as an induced subgraph in G and all occurrences of H_2 as an induced subgraph of G and then setting an occurrence of H_1 to adjacent to an occurrence of H_2 if and only if the occurrence of H_1 is an induced subgraph of the corresponding occurrence of H_2. The graph $G(H_1, H_2)$ will be called the *Grassmannian* of G with respect to H_1, H_2. The name is chosen by analogy with Grassmannians traditionally defined on the collection of k-dimensional vector subspaces of an n-dimensional vector space.

Consider the following constructions which were inspired by Bouwer [11, 12]:

Construction 1. Let us consider three disjoint sets on 9 elements: $X = \{x_1, x_2, x_3\}$, $Y = \{y_1, y_2, y_3\}$, $Z = \{z_1, z_2, z_3\}$ and their union $V = X \cup Y \cup Z$. Let E denote the set of 27 pairs $\{x_i, y_j\}, \{x_i, z_k\}, \{y_j, z_k\}$, and let T denote the set of 27 triples $\{x_i, y_j, z_k\}$. Define the bipartite graph B_1 with bipartition (E, T) in which a pair $e \in E$ is adjacent to the triple $t \in T$ if and only if $e \subset t$.

Construction 2. Consider the complete tripartite graph $K_{3,3,3}$. Let B_2 be the graph $K_{3,3,3}(K_2, K_3)$.

Construction 3. Define B_3 to be the graph $K_3 \square K_3 \square K_3(K_1, K_3)$, recalling the shorthand notation: $K_{3,3,3} = K_{3(3)}$ and $K_3 \square K_3 \square K_3 = K_3^3$.

Using the spatial configuration of Fig. 6.66, it is not hard to show the following:

Theorem 6.23. *The graphs B_1, B_2, B_3 are all isomorphic to the Gray graph.*

Proof. To see that B_2 is isomorphic to B_1, label the vertices in each color class with labels from X, Y, and Z, respectively. The 27 edges of $K_{3,3,3}$ are naturally labeled by labels from E. Furthermore, the 27 cycles of length 3 are naturally labeled by T, and the incidences of B_1 correspond precisely to the incidences in B_2. Consider the vertex set of $K_3 \square K_3 \square K_3$ to be $X \times Y \times Z$. The labeling can be carried on to B_3. The vertices of B_3 can be equivalently labeled as sets $\{x_i, y_j, z_k\}$ rather than triples (x_i, y_j, z_k). Furthermore, a triangle in $K_3 \square K_3 \square K_3$ is determined by two coordinates: $(x_i, y_j, *), (x_i, *, z_k), (*, y_j, z_k)$, or, equivalently, by an unordered pair $\{x_i, y_j\}, \{x_i, z_k\}, \{y_j, z_k\}$. This establishes the isomorphism between B_1 and B_3 where a pair $e \in E$ is mapped to a triangle of B_3 and a triple $t \in T$ is mapped to a vertex of B_3. In [66], it is shown that the Gray graph is the Levi graph of the Gray configuration, the configuration of 27 points and 27 lines as depicted in Fig. 6.66. This identifies B_3 as the Gray graph. An alternative argument that the three graphs B_i, $i = 1, 2, 3$ are isomorphic to the Gray graph follows from the fact that the

6.7 Realizing the Gray Graph and Bouwer Graph

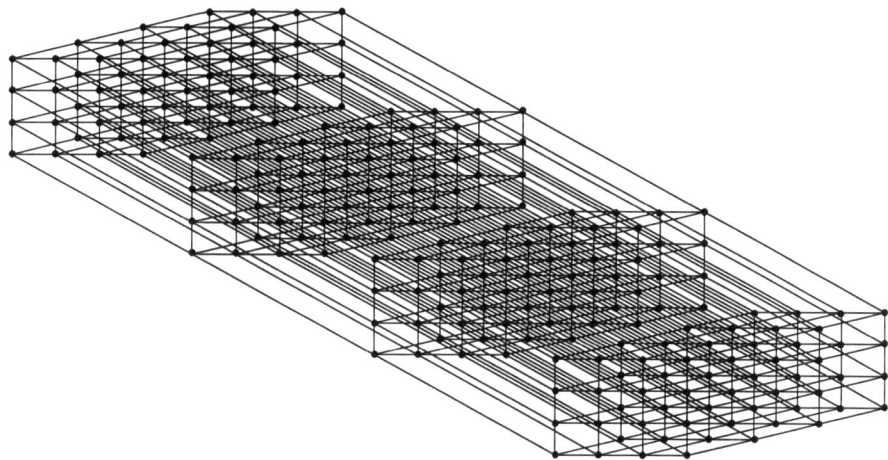

Fig. 6.67 The generalized Gray configuration (256_4). It is composed of four clearly visible $(64_3, 48_4)$ subconfigurations

graphs are trivalent and semisymmetric on 54 vertices, see [62]. They are clearly trivalent. The construction of B_1 shows that the graph is edge transitive. The graph $B_i^{(2)}$ consists of two connected components. One is isomorphic to $K_3 \square K_3 \square K_3$ in which each edge belongs to exactly one triangle, while the other one does not have this property. Hence, B_i is not vertex transitive.

This spatial configuration casts a new light on the Menger graph M and the dual Menger graph D of the Gray configuration. Graph M, isomorphic to $K_3 \square K_3 \square K_3$, is defined on the vertex set T with triples t, s adjacent if and only if $|s \cap t| = 2$. The graph D is defined on the set E with two pairs e, f adjacent if and only if $|e \cap f| = 1$.

There is a generalization of these constructions from the case $n = 3$ to general n. Instead of triples, define n-tuples. Let $V(n)$ be a set $\{x_{ij} | i, j \in \mathbb{Z}_n\}$ of n^2 elements; let $T(n)$ be the set of all n-tuples whose i'th entry is x_{ij} for some j. Similarly, let $E(n)$ denote the set of n^n $(n-1)$-tuples that are obtained from the from $T(n)$ by deleting any entry in any n-tuple. Note that each n-tuple from $T(n)$ gives rise to n $(n-1)$-tuples from $E(n)$. However, each $(n-1)$-tuple is obtained n times. Then define a graph $B_1(n)$ on $2n^n$ vertices or, equivalently, a triangle-free (n_n^n) configuration that is both combinatorial and geometric. Any n-tuple v from $T(n)$ is adjacent to an $(n-1)$-tuple u from $E(n)$ if and only if u can be obtained from v by deleting an entry from v. Figure 6.67 shows how to construct this geometric configuration in the case $n = 4$. In addition, the theorem generalizes. If we define $B_2(n) = K_{n(n)}(K_{n-1}, K_n)$ and $B_3(n) = K_n^n(K_1, K_n)$, then the graphs $B_1(n)$, $B_2(n)$, and $B_3(n)$ are isomorphic. This generalization implies the existence of geometric triangle-free point-, line-, and flag-transitive non-self-dual configurations (v_k) for any value of k; see also [43, 44]. It also implies the

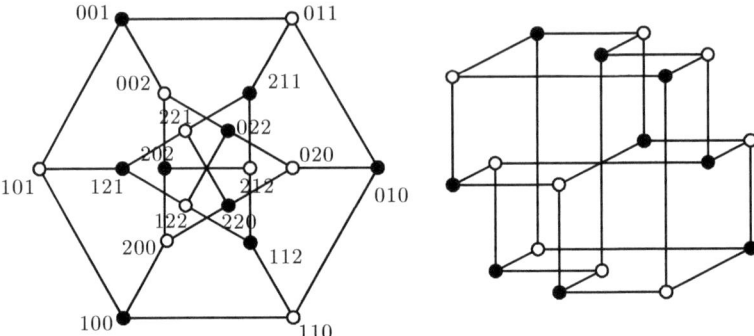

Fig. 6.68 A spatial drawing of the Pappus graph

existence of semisymmetric graphs of large valence $n > 2$. Namely, the graph $B(n)$ on $2n^n$ vertices of valence n has girth 8, diameter $2n$, each vertex corresponding to an n-tuple has $(n-1)^n$ antipodal vertices, while each vertex corresponding to an $(n-1)$-tuple has $(n-1)^{n-1}$ antipodal vertices. The numbers of vertices at distances 0, 1, 2, 3, and 4 from a given n-tuple are $1, n, n(n-1), n(n-1)^2$, and $n(n-1)^3/2$, respectively, while the corresponding numbers from a given $(n-1)$-tuple are $1, n, n(n-1), n(n-1)^2$, and $(n-1)^4$.

The construction of the configurations naturally generalizes not only to (n_n^n) but to configurations (p_q, n_k) for all q and k (with appropriate, easily computable p and n): In q-dimensional space, take a lattice hypercube with k points on each side; see [12]. By a suitable projection in the plane, we obtain a geometric configuration of points and lines.

Theorem 6.24. *For any values of $q, k \geq 1$, there exists a point-transitive, line-transitive, flag-transitive, triangle-free geometric configuration (p_q, n_k), where $p = k^q$ and $n = qk^{q-1}$. The configuration is self-dual if and only if $q = k = 2$.*

The Levi graphs $B(n, k)$ in this case admit the following equivalent descriptions: $B(n, k) = K_n^k(K_1, K_n) = K_{k(n)}(K_{n-1}, K_n)$. These configurations and hence their Levi graphs were first considered by I. Bouwer, and we therefore denote them *Bouwer graphs*. The Gray graph is the Bouwer graph $B(3, 3) = B(3)$.

It is perhaps of interest to note another relationship between the Gray graph and the Pappus graph. Namely, the edges of the subdivision of the Pappus graph can be interpreted as lines of the Gray configuration, and the 18 vertices of the Pappus graph are distinguished points of the Gray configuration. By deleting 9 points of the Gray configuration, we obtain a geometric $(18_3, 27_2)$ configuration that produces an unusual drawing of the Pappus graph in 3-space. The Pappus graph is a skeleton of a body that is obtained from a 2 by 2 cube with two centrally symmetric cubes removed. The vertex labeling of the Pappus graph in the left side of Fig. 6.68 defines a representation depicted in the right side of the figure. The vertices of the Pappus graph are labeled by triples i, j, k that represent triples (x_i, y_j, z_k). Note that there are 18 triples with the property $i + j + k \not\equiv 0 \mod 3$. Two triples are adjacent if and only if they agree in two coordinates. Hence, each of the 27 edges can be uniquely labeled by a pair $(x_i, y_j), (x_i, z_k)$, or (y_j, z_k) inducing a natural 3-edge coloring of the Pappus graph. The triples are the coordinates of the vertices in the spatial drawing. This drawing can also be interpreted as a spatial $(18_3, 27_2)$ subconfiguration of the Gray configuration whose Menger graph is the Pappus graph.

6.8 The Zindler Degree of Regularity of an Incidence Structure

Suppose we have an incidence structure with p points and g blocks. Let the degrees of the points as vertices of the incidence graph be given by γ_i, $\gamma_1 < \gamma_2 < \cdots < \gamma_k$, and let there be p_i elements of degree γ_i, so $p = p_1 + \cdots + p_k$. Similarly let $\{\pi_i\}$ be the set of degrees of vertices in the incidence graph corresponding to blocks, $\pi_1 < \cdots < \pi_l$, and suppose there are g_i blocks containing π_i points. Counting incidences, we have immediately

$$\sum_{i=1}^{k} p_i \gamma_i = \sum_{i=1}^{l} g_i \pi_i$$

For example, the incidence structure in Fig. 6.69 has vertex degrees $\gamma_1 = 3$ and $\gamma_2 = 4$ with $p_1 = 6$ white vertices of degree 3 and $p_2 = 8$ blue vertices of degree 4. For lines, the structure has $\pi_1 = 3$, $\pi_2 = 4$, and $\pi_3 = 6$, with the $g_1 = 3$ green lines containing $\gamma_1 = 3$ points, the $g_2 = 2$ black lines containing $\pi_2 = 4$ points, and the $g_3 = 1$ red line each containing $\pi_3 = 6$ points. The 50 incidences are filtered via

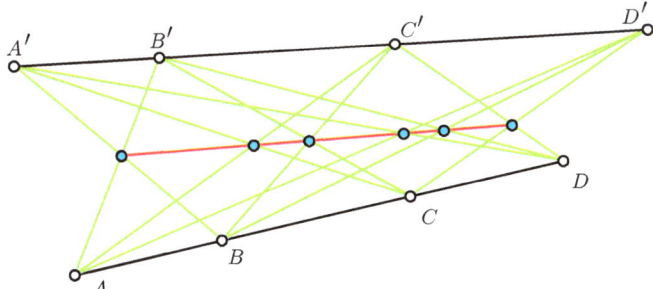

Fig. 6.69 The $\binom{4}{2}$ crosswise intersection points, in *blue*, are collinear if the two *black lines* and their points are parallel translates or, more generally, if the points $[A, B, C, D]$ and $[A', B', C', D']$ have the same cross ratio

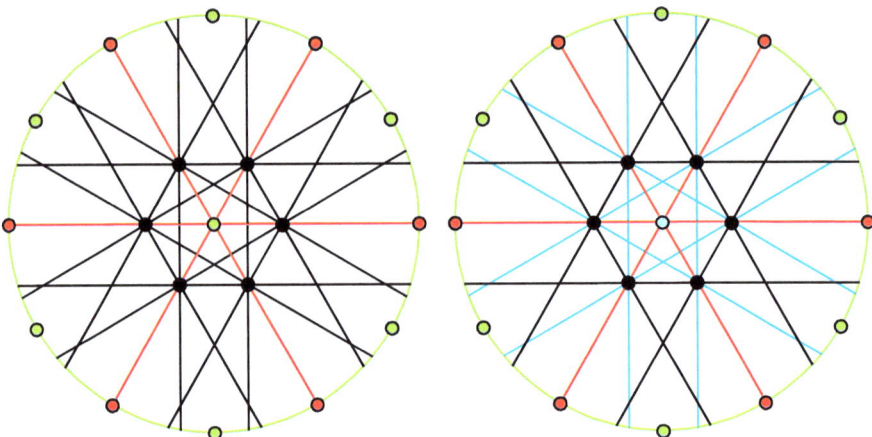

Fig. 6.70 An incidence structure generated by a regular hexagon. The *green circle* is meant to represent the line at infinity. This structure is regular to third order

$$p_1\gamma_1 + p_2\gamma_2 = 6 \cdot 3 + 8 \cdot 4 = 50 = g_1\pi_1 + g_2\pi_2 + g_3\pi_3 = 12 \cdot 3 + 2 \cdot 4 + 1 \cdot 6$$

Since there is more than one γ_i and π_i, the incidence structure of Fig. 6.69 is not sufficiently regular to be considered a configuration; however, it does exhibit more regularity than expected simply from the set of numbers $\{\gamma_i\}$ and $\{\pi_i\}$. Specifically, the two colors of vertices are regular with respect to the three colors of lines: Each green vertex is incident to two green lines and one red line, and each white vertex is adjacent one black line and four green lines. Dually, the lines of the same color each contain the same number of vertices of the same color. Following Zindler [110], we say that this structure is regular to the second order.

The incidence structure of Fig. 6.70 is not regular to second order since, for example, there are two different types of vertices of degree 3: three which are incident to two black lines and one green line and one which is incident to three red lines. In this case, we can carry out the same filtering procedure used above, but instead of starting with the degree of each vertex and block, use the set of colored degrees. The new, finer coloring does have the property that each vertex and line of the same color has the same colored degree for each color (see the structure on the right of Fig. 6.70), and we say it has regularity of third order. In general, this recursive procedure can be continued until it eventually results in an order of regularity, although it may happen that each vertex and each line is colored differently; in which case, the structure is maximally irregular. Every automorphism of the incidence graph must preserve these color sets, so a maximally irregular structure must have a trivial automorphism group.

The Zindler degree of regularity can be arbitrarily high even if there are very few different valences. For example, consider the treelike constructions described in Sect. 6.6. One can construct a treelike k-valent incidence graph of any desired diameter and cap the leaves off as indicated. If, however, instead of starting with a point of valence k, we take one of valence $k + 1$ instead, we will have an incidence graph whose order of regularity is of the order of the radius of the tree.

6.9 Polycirculants and Polycyclic Configurations

Let G be a connected graph on n vertices, and let $\alpha \in \text{Aut}(G)$ be a semiregular automorphism,

$$\alpha = (v_{11}, v_{12}, \ldots, v_{1k})(v_{21}, v_{22}, \ldots, v_{2k}) \cdots (v_{m1}, v_{m2}, \ldots, v_{mk})$$

i.e. α can be written as a product of m disjoint cycles, each of length k. Such a graph G is said to be *polycirculant with respect to α* or just *polycirculant*. The vertex set of a graph polycirculant with respect to α above graph satisfies $|V(G)| = n = m \cdot k$ and is also specifically said to be m-circulant. It is possible for a given G to be polycirculant with respect to more than one automorphism. If necessary to specify the automorphism, we will write $G = (G, \alpha)$.

Every graph is trivially polycyclic with respect to the identity. At the other extreme, if the order of α is $|V|$, so there is exactly one cycle, the graph G is member of the well-studied class of *circulant* graphs. Since the automorphism α acts freely and transitively on the vertex set of a circulant graph, a circulant graph is the Cayley graph of the cyclic group $\langle \alpha \rangle$.

In 1981, Dragan Marušič [68] raised the following question:

Does there exist a vertex-transitive graph which is not a polycirculant?

More precisely, he conjectured:

Conjecture 6.25 (Marušič, 1981). Every vertex-transitive graph is polycirculant with respect to a nontrivial automorphism.

If to each graph G we may assign the number $m(G)$, the maximal m for which G is m-circulant, the polycirculant conjecture would state that each vertex-transitive graph G satisfies $m(G) > 1$.

Question. What is the value of $m(G)$ for all polycirculant graphs up to, say, ten vertices?

For bipartite graphs, we say a polycirculant graph (G, α) is *strong* with respect to α if α respects the bipartition.

We may ask the following question:

Question. Is there any bipartite vertex-transitive graph which is not strongly polycirculant?

6.9.1 Cubic Circulants

Circulants of valence 3 have been characterized as being odd prism graphs, or Möbius ladders, of which only two are arc transitive, K_4 and $K_{3,3}$.

6.9.2 Bicirculants

A polycirculant graph with two cycles is said to be *bicirculant*. The quotient graph of a bicirculant graph (G, α) with respect to α has only two vertex orbits, so the quotient graph is either a theta graph, in which case G is a cyclic Haar graph, or the quotient group also has loops or semiedges. It is not hard to see that there may be at most one semiedge per vertex, which is true for the quotient of any polycirculant (G, α) with respect to α. Cubic bicirculants have also been characterized via their quotients as follows:

1. The dipole, giving rise to a cyclic Haar graph.
2. The handcuff graph GP(1, 1), giving rise to I-graphs and in particular generalized Petersen graphs.
3. The double edge with a pending edge at each of its two vertices, giving rise to a Möbius ladder graph or prism.

It is not hard to classify vertex-transitive trivalent bicirculants.

Theorem 6.26. *A connected trivalent bicirculant is vertex transitive if and only if it is a cyclic Haar graph or a vertex-transitive generalized Petersen graph or K_4.*

For arc-transitive graphs, we have the following:

Theorem 6.27. *A connected trivalent bicirculant graph is arc transitive if and only if it is K_4, a cyclic Haar graph with voltages $0, 1, r+1$ such that $1 + r + r^2 \cong 0 \mod n$, or it is one of the seven arc-transitive generalized Petersen graphs* GP(4, 1), GP(5, 2), GP(8, 3), GP(10, 2), GP(10, 3), GP(12, 5), *or* GP(24, 5).

There are infinitely many graphs satisfying the above theorem. Only three of them are not bipartite; only two have girth 4, and only two have girth 5. The rest are all Levi graphs of combinatorial configurations. Are all configurations of this type except the Fano plane and the Möbius-Kantor configuration geometrically realizable?

Cubic trivalent arc-transitive bicirculants were classified in [78].

Table 6.1 contains a list of small graphs on $v < 100$ vertices from the above theorem. The graph $\text{Ha}(n, \{0, 1, r+1\})$ has LCF notation $(2r+1, -2r-1)^n$. For more information, see [63].

Question. Are there any bipartite bicirculants which are not strongly bicirculant?

Table 6.1 Arc-transitive cubic bicirculants on at most 100 vertices

Id	v	Graph	Girth	Special name
0	4	K_4 3		Tetrahedron
1	6	$\mathrm{Ha}(3, (0, 1, 2)) = \mathrm{LCF}(3^6)$	4	$K_{3,3}$
2	8	$\mathrm{GP}(4, 1)$	4	Q_3
3	10	$\mathrm{GP}(5, 2)$	5	Petersen
4	14	$\mathrm{Ha}(7, (0, 1, 3)) = \mathrm{LCF}(\{5, -5\}^7)$	6	Heawood
5	16	$\mathrm{GP}(8, 3)$	6	Möbius–Kantor
6	20	$\mathrm{GP}(10, 2)$	5	Dodecahedron
7	20	$\mathrm{GP}(10, 3)$	6	Desargues
8	24	$\mathrm{GP}(12, 5)$	6	Nauru
9	26	$\mathrm{Ha}(13, (0, 1, 4)) = \mathrm{LCF}(\{7, -7\}^{13})$	6	
10	38	$\mathrm{Ha}(19, (0, 1, 8)) = \mathrm{LCF}(\{15, -15\}^{19})$	6	
11	42	$\mathrm{Ha}(21, (0, 1, 5)) = \mathrm{LCF}(\{9, -9\}^{21})$	6	
12	48	$\mathrm{GP}(24, 5)$	6	
13	62	$\mathrm{Ha}(31, (0, 1, 6)) = \mathrm{LCF}(\{11, -11\}^{31})$	6	
14	74	$\mathrm{Ha}(37, (0, 1, 11)) = \mathrm{LCF}(\{21, -21\}^{37})$	6	
15	78	$\mathrm{Ha}(39, (0, 1, 17)) = \mathrm{LCF}(\{33, -33\}^{39})$	6	
16	86	$\mathrm{Ha}(43, (0, 1, 7)) = \mathrm{LCF}(\{13, -13\}^{43})$	6	
17	98	$\mathrm{Ha}(49, (0, 1, 19)) = \mathrm{LCF}(\{37, -37\}^{49})$	6	

We have seen that a bicirculant, together with a semiregular automorphism, defines a quotient graph. For connected simple bicirculants, only three quotients are possible. However, the same graph may have more than one quotient. Besides the prism graphs $\mathrm{GP}(n, 1)$, the graph $\mathrm{GP}(8, 3)$ is the only graph in the intersection of the classes of trivalent cyclic Haar graphs and I-graphs.

6.10 Exercises

Exercise 6.1. Given a combinatorial incidence structure $\mathcal{C}$ and one of its points P, explain how to obtain the Levi graph of $\mathcal{C} - P$ from the Levi graph of $\mathcal{C}$. In particular, draw the Levi graph of the Desargues configuration with one point removed.

Exercise 6.2. Give a geometric proof that the Fano plane (7_3) is not realizable as a configuration of points and lines in the Euclidean plane.

Exercise 6.3. Give a geometric proof that the Möbius–Kantor configuration (8_3) is not realizable as a configuration of points and lines in the Euclidean plane.

Exercise 6.4. Let $\mathcal{A}$ be an affine plane, and let $\mathcal{P}(\mathcal{A})$ be its extended plane. Show that there exist at least four points P, Q, R, S such that no three of them are collinear.

Exercise 6.5. Show that the near pencil is not the extended plane of any affine plane.

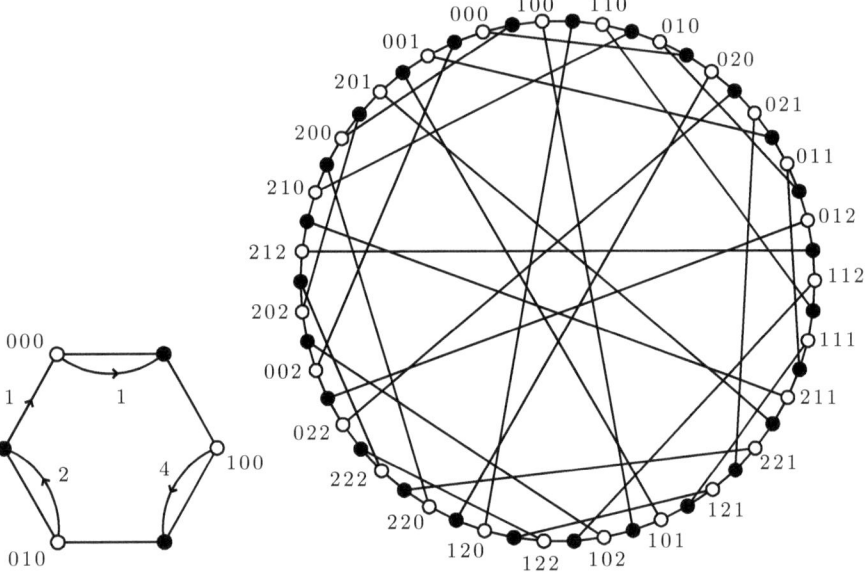

Fig. 6.71 A voltage graph and the Gray graph

Exercise 6.6. Show that the Fano plane satisfies the axioms of the projective plane.

Exercise 6.7. Show that the $z = 1$ model of the real projective plane is disconnected as a subset of $\mathbb{R}^3$. How many connected components does it have?

Exercise 6.8. Prove Lemma 6.5.

Exercise 6.9. Show that the vector space configuration $\mathbb{Z}_2^n$ is that of the complete graph on 2^n vertices.

Exercise 6.10. Show that the graph constructed in Sect. 6.2 from $n - 1$ mutually orthogonal $n \times n$ Latin squares satisfies the requirements for the Levi graph of an order n projective plane.

Exercise 6.11. Determine the line-plane incidence table of the Desargues incidence structure in space.

Exercise 6.12. Show that the six points of similitude of three circles with distinct radii in the Euclidean plane form the points of a complete quadrangle.

Exercise 6.13. Find the Petrie dual of the regular map in Fig. 6.39 of the hypercube onto the torus.

Exercise 6.14. Try to find realizations of point-plane incidence structures for the other three Möbius hypercubes.

6.10 Exercises

Exercise 6.15. Show that the dual dimension of the configuration of the complete graph K_n is exactly 2.

Exercise 6.16. Show that Cremona–Richmond configuration has dimension four.

Exercise 6.17. Prove that there is a unique $(5_4, 10_2)$ combinatorial configuration.

Exercise 6.18. Prove that there is a unique $(5_4, 10_2)$ geometric configuration, and find a symmetric drawing for it.

Exercise 6.19. Prove that there is a unique $(10_2, 5_4)$ geometric configuration, and find a symmetric drawing for it.

Exercise 6.20. Prove that there is a unique (13_4) combinatorial configuration. Show that it is lineal and cyclic.

Exercise 6.21. Prove that the following matrix over the field $\mathbf{F}_3$ is a projective coordinatization of a (13_4) configuration.

$$\begin{bmatrix} 1 & 0 & 0 & 0 & 0 & 1 & -1 & 1 & 1 & 1 & -1 & 1 & 1 \\ 0 & 1 & 0 & 1 & -1 & 0 & 0 & 1 & -1 & 1 & 1 & -1 & 1 \\ 0 & 0 & 1 & 1 & 1 & 1 & 1 & 0 & 0 & 1 & 1 & 1 & -1 \end{bmatrix}$$

Exercise 6.22. How many ways can you construct the Desargues configuration using the Grünbaum calculus?

Exercise 6.23. Show that voltage assignment of Fig. 6.71 yields the Gray graph.

Exercise 6.24. What is the dimension of the generalized Gray configuration?

Exercise 6.25. Find a graph which is polycyclic with respect to two different automorphisms whose orders are coprime.

References

1. K. Appel, W. Haken, Every planar map is four colorable. Bull. Am. Math. Soc. **82**(5), 711–712 (1976)
2. P.K. Aravind, The generalized kochen-specker theorem. Phys. Rev. A **68**, 052104 (2003)
3. D.S. Archdeacon, R.B. Richter, The construction and classification of self-dual polyhedra. JCTB **54**(1), 37–48 (1992)
4. J. Ashley, B. Grünbaum, G.C. Shephard, W. Stromquist, Self-duality groups and ranks of self-dualities. in *Applied Geometry and Discrete Mathematics*, vol. 4 of DIMACS Series in Discrete Mathamatics and Theoretical Computer Science (American Mathematical Society, Providence, 1991), pp. 11–50
5. L. Babai, C.D. Godsil, On the automorphism groups of almost all Cayley graphs. Eur. J. Combin. **3**(1), 9–15 (1982)
6. D.W. Barnette, B. Grünbaum, On Steinitz's theorem concerning convex 3-polytopes and on some properties of planar graphs. in *The Many Facets of Graph Theory (Proc. Conf., Western Mich. Univ., Kalamazoo, Mich., 1968)* (Springer, Berlin, 1969), pp. 27–40
7. A. Betten, G. Brinkmann, T. Pisanski, Counting symmetric configurations v_3. in *Proceedings of the 5th Twente Workshop on Graphs and Combinatorial Optimization (Enschede, 1997)*, Discrete Appl. Math. **99**(1–3), 331–338, (2000)
8. N. Biggs, *Algebraic Graph Theory*, 2nd edn. (Cambridge Mathematical Library, Cambridge University Press, Cambridge, 1993)
9. M. Boben, T. Pisanski, A. Žitnik, I-graphs and the corresponding configurations. J. Combin. Des. **13**(6), 406–424 (2005)
10. J. Bokowski, B. Sturmfels, *Computational Synthetic Geometry*, vol. 1355 of Lecture Notes in Mathematics (Springer, Berlin, 1989)
11. I.Z. Bouwer, An edge but not vertex transitive cubic graph. Bull. Can. Math. Soc. **11**, 533–535 (1968)
12. I.Z. Bouwer, On edge but not vertex transitive regular graphs. J. Combin. Theor. B **12**(1), 32–40 (1972)
13. G. Brinkmann, B.D. McKay, C. Saager, The smallest cubic graphs of girth nine. Combin. Probab. Comput. **4**(4), 317–329 (1995)
14. R.H. Bruck, H.J. Ryser, The nonexistence of certain finite projective planes. Can. J. Math. **1**, 88–93 (1949)
15. S. Chowla, H.J. Ryser, Combinatorial problems. Can. J. Math. **2**, 93–99 (1950)
16. H.S.M. Coxeter, Self-dual configurations and regular graphs. Bull. Am. Math. Soc. **56**, 413–455 (1950)
17. H.S.M. Coxeter, *Twelve Geometric Essays* (Southern Illinois University Press, Carbondale, 1968)

18. H.S.M. Coxeter, *Projective Geometry* (Springer, New York, 1994); Revised reprint of the second (1974) edition
19. H.S.M. Coxeter, in *Coloured Symmetry*, ed. by H.S.M. Coxeter et al. M. C. Escher: Art and Science (Elsevier, Amsterdam, 1986), pp. 15–33
20. H.S.M. Coxeter, R. Frucht, D.L. Powers, *Zero-Symmetric Graphs* (Academic [Harcourt Brace Jovanovich Publishers], New York, 1981), Trivalent graphical regular representations of groups
21. H.S.M. Coxeter, W.O.J. Moser, in *Generators and Relations for Discrete Groups*, vol. 14 of Ergebnisse der Mathematik und ihrer Grenzgebiete [Results in Mathematics and Related Areas], 4th edn. (Springer, Berlin, 1980)
22. P.R. Cromwell, *Polyhedra* (Cambridge University Press, Cambridge, 1997)
23. W.H. Cunningham, J. Edmonds, A combinatorial decomposition theory. Can. J. Math. **32**(3), 734–765 (1980)
24. R.D. von Sterneck, Die configurationen 11_3. Monatsh. Math. Phys. **5**(1), 325–330 (1894)
25. R. von Sterneck, Die configurationen 12_3. Monatsh. Math. Phys. **6**(1), 223–254 (1895)
26. A. Deza, M. Deza, V. Grishukhin, Fullerenes and coordination polyhedra versus half-cube embeddings. Discrete Math. **192**(1–3), 41–80 (1998). Discrete metric spaces (Villeurbanne, 1996)
27. C. Droms, B. Servatius, H. Servatius, The structure of locally finite two-connected graphs. Electr. J. Comb. **2** (1995)
28. C. Droms, B. Servatius, H. Servatius, Connectivity and planarity of Cayley graphs. Beiträge Algebra Geom. **39**(2), 269–282 (1998)
29. D. Eppstein, Finding large clique minors is hard. J. Graph Algorithms Appl. **13**(2), 197–204 (2009)
30. L. Euler, Recherches sur une nouvelle espece de quarres magiques, in *Verhandelingen Uitgegeven Door het Zeeuwsch Genootschap der Wetenschappen te Vlissingen 9*, Middelburg, 1782, pp. 85–239
31. J. Folkman, Regular line-symmetric graphs. J. Combin. Theor. **3**, 215–232 (1967)
32. R.M. Foster, *The Foster Census* (Charles Babbage Research Centre, Winnipeg, 1988); R. M. Foster's census of connected symmetric trivalent graphs, With a foreword by H. S. M. Coxeter, With a biographical preface by Seymour Schuster, With an introduction by I. Z. Bouwer, W. W. Chernoff, B. Monson and Z. Star, Edited and with a note by Bouwer
33. R. Frucht, A canonical representation of trivalent Hamiltonian graphs. J. Graph Theor. **1**(1), 45–60 (1977)
34. R. Frucht, J.E. Graver, M.E. Watkins, The groups of the generalized Petersen graphs. Proc. Camb. Philos. Soc. **70**, 211–218 (1971)
35. C. Godsil, G. Royle, in *Algebraic Graph Theory*, vol. 207 of Graduate Texts in Mathematics (Springer, New York, 2001)
36. M. Goldberg, A class of multi-symmetric polyhedra. Tohoku Math. J. **43**, 104–108 (1937)
37. H. Gropp, in *On the History of Configurations*, ed. by A. Deza, J. Echeverria, A. Ibarra. International Symposium on Structures in Mathematical Theories (University del Pais Vasco, Bilbao, 1990), pp. 263–268
38. H. Gropp, On the existence and nonexistence of configurations n_k. J. Combin. Inform. Syst. Sci. **15**(1–4), 34–48 (1990); Graphs, designs and combinatorial geometries (Catania, 1989)
39. H. Gropp, Configurations and graphs. Discrete Math. **111**(1–3), 269–276 (1993); Graph theory and combinatorics (Marseille-Luminy, 1990)
40. H. Gropp, The construction of all configurations $(12_4, 16_3)$, in *Fourth Czechoslovakian Symposium on Combinatorics, Graphs and Complexity (Prachatice, 1990)*, vol. 51 of Annals in Discrete Mathetics (North-Holland, Amsterdam, 1992), pp. 85–91
41. H. Gropp, Configurations and their realization. Discrete Math. **174**(1–3), 137–151 (1997); Combinatorics (Rome and Montesilvano, 1994)
42. J.L. Gross, T.W. Tucker, *Topological Graph Theory* (Dover, Mineola, 2001); Reprint of the 1987 original [Wiley, New York; MR0898434 (88h:05034)] with a new preface and supplementary bibliography

43. B. Grünbaum, (1–2–3)-complexes. Geombinatorics **13**(2), 65–72 (2003)
44. B. Grünbaum, in *Configurations of Points and Lines*, vol. 103 of Graduate Studies in Mathematics (American Mathematical Society, Providence, 2009)
45. B. Grünbaum, J.F. Rigby, The real configuration (21_4). J. Lond. Math. Soc. (2) **41**(2), 336–346 (1990)
46. B. Grunbaum, G.C. Shephard, Is selfduality involutory? Am. Math. Mon **95**(8), 729–733 (1988)
47. B. Grünbaum, G.C. Shephard, *Tilings and Patterns*. A Series of Books in the Mathematical Sciences (W. H. Freeman and Company, New York, 1989); An introduction
48. B. Grünbaum, G.C. Shephard, Isohedra with dart-shaped faces. Discrete Math. **241**(1–3), 313–332 (2001); Selected papers in honor of Helge Tverberg
49. W.H. Haemers, D.G. Higman, S.A. Hobart, in *Strongly Regular Graphs Induced by Polarities of Symmetric Designs*. Advances in Finite Geometries and Designs (Chelwood Gate, 1990) (Oxford Sci. Publ., Oxford University Press, New York, 1991), pp. 163–168
50. D. Hilbert, S.Cohn-Vossen, *Geometry and the Imagination* (Chelsea Publishing Company, New York, 1952); Translated by P. Neményi
51. D. Hilbert, S. Cohn-Vossen, *Anschauliche Geometrie* (Wissenschaftliche Buchgesellschaft, Darmstadt, 1973); Mit einem Anhang: "Einfachste Grundbegriffe der Topologie" von Paul Alexandroff, Reprint der 1932 Ausgabe
52. M. Hladnik, D. Marušič, T. Pisanski, Cyclic Haar graphs. Discrete Math. **244**(1–3), 137–152 (2002); Algebraic and topological methods in graph theory (Lake Bled, 1999)
53. J.E. Hopcroft, R.E. Tarjan, Dividing a graph into triconnected components. SIAM J. Comput. **2**, 135–158 (1973)
54. S. Jendroľ, On symmetry groups of self-dual convex polyhedra, in *Fourth Czechoslovakian Symposium on Combinatorics, Graphs and Complexity (Prachatice, 1990)*, vol. 51 of Annals of Discrete Mathematics (North-Holland, Amsterdam, 1992), pp. 129–135
55. T.P. Kirkman, On autopolar polyhedra. Philos. Trans. Roy. Soc. Lond. **147**, 183–215 (1857)
56. H.S. Koike, I. Kovács, T. Pisanski, Enumeration of cyclic n_3 and n_4 configurations. Isomorphic tetravalent cyclic Haar graphs to appear in Ars Mathematica Contemporanea, (2014)
57. I. Kovács, M. Servatius, On cayley digraphs on nonisomorphic 2-groups. J. Graph Theor. **70**(4), 435–448 (2012)
58. C.W.H. Lam, G. Kolesova, L. Thiel, A computer search for finite projective planes of order 9. Discrete Math. **92**(1–3), 187–195 (1991)
59. C.W.H. Lam, L. Thiel, S. Swiercz, The nonexistence of finite projective planes of order 10. Can. J. Math. **41**(6), 1117–1123 (1989)
60. F.W. Levi, *Geometrische Konfigurationen. Mit einer Einführung in die Kombinatorische Flächentopologie* (S. Hirzel, Leipzig, 1929)
61. W. Magnus, A. Karrass, D. Solitar, *Combinatorial Group Theory*, 2nd edn. (Dover, Mineola, 2004); Presentations of groups in terms of generators and relations
62. A. Malnič, D. Marušič, P. Potočnik, C. Wang, An infinite family of cubic edge- but not vertex-transitive graphs. Discrete Math. **280**(1–3), 133–148 (2004)
63. D. Marušič, T. Pisanski, Weakly flag-transitive configurations and half-arc-transitive graphs. Eur. J. Combin. **20**(6), 559–570 (1999)
64. D. Marušič, T. Pisanski, The remarkable generalized petersen graph $gp(8, 3)$. Math. Slovaca **50**, 117–121 (2000)
65. D. Marušič, T. Pisanski, The Gray graph revisited. J. Graph Theor. **35**(1), 1–7 (2000)
66. D. Marušič, T. Pisanski, S. Wilson, The genus of the gray graph is 7. Eur. J. Combin. **26**(3–4), 377–385 (2005)
67. D. Marušič, M. Y. Xu, A $\frac{1}{2}$-transitive graph of valency 4 with a nonsolvable group of automorphisms. J. Graph Theor. **25**(2), 133–138 (1997)
68. D. Marušič, On vertex symmetric digraphs. Discrete Math. **36**, 69–81 (1981)
69. B. Mohar, P. Rosenstiehl, Tessellation and visibility representations of maps on the torus. Discrete Comput. Geom. **19**(2), 249–263 (1998)

70. B. Monson, T. Pisanski, E. Schulte, A.I. Weiss, Semisymmetric graphs from polytopes. J. Combin. Theor. Ser. A **114**(3), 421–435 (2007)
71. F.R. Moulton, A simple non-Desarguesian plane geometry. Trans. Am. Math. Soc. **3**(2), 192–195 (1902)
72. R. Nedela, M. Škoviera, Which generalized Petersen graphs are Cayley graphs? J. Graph Theor. **19**(1), 1–11 (1995)
73. P.M. Neumann, A lemma that is not Burnside's. Math. Sci. **4**(2), 133–141 (1979)
74. A. Orbanić, M. Petkovšek, T. Pisanski, P. Potočnik, A note on enumeration of one-vertex maps. Ars Math. Contemp. **3**(1), 1–12 (2010)
75. M. Petkovšek, H. Zakrajšek, Enumeration of I-graphs: Burnside does it again. Ars Math. Contemp. **2**(2), 241–262 (2009)
76. T. Pisanski, M. Boben, D. Marušič, A. Orbanić, A. Graovac, The 10-cages and derived configurations. Discrete Math. **275**(1–3), 265–276 (2004)
77. T. Pisanski, A. Žitnik, A. Graovac, A. Baumgartner, Rotagraphs and their generalizations. J. Chem. Inform. Comput. Sci. **34**(5), 1090–1093 (1994)
78. T. Pisanski, A classification of cubic bicirculants. Discrete Math. **307**(3–5), 567–578 (2007)
79. T. Pisanski, Yet another look at the Gray graph. New Zealand J. Math. **36**, 85–92 (2007)
80. T. Pisanski, M. Randić, Bridges between geometry and graph theory, in *Geometry at Work*, vol. 53 of MAA Notes (Mathematical Association of America, Washington, DC, 2000), pp. 174–194
81. T. Pisanski, D. Schattschneider, B. Servatius, Applying Burnside's lemma to a one-dimensional Escher problem. Math. Mag. **79**(3), 167–180 (2006)
82. G. Ringel, *Map Color Theorem* (Springer, New York, 1974); Die Grundlehren der mathematischen Wissenschaften, Band 209
83. J.J. Rotman, in *An Introduction to the Theory of Groups*, vol. 148 of Graduate Texts in Mathematics, 4th edn. (Springer, New York, 1995)
84. C.P. Rourke, B.J. Sanderson, *Introduction to Piecewise-Linear Topology* (Springer, New York, 1972)
85. G. Sabidussi, Graphs with given group and given graph-theoretical properties. Can. J. Math. **9**, 515–525 (1957)
86. G. Salmon, *A Treatise on Conic Sections*. 6'th ed. New York Chelsea Pub., (1954) Reprinted by the American Mathematical Society (Providence, Rhode Island, 2005)
87. D. Schattschneider, Escher's combinatorial patterns. Electron. J. Combin. **4**(2) (1997); Research Paper 17, approx. 31 pp. (electronic). The Wilf Festschrift (Philadelphia, PA, 1996)
88. A.E. Schroth, How to draw a hexagon. Discrete Math. **199**(1–3), 161–171 (1999)
89. B. Servatius, H. Servatius, The 24 symmetry pairings of self–dual maps on the sphere. Discrete Math. **140**, 167–183 (1995)
90. B. Servatius, H. Servatius, Self-dual graphs. Discrete Math. **149**(1–3), 223–232 (1996)
91. B. Servatius, H. Servatius, in *Symmetry, Automorphisms, and Self-duality of Infinite Planar Graphs and Tilings*, ed. by V. Balint. Proceedings of the International Geometry Conference in Zilina, 1998, pp. 83–116
92. R.P. Stanley, in *Enumerative Combinatorics. Vol. 2*, vol. 62 of Cambridge Studies in Advanced Mathematics (Cambridge University Press, Cambridge, 1999); With a foreword by Gian-Carlo Rota and appendix 1 by Sergey Fomin
93. A. Steimle, W. Staton, The isomorphism classes of the generalized Petersen graphs. Discrete Math. **309**(1), 231–237 (2009)
94. E. Steinitz, Über die construction der configurationen n_3. Ph.D. thesis, Kgl. Universität Breslau, 1894
95. E. Steinitz, H. Rademacher, *Vorlesungen über die Theorie der Polyeder unter Einschluss der Elemente der Topologie* (Springer, Berlin, 1976); Reprint der 1934 Auflage, Grundlehren der Mathematischen Wissenschaften, No. 41
96. I. Stewart, *Galois Theory*, 3rd edn. (Chapman & Hall/CRC Mathematics. Chapman & Hall/CRC, Boca Raton, 2004)

97. B. Sturmfels, N. White, All 11_3 and 12_3-configurations are rational. Aequationes Math. **39**(2–3), 254–260 (1990)
98. W.T. Tutte, A family of cubical graphs. Proc. Camb. Philos. Soc. **43**, 459–474 (1947)
99. W.T. Tutte, A census of planar maps. Can. J. Math. **15**, 249–271 (1963)
100. W.T. Tutte, How to draw a graph. Proc. Lond. Math. Soc. (3) **13**, 743–767 (1963)
101. W.T. Tutte, *Connectivity in Graphs*. Mathematical Expositions, No. 15 (University of Toronto Press, Toronto, 1966)
102. V.G. Vizing, On an estimate of the chromatic class of a p-graph. Diskret. Analiz No. **3**, 25–30 (1964)
103. D. Wells, *The Penguin Dictionary of Curious and Interesting Geometry* (Penguin Books, New York, 1991)
104. A.T. White, in *Graphs, Groups and Surfaces*, vol. 8 of North-Holland Mathematics Studies, 2nd edn. (North-Holland Publishing Co., Amsterdam, 1984)
105. W. Whiteley, A matroid on hypergraphs, with applications in scene analysis and geometry. Discrete Comput. Geom. **4**(1), 75–95 (1989)
106. H. Whitney, Congruent graphs and the connectivity of graphs. Am. J. Math. **54**(1), 150–168 (1932)
107. H. Whitney, 2-Isomorphic Graphs. Am. J. Math. **55**(1–4), 245–254 (1933)
108. P.K. Wong, Cages—a survey. J. Graph Theor. **6**(1), 1–22 (1982)
109. E.M. Wright, Burnside's lemma: A historical note. J. Combin. Theor. Ser. B **30**(1), 89–90 (1981)
110. K. Zindler, Zur Theorie der Netze und Configurationen. [J] Wien. Ber., Math. Naturw. Kl. **98**, 499–519 (1888)

Index

Symbols
I-graph, 261
g-cage, 28, 29
k-factor, 31
n-connected, 39
t-arc, 78
t-arc transitive, 79
1-factor, 31, 33, 34
1-skeleton, 21
1/2-arc transitive, 79
10-cage, 29
2-cage, 89
2-factor, 154
3-block tree, 151
3-cage, 28, 89
4-cage, 28, 89
5-cage, 28, 89
5-cube graph, 193
6-cage, 29
7-cage, 29, 89
8-cage, 29, 89
9-cage, 29

A
abelian group, 56
    Fundamental theorem, 64
absolute line, 185
absolute point, 185
action
    transitive, 67
acyclic, 18
adjacency matrix, 15
adjacency relation, 15
affine configuration, 219
affine plane, 197, 198, 261
alternating group, 65, 98

antipodal dual, 117
antiprism, 22
    symmetry, 99
arc, 37, 83
Archimedean graph, 23, 49
Archimedean polyhedron, 154
Archimedean solid, 76
arrangement of lines, 191
automorphism, 20, 55
automorphism group, 194, 195, 232
autopolar, 159
autopolar configuration, 175, 184
autopolar decomposition, 186
axioms for the projective plane, 198

B
Balaban 10-cage, 50, 195
balanced configuration, 216
ball
    closed, 44
barycentric subdivision, 109, 232
base graph, 89
benzene, 52
benzenoid graph, 52
bi-adjacency matrix, 185
bicirculant, 260
binary tree, 29
bipartite graph, 32, 195
    bipartite complement, 196
    semi-regular, 35
    strongly polycirculant, 259
bipartition, 32
bipartition set, 77
block, 157, 193
block cutpoint tree, 151
Borel subgroup, 162

bouquet of circles, 84
Bouwer graph, 256
Brooks' theorem, 42
Burnside's theorem, 72

C
cage, 27, 89
cartesian coordinates, 199
Cauchy–Frobenius theorem, 72
Cayley graph, 62, 98, 99, 101, 103, 155, 232, 259
Cayley line, 4
Cayley map, 138
center of similitude, 228
chamfering operation, 126
chromatic index, 42
chromatic number, 41
circulant
    cubic
        arc-transitive, 260
circulant graph, 63, 195
closed ball, 44
combinatorial configuration, 164, 193, 263
combinatorial incidence structure, 261
commutative group, 56
commute, 56
complete bipartite graph, 19, 193
complete graph, 16
complete multipartite graph, 19
complete quadrangle, 11, 12, 163, 225, 228, 237, 252, 262
complete quadrilateral, 12
complex number, 60
concurrence, 240
concurrent lines, 210, 240
cone, 38
configuration, 193
    automorphism, 194
    automorphism group, 193
    combinatorial, 164
    defect, 239
    dimension, 263
    Reye, 181
configuration table, 8, 13, 36
conic, 203, 204
conjugate, 59
connected, 18
connected sum, 111
convex closure, 45
convex hull, see convex closure, 45
convex polyhedron, 105
convex set, 45
coordinatization matrix, 237

coset incidence structure, 103, 162
    Borel subgroup, 162
    parabolic subgroup, 162
cover
    Kronecker, 85
covering graph, 89
    cyclic, 188
covering projection, 87
covering transformation, 103
Coxeter graph, 195
Cremona–Richmond configuration, 179, 193, 226, 263
Cremona–Richmond graph, see Tutte 8-cage, 29
cross polytope, 223
cross product, 98
cross ratio, 227, 257
crosscap, 122
crystallographic group, 155
cube, 22, 57, 99, 154
    colored, 99
    symmetry, 99
        color-preserving, 99
cube graph, 45, 79, 87, 154, 193
cubeoctahedron graph, 53
cubic graph, 32
    LCF notation, 32
cycle, 18
cyclic configuration, 191, 263
cyclic group, 62, 63, 259
cyclic Haar graph, 188, 195, 196, 260
cyclic permutation, 36

D
Dürer graph, 26, 47
dart, 83
degenerate realization, 221
degree, 15, 36
degree of freedom, 14
derangement, 36, 51
derived graph, 85
Desargues configuration, 170, 177, 193, 236, 261, 263
Desargues graph, 87, 103, 187, 193
Desargues theorem, 3, 205
dihedral group, 64, 176
dilation coefficient, 47, 53
dimension, 236
dipole, 85, 188, 260
direct product of groups, 64
directed graph, 36
discrete graph, 16
distance, 18, 44

Index

distance function, 43
distance matrix, 44
distance sequence, 44, 77
division ring, 205
divisor graph, 15, 48
dodecahedron, 23
dodecahedron graph, 41
double pyramid, 99
drawing
　energy, 47
　symmetry, 55
dual dimension, 237, 263
dual incidence structure, 160
dual Menger graph, 192, 255
dual pencil, 252
dual polyhedron, 11
dual statement, 210

E
edge
　endvertex, 32
edge coloring, 42
edge join, 134
edge length, 46
edge permutation, 118
edge-transitive, 154
endpoint, 15
endvertex, 15, 17
Escher problem, 73, 100, 155
Escher stamp, 99
Euclidean plane, 197
Euler characteristic, 31, 111
Euler formula, 153
Euler totient function ϕ, 63
Euler's 36 officers problem, 213
extended Euclidean plane, 199
extended plane, 198
extended real plane, 205
exterior face, 105

F
face, 30
face permutation, 153
Fano configuration, 5, 12, 46, 163, 184, 193, 199
Fano plane, 170, 199, 217, 261, 262
Fary's Theorem, 143
field, 5, 199, 205
finite group, 98
finite projective plane, 209
fixed point, 36, 68
fixed-point free involution, 36

fixed-point free permutation, 36
flag graph, 118
flags, 113
Folkman graph, 77
forest, 48
Four Color Theorem, 41
four dimensional cube Q_4, 184
free action, 136
free group, 66
frieze group, 155
fullerene, 49, 153

G
general linear group, 218
generalized n-gon, 161
generalized digon, 193
generalized Gray configuration, 263
generalized Petersen graph, 49, 86, 101, 174, 175, 260
　arc-transitive, 78, 260
generalized quadrangle, 193
generating set, 65
　redundant, 65
generator, 62
generic choice, 225
generic realization, 239
genus, 124
geometric configuration, 263
Goldberg operations, 126
Grünbaum calculus, 251, 263
Grünbaum graph, 159
Graeco-Latin square, 212
graph, 15, 192, 259
　k-regular, see k-valent, 31
　1/2-arc transitive, 79
　acyclic, 18
　antiprism, 21
　arc, 37
　arc-transitive, 78
　automorphism, 55, 101, 102, 196
　automorphism group, 99, 193
　bipartite, 32, 50, 154
　　semi-regular, 50
　cage, 27
　cartesian product, 39, 51
　circulant, 63
　clique, 40
　complement, 52
　cone, 38, 51
　　apex, 38
　connected, 196
　connectivity, 39
　cover, 100

graph (cont.)
  cut vertex, 38
  diameter, 102
  disconnected, 196
  distance, 40, 44
  distance sequence, 44, 102
  double cover, 100
  edge coloring, 42
  edge set, 15
  edge-transitive, 76
  general, 84
  generalized Petersen graph, 25
  girth, 28, 49, 102, 193, 194
  Hamiltonian, 32
  hamiltonian, 196
  independent set of vertices, 196
  isomorphism, 51, 55, 154
  Möbius ladder, 48
  maximum valence, 42
  multigraph, 37
  one-point union, 38
  one-regular, 80
  polyhedral, 105
  pre-graph, 37
  prism, 21
  pyramid, 20
  regular, 31
  regular cover, 100
  semi-edge, 37
  semi-symmetric, 77
  simple, 15
  square, 52
  strong product, 40
  suspension, 38, 51
  tensor product, 40
  vertex coloring, 41
  vertex set, 15
  vertex transitive, 101
  vertex-transitive, 76
  wheel, 20
graph complement, 38
graph drawing, 30, 47
graph induced by autopolarity, 185
graph invariant, 28, 39
graph isomorphism, 37
graph join, 38
graph operation, 38
graph representation, 46
graph square, 40
graph union, 38
graphene, 52
Graphical regular representation, 71
Grassmannian, 218, 254
Grassmannian configuration, 219

Gray configuration, 253, 257
Gray graph, 77, 102, 103, 252, 254, 256, 263
group, 55, 97, 101
  action, 67
  additive, 98
  automorphism, 100
  cardinality, 57
  element, 55
    order, 57, 98
  elementary abelian, 100
  finitely presented, 66
  generator, 62, 100, 193
    redundant, 193
  infinite, 98
  integers, 56
  inverse, 57, 97
  law of composition, 55
  multiplicative, 98
  non-commutative, 60
  normal subgroup, 60
  order, 55, 98
  real numbers, 56
  representation
    faithful, 193
  set of generators, 62
  subgroup
    conjugate, 64
    coset, 60
    index, 60
    normal, 59, 60, 98
  unit element, 56
  unit quaternions, 68
group action, 67
  imprimitive, 71
  primitive, 71
  regular, 70
group axioms, 56, 97
group homomorphism, 61, 98
  kernel, 61, 98
group isomorphism, 61, 98
group of covering transformations, 103
group of rotations, 57
GRR, 71

H
Haar graph, 89
hairy hexagon, 166
half edge, 37
half-edge, 84, 85
Hamilton cycle, 19, 32, 77, 172
handcuff graph, 86, 260
Handshaking lemma, 15
harmonious map, 142

Index

Heawood graph, 29, 50, 89, 101, 171, 196
heptahedron (see tetrahemihexahedron), 107
hexagonal tiling, 193
hexagrammum mysticum, 1, 29
hexahedron, 22
homeomorphism, 109
homogeneous coordinates, 199
    incidence, 201
    polarity, 202
hypercube, 182
hypercube graph, 39, 51

I

icosahedron, 23, 93, 164
icosahedron graph, 41, 52
ideal point (see point at infinity), 198
identity permutation, 36
imprimitive group action, 71
in perspective form a point, 176
incidence, 15
    undesired, 239
incidence graph, 9, 35, 36, 103, 158, 191, 192, 209, 210
incidence matrix, 16, 160
incidence structure, 12, 42, 157, 191, 192, 194, 197
    balanced, 191
    block regular, 164
    co-simple, 158, 192
    combinatorial configuration, 164
    connected, 158
    desarguesian, 206
    dimension, 236
    disconnected, 158
    dual dimension, 237
    dual structure, 160
    Levi graph, 193
    lineal, 163, 191
    operation
        cartesian product, 247
        disjoint union, 245
        parallel replication, 245
    pappian, 206
    parallel blocks, 160
    parallel class, 160
    parallel switch, 247
    picture, 239
        lift, 239
    point regular, 164
    rank d incidence structure, 193
    rank 3 incidence structure, 193
    regular, 164
    resolvable, 160
    self-duality, 158
    self-polar, 191
    simple, 157, 192
    square, 164
    substructure, 158
        strong, 158
        weak, 158
    symmetric, 164
    tetrahedron, 14
    valence of a block, 164
    valence of a point, 164
incidence switch, 243
incidence table, 8, 158, 214
independent, 15
index, 60
induced subgraph, 17, 193
infinite outer face, 30
inner automorphism, 91, 97, 101
integer
    Haar graph, 196
integers
    cyclic equivalent, 196
integers mod n, 57
internal vertex, 17
intersection graph, 44, 53
interval graph, 44
inverse, 57
involution, 36, 37, 50
    fixed-point free, 36
isometry, 47
isomorphism, 8, 26, 55

K

Kekulé structure, 53
kernel, 61
Kirkman point, 4
Klein 4-group, 62
Klein configuration, 194
Kronecker double cover, 100, 178

L

Latin square, 211, 262
    orthogonal, 212
LCF, 32, 50
leapfrog, 130
left coset, 60
left regular action, 68
Levi graph, 9, 164, 191, 193, 194, 261
lifting, 238
line arrangement, 159
line at infinity, 198
line graph, 40, 80, 102, 103, 154

lineal incidence structure, 163
linear arrangement
    Euclidean, 163
local isomorphism, 87
loop, 37, 84

M
Möbius band, 153
Möbius hypercube, 184, 194, 195, 262
    of dimension 4., 184
Möbius ladder, 48, 196, 260
Möbius–Kantor configuration, 174, 175, 191, 261
Möbius–Kantor graph, 174, 186, 188, 190, 195
map, 114, 153
    1-skeleton, 153
    axioms, 153
    combinatorial map, 154
    crosscap, 153
    dual, 153, 154
    edge-contraction, 153
    edge-transitive, 154
    face permutation, 153
    flag graph, 153
    gothic operation, 154
    handle, 153
    isomorphism, 154
    medial, 154
    non-orientable, 153
    one-dimensional subdivision, 153
    operation, 154
    orientable, 118, 153
    Petrie dual, 154
    self-dual, 155
    skeleton, 154
    snub, 154
    topological type, 153, 154
    vertex split, 153
    vertex splitting, 153
    vertex-transitive, 154
map projection, 119
matrix multiplication, 57
matrix of transformation, 64
maximum valence, 42
medial, 129
Menger graph, 158, 172, 192, 255, 257
Menger graph of a given type, 158
metric space, 43
Miquel configuration, 6, 13, 163
morphism
    color reversing, 158
Moser graph, 53

Moulton plane, 199
multigraph, 37

N
near-pencil, 159, 191, 199, 261
non-commutative, 60
nonorientable genus, 124
normal subgroup, 59, 60, 98, 100
number
    Haar graph, 196

O
octahedron, 22, 153
octahedron graph, 40, 41, 153
octahemioctahedron, 154
one-dimensional subdivision map, 153
one-point union, 38
one-skeleton, 105
opposite arc, 37
orbit, 36, 37, 67
orbit graph, 83
order, 36, 57
ordinary covering graph, 88
orientability, 112
orientable genus, 124
orientation classes, 118
outer automorphism, 91, 97

P
Pappus configuration, 7, 13, 36, 170, 176, 236
Pappus graph, 195, 235, 256, 257
    spatial drawing, 256
Pappus theorem, 6
Pappus's Theorem, 176, 205
Pappus's theorem, 235
parabolic subgroup, 162
parallel edges, 37, 84
Pascal configuration, 3
Pascal line, 1, 4, 12
Pascal theorem, 13, 203, 204
path, 17
    endvertex, 17
    internal vertex, 17
pencil, 159, 198
pentad, 97
permutation, 34, 35, 50, 98
    cycle notation, 36, 153
    cycle structure, 51
    degree, 36
    even, 65
    fixed-point free, 36, 50

Index 277

odd, 65
orbit, 36
order, 36
polycyclic, 36
semi-regular, 51
transposition, 65, 100
permutation derived graph, 88
permutation group, 67
permutation voltage assignment, 89
permutation voltage graph, 88
Petersen graph, 25, 28, 78, 79, 192, 195
Petrie dual, 117, 262
Petrie walk, 117, 153
phenalene, 53
Plücker line, 4
planar hexagonal lattice, 52
planar incidence structure, 163
planar representation, 47
plane
    affine, 198
plane embedding, 30
Platonic graph, 23, 45, 49, 52
Platonic solid, 76
Platonic solids, 23
point at infinity, 198
point of perspectivity, 243
point of similitude, 262
polar duality, 245
polar mapping, 244
polar structure, 185
polarity, 202
polycirculant, 260
polycirculant conjecture, 259
polycirculant graph, 259
polycyclic permutation, 36
polygonal surface, 107, 108
polyhedral graph, 23
polyhedron, 23, 105
    convex
        cubic, 49
    polyhedral graph, 50
    skeleton, 43
pre-graph, 37
pregraph, 83
    cubic, 84
prime field, 215
prism, 22, 68
    rotational group, 68
    symmetry, 99
prism graph, 44, 100, 196, 260
projective line, 46
projective linear group, 218
projective plane, 193, 198
    axioms, 198, 209

    finite, 209
        of order five, 214
        of order four, 214
        of order six, 213
    incidence graph, 209
    non-desarguesian, 205
    non-pappian, 205
    order, 211
    order 2, 199
    order 3, 212
    order 5, 214
    rational, 199
    subspace model, 201
    topological, 108
    unit sphere model, 201
projective space, 216
pure graph square, 40
pyramid
    symmetry, 99
pyramid graph, 20
pyramid graph, see wheel graph, 21
Pyramid of Cheops, 58

Q

quaternion, 60
quaternion group, 60
quaternions, 68
quotient graph, 83
quotient group, 60

R

rank, 158
rationals, 56
real numbers, 56
real projective plane
    model, 262
realizable, 6
realization, 235
    dimension, 236
    generic, 239
    spatial, 253
realization defect, 235
regular, 136
regular autopolar decomposition, 186
regular graph
    k-valent, 31
regular map, 136, 232, 262
regular pentagon, 12
regular polygon, 64
    center, 64
    line of symmetry, 64
    reflection, 64

regular polygon (*cont.*)
  rotation, 64
regular prism
  full symmetry group, 64
  rotational symmetry group, 64
relation
  across, 49
  incidence, 193
representation, 235
  lifting, 238
Reye configuration, 181, 193, 194, 226, 240
right coset, 60
right regular action, 68
rotagraph, 87, 101
rotational symmetry, 100

S

Salmon point, 4
Schlegel diagram, 105
self-dual, 145
self-dual pairing, 146
self-polar, 159
semi-edge, 37, 83
semi-regular permutation, see polycyclic, 36
semilinear incidence structure, 163
sharp lift, 239, 240
simple graph, 15, 84
skeleton, 43
skeleton, see 1-skeleton, 21
spacial representation, 47
spanning subgraph, 19
  1-factor, 31
  k-factor, 31
spanning tree, 31
sphere, 193
stabilizer, 67, 176
Steiner point, 3, 4
straight-line embedding, 143
strong polycirculant, 259
strong product, 40
strong realization, 235, 239
strongly regular autopolar decomposition, 186
subconfiguration, 239
subdivision graph, 40
subgraph, 16
subgroup, 58, 98
  orbit, 196
substructure, 158
surface
  Euler characteristic, 153
  non-orientable, 112
  orientable, 112
  topological type, 153

suspension, 38
symmetric group, 36, 65, 100
symmetry
  orientation preserving, 58, 65
  orientation reversing, 65
  reflective, 58
syntheme, 97

T

tensor product, 40
tetrahedron, 10, 22, 66, 154
  rotational group, 66
tetrahedron graph, 41, 79, 87
tetrahemihexahedron, 107
tetrahexahedron, 153
tetravalent, 80
theta graph, 85, 89
tic-tac-toe structure, 164
tiling
  self-dual, 155
  square grid, 154
topological type, 153
torus, 153, 193, 194
transformation group, 57
  representable, 57
transposition, 65
tree, 18, 30, 48
trivalent graph (see cubic graph), 260
truncated hexagonal prism, 59
truncation, 125
Tutte 8-cage, 29, 50, 193

U

uniform polyhedron, 154
unit distance graph, 47
unit element, 56
unit sphere graph, 45, 52
unit sphere model, 201
unitary map, 121, 153
universal graphs, 44
unsplittable incidence structure, 159

V

valence, 15, 84
vector product, 202
vector representation, 46
vector space configuration, 216, 218, 262
vertex coloring, 41
vertex join, 133
Vizing's theorem, 42
voltage graph, 84, 89, 102, 263

W

wallpaper group, 76, 99
weak realization, 221, 235
wheel graph, 20, 51
Whitney flip, 135
Whitney twist, 135

Whitney's Theorem, 150, 153
word, 66, 98

Z

Zindler degree of regularity, 258

MIX
Papier aus verantwortungsvollen Quellen
Paper from responsible sources
FSC® C105338

If you have any concerns about our products,
you can contact us on
ProductSafety@springernature.com

In case Publisher is established outside the EU,
the EU authorized representative is:
**Springer Nature Customer Service Center GmbH
Europaplatz 3, 69115 Heidelberg, Germany**

Printed by Libri Plureos GmbH
in Hamburg, Germany